MICHAEL FRAYN

The Human Touch

Our part in the creation
of a universe

faber and faber

First published in 2006
by Faber and Faber Limited
3 Queen Square London WC1N 3AU
This paperback edition first published in 2007

Typeset by Faber and Faber Limited
Printed in England by Mackays of Chatham plc, Chatham, Kent

A CIP record for this book
is available from the British Library

ISBN 978-0-571-23218-5

Contents

V Homewards

The Human Touch

Prospectus

You look up at the stars on a calm, clear night, and you're awed by the tranquil vastness of it all. You turn to a book on cosmology and you find it impossible to take in how much vaster still the universe is than you could ever guess even from that visible skyful of sparkling darkness.

And then, when you think about it quietly, you can't help being struck by something even more surprising: what a muddle it all is. Nothing's quite straight, nothing's quite circular – it's like an old cottage, built without a plumb line or a spirit level. Stars and galaxies are splashed about all anyhow. Some things are heating up, some things are cooling down. There are things exploding, there are things crashing into other things. And everything is inextricably mixed up with everything else, as if the cottage were lived in by some crazy old eccentric.

Has anyone ever seen a universe like it? Well, no, because this is the first of its many idiosyncrasies: its uniqueness. There aren't any other universes to compare it with – not ones that anyone has ever seen, at any rate. Of almost everything else there is more than one example of the same kind of thing, so that we can make comparisons and judge how things stand. Of universes we have just the one, so we don't know whether this is the way you'd expect a universe to be. I can't help feeling, though, that if someone had asked me before the universe began how it would turn out, I should have guessed something a bit less like an old curiosity shop and a bit more like a formal French garden – an orderly arrangement of straight avenues, circular walks, and geometrically shaped trees and hedges.

Our own particular speck of this universe, the planet we live on, is as irregular as everything else. A sphere, which seems a neat enough idea – but a sphere that isn't exactly spherical, wobbling a little on its axis, spinning not quite regularly. With a surface as rumpled as an unmade bed, splashed with seas and lakes as haphazard as the spills

3

on a bar, under a shifting blanket of air and water vapour as confused as a drawerful of tangled string.

The oddest feature of this wobbly spheroid, though, is one particular class of things scattered about amidst the rest: a range of entirely anomalous objects that construct themselves out of the material around them, and then replicate themselves – perhaps the only objects of this sort in the entire universe. Among these weird anomalies is a sub-group with a few thousand million members that are even odder, because they also have some inkling of just how odd they are.

And then among these few thousand million is an even smaller, even odder group: you and me. You reading this, and me writing it. What distinguishes us from all the rest is that we are currently communicating with each other. I don't understand much about you – just enough to know that you see the universe around you in much the same way (but not exactly the same way) as I do. I know this because otherwise you wouldn't understand a word I'm saying, and you wouldn't have got this far. And I know for the same reason that you are aware of being a member of an even smaller and stranger class than the two of us make up together: the class of your own self, which has only one member, just as I am the sole member of the class of my own self. Each of us is for himself what cosmologists, when they are talking about black holes and the moment when the universe began (if it did), call a singularity – the only real singularity in our own experience apart from the totality of the universe itself, the Big Bang and black holes not being things of which we have any personal acquaintance.

So you have something in common with the universe as a whole, and so do I. It's odd, though, how out of scale we both are with our great fellow singularity. Our metre or two of flesh doesn't occupy much of the distance between the furthest observable galaxies in the universe above our heads and the furthest observable below our feet. The twenty-four hours of our day, and the three score and ten years of our natural span, make only a brief interval in the fourteen billion years lifetime of the universe so far,[1] not to mention the more than fourteen billion yet to come.

And yet there's something odder still about where we're standing in relation to all this. We have the good luck to be, not on the outside of all this, looking in, but right in the thick of things. We seem to be quite literally in the middle of the universe, with its ever-expanding boundaries marked only because they are receding from us as rapidly as the light from them can travel back to us, at a distance of about fourteen billion light years in every direction.[2] We also seem to be somewhere in the middle of the duration of the universe. Somewhere, perhaps, between the start of things and the end of them; or somewhere in the first half of an enormous cycle, before the boundaries stop moving away from us and begin to move in upon us again; or else somewhere infinitely far along the road from no beginning, with infinitely far left to travel to reach no end.

We're also somewhere in the middle of the scale of things. Small as we are by comparison with the universe as a whole, if we look into the details of the world around us we become giants – more than a billion times as tall as an atom, living perhaps two hundred billion times as long as a relatively long-lived meson. Where exactly are you located inside your moderate couple of metres, for the length of your middling seventy years? Where am I, in mine? Lodged somehow, never mind for the moment how, in a brain whose hundred billion neurons are connected in one million billion ways, a brain which is one sub-system in a body built upon hierarchies of sub-subsystems and cells – of a hundred trillion cells, each built from something like sixty-seven trillion atoms; not to mention the hierarchy of unobservable quarks and strings postulated within each atom.[3]

Another odd thing about our situation: how handily placed we are to think about all this. Not only are there six billion of us talking away to each other, and exercising the language and conceptual processes we need, but we have inherited the handiwork of another hundred billion or so human beings, spread out over the last fifty thousand years.[4] Slowly, slowly, they have shaped our immediate surroundings to our needs, and invented the words and numbers and concepts to represent what we are looking at in a form that enables us to think about it.

Are we in the middle of human history, as of everything else? When we turn away from the faceless hundred billion who preceded us (not to mention the forms of life that preceded *them*, gradually developing over the previous three or four billion years), and look forward into the future, it's not just the individuals who are difficult to distinguish, but even the haziest approximation of their numbers. Are we looking at billions, or millions of billions? Or do we see them dwindling away again in the distance, to hundreds, to handfuls, to two, to one? Are we looking at no one at all, from 10.35 tomorrow

5

morning? Are we in the oddest situation of all – of being among the last ones to have all the luck?

The sheer strangeness of our situation, when we look at it in this way, threatens to swamp us. No one could ever have predicted any of this! But then no one could ever have predicted anything. This is what sticks in so many people's throats about evolution. They look at the sheer idiosyncrasy of the spider and the bower bird, of the lymph gland and the enzymes of the digestive system, and they think, 'But no one could have seen any of this coming! You can't start with a few general principles and end up with all *this*!' Nor could anyone have predicted, from a study of genetics and child psychology, that you would be you and I should be I.

But then we're not *trying* to predict the spider or your identity. We've *got* spiders, we've *got* you. All these things are the given. The starting point of our speculations about the universe and our place in it is not Day Zero of creation, or even Day One – it is the evening of Day Six, after the creation of man; or, more precisely, the situation that has developed out of all this since, including you and me as we now are. And if we and our fellows were not here to ask why things are as they are, if we were not in the situation and endowed with the faculties that enable us to ask the question, then no question would arise. If the question can be asked, the oddity felt, then (paradoxically) there *is* no question, there *is* no oddity.[5]

And if we *are* the last, likewise, then we are. Does it matter much, in the great scheme of things, if you and I close our eyes, and close them for good? If the particles that comprise us, supporting our tiny sense of being here at the centre of space and time, are scattered to wherever they came from in the first place, among the indifferent particles around? In this vast organisation we're all nobodies. We shall all of us go in the end – and what will have changed? The number of quarks in the universe will still be the same. A curious little fleck of foam that wind and tide raised for a few moments on a few square centimetres of the great ocean will have vanished back into the waters. The universe will go on as if we had never been. We both know that as surely as we know anything.

But now we come to the oddest thing of all about our relationship with the universe: we both also know that the opposite of that is true. We know that if we go, so will everything.

What *is* the universe, after all? Vastly big in its totality – and vastly small in its details. What makes it big, what makes it small? You and I do, by standing where we stand, by being 1.80 metres tall, by living for three score years and ten. We have some notion of how vastly bigger it is than ourselves, how vastly smaller, only because we've laid our metre-rule against it. Only because we've put our stopwatches on it.

It's big, it's small, it's so many billion light years across and so many years old because you and I and some of our friends say it is. If we weren't here in the audience, comparing and measuring, gasping and applauding, the whole show would have gone for nothing. The universe would not be several billion light years across, or four centimetres, or any other distance. It would not be odd or awe-inspiring – or even banal. It would have no characteristics at all. And if it had no characteristics then in what sense would it be *anything*? In what sense would it exist?

The lifespan of the omega meson would not be 10^{-22} seconds or twelve centuries or any other length of time. *Here* would not be here. *Now* would not be now. And if *here* is not here, nor *now* now, *there* is not there, nor *then* then. There would be no *is*, no *was*, no *will be*. If no *is*, no *was*, no *will be*, then no passage of time.

So we are perhaps not after all such nobodies. We are not for nothing. The middle of things is not an entirely inappropriate place for us to be.

We seem to know two mutually contradictory things to be true. This paradox is something that I have been puzzling about for most of my adult life. So have many other people, in many generations before me. It's the world's oldest mystery, and it has taken many different forms. Are the qualities (physical, moral, aesthetic) that distinguish one thing from another objective realities, or are they our subjective imposition upon things? Can we have any real acquaintance with things outside ourselves at all, or does the knowable world consist purely of our experiences? Is the world in one way or another *out there*, or is it *in here*? Succeeding schools of philosophers have reached for one horn of the dilemma or the other, but it's impossible to seize both horns equally securely at the same time. This is one of the reasons why philosophy has never culminated in any generally accepted body of doctrine, or even seemed to be on the road towards it, as science is thought to be by many scientists.

If we go over the ground together once again, can we at any rate get the problems into a clearer perspective? Perhaps we can help each

7

other. The thoughts that one thinks in the privacy of one's own head tend to be elusive – and often prove nugatory or false when one attempts to bring them into the light of day. The possibility that some-one might be listening (and you must have got this far, at any rate!) makes us all more coherent. After all, we learnt all the words we know in the first place only from talking to each other.

I'm not proposing anything that requires specialised knowledge (you may have some but I don't) – only that we keep our eyes open, like any thoughtful tourists visiting a strange place (and the place we find ourselves in is *very* strange). We can't look at everything, but we can choose a few particular sites, a few vantage points with wide views. What I'm proposing is that we should go this way and that, without any particular system, wherever a path seems to offer, to get the lie of the land. It may mean that we find ourselves crossing paths we followed earlier – perhaps even covering the same ground again for a while – or coming by different paths upon the same objective.

Is this excursion yet another attempt at philosophy? Not really. I shouldn't have the courage to make any such claim, because I can imagine how scornfully it would be dismissed by most professional philosophers. Philosophy these days is a very specialised and technical discipline, comprehensible only to the initiate, and undertaken almost entirely for their benefit. Reading a philosophical journal, I'm some-times reminded of Dawkins's Law of the Conservation of Difficulty,[6] which states that obscurantism in an academic subject expands to fill the vacuum of its intrinsic simplicity.[7]

Am I claiming that this proposed joint expedition is something *important*? I don't know what to say. Even if we actually manage to resolve the great paradox that has been teasing mankind for the last few thousand years, it won't help with any of the world's practical problems, or make us better people. But then nor will the stories we read, or the pictures we look at, or the music we listen to. All these things are as important as you personally happen to find them, no less and no more. A mere diversion, perhaps, if that's what they seem, or entirely pointless. Or just possibly keys that unlock something in the thoughts and feelings of which our life is composed.

And whether we do in fact get any nearer to resolving the paradox . . . we shall see.

I: Principles

Traffic

Fleeting glimpses of a fleeting world

When you look up at the great stillness of the night sky, one thing seems sure. You can't help agreeing with Robert Frost, at the same ever-recurring moment of human contemplation: that calm seems certainly safe to last tonight. Whatever the discrepant scale and lop-sidednesses of the universe, at least it's *there*. Give or take the odd supernova (none visible tonight, at any rate to the naked eye), and perhaps a racing, vanishing meteorite to set off the permanence around it, there is a fundamentally stable array of *things* out there by which you can set your watch and direct your steps.

And yet we know that the whole display is stirring. Not only is the Earth moving round the Sun at thirty kilometres per second, but the Sun is moving round in the Milky Way at 250 km per second – and the whole of the Milky Way is moving at some 600 km per second towards the Virgo cluster, which is itself in motion.[1] Take a photograph of it, and the whole universe is revealed to be a blur of movement. However high you make the shutter speed, and however sharp the photograph looks at first sight, if you enlarge it sufficiently you'll see that the subject has changed during the time that the shutter was open. If you go close enough to the object to see its sub-atomic structure, you know that the movements of its parts are so fast and so ambiguous that no shutter speed could ever be fast enough to resolve them.

Everything is relative, even permanence. We are struck by the permanence of the sand in comparison with the transience and insubstantiality of the footprints we have left in it. They weren't there before we walked across the beach – they won't be there tomorrow. The sand in which they are pressed, however, was and will be. When we look more closely even at the sand, though, its persistence and solidity begin to fade a little as well. The footprints will vanish by tomorrow because the grains of sand in which they're written are rolled away every day by the incoming tide and replaced with fresh grains. The eternal ocean that shifts the sand is even less enduring. It evaporates and falls as rain that scours the ancient rocks of the land to the shifting sand of the beach. The sun that heats the sea that grinds the rocks that make the sand was born, grows old, and dies. The world, as we have all understood since Heraclitus, is a process, in which the persistent shape and identity of oceans and rivers are maintained by the endless flux of the particles that arrange themselves to compose them.

We look closer, and see that each particle is in itself a world in flux, a hierarchy of still smaller particles – of particles that are not precisely particles, but additionally and alternatively wave formations, fluctuations in probability, whose precise state can never be fully expressed. One of the implications of indeterminacy is that there are no *things* – no fixed, fully specifiable entities. These gritty grains of sand, so eminently and geometrically and tangibly *there*, are analysable into constituents whose defining characteristics can never be completely and precisely determined.[2]

The persistence of the sand and the stars is a quality with which we have endowed them, to distinguish them from the meteorites and the footprints, and to enable us to enter into practical relationships with them. It arises from our traffic with them. And traffic involves action and change. Without perpetual change we could not see the world in front of our eyes – the stars and the sand any more than anything else. Just as the passenger in a car notices the speed only when it goes faster or slower, so we notice the static world only when something changes in it, or when we shift the focus of our attention. Our eyes, and all the neurological software that supports them, are evolved to see movement. Even if the subject doesn't move, the eye does – continuously. The image on the retina fades almost immediately if it is not perpetually refreshed, and to keep it there at all the eye moves as restlessly as the electron around the nucleus, scanning and re-scanning the scene hundreds of times a second.

This isn't simply a limitation of the human apparatus – it is a limitation that applies to *all* relations between objects. A rock lies inertly upon a rock. But the pressure that it exerts is the manifestation of a continuous stream of events – the gravitational force which one mass exerts upon another, whether by some as yet undetected particles or by some as yet undetected waves (or both). Even the most static aspects of the world, as we now understand from digitalised information technology, can be represented as a stream of events, a flow of fragments of information (as when a stream of electrons lights one by one the pixels that compose a television image). Actually Locke proposed a prototype of this idea three hundred years ago, in relation to the most elusive and controversial entities of all – human beings. What differentiated one person from another, he suggested, was simply the differing memories they had, just as what differentiates one CD from another is the information encoded on it; and to be realised as actual rather than potential memories this information would have to be recalled, by way of a series of experiences extended in time. I once proposed using this radical and modern perception to solve all the problems created by commuter traffic, overcrowded air lanes, decaying rail networks, limited fuel resources, etc. If Locke is right everyone could stay exactly where they are. What would be moved from place to place – cleanly, safely, electronically – would be all the information in their memory, to be unloaded onto the people blanks waiting like blank discs at their destination. Our personality, our self, is the virtual reality that is stored in our physical body. If we could find some way to transfer that reality to another machine, we could be immortal.

This proposal seems to be a little nearer to being adopted. At any rate biologists can now envisage the possibility of producing identical blanks, clones, which would truly be differentiated only by the information they carried. But let's be more radical. After all, the roads are jammed with lorries as well as cars. So let's send all our freight the same way. Everything, from frozen pizza to the blank human beings who will be stacked up for imprinting, is an arrangement of electrons around the nucleus, of protons and neutrons inside it. So why not keep the particles where they are, and merely email the arrangement of them, to be imposed upon the supply of particles waiting at the receiving end?

And yet we feel somehow cheated by the idea. It wouldn't be quite the same, that longed-for visit from the one we love, if we knew that she was physically just a blank from the stack in the post office, however well she recalled every kiss and caress from our last meeting. I remember the indignant disappointment expressed by my five-year old grandsons when their father was away in America and they faxed off the loving drawings they had made for him only to see the pages emerging from the back of the machine – not in America at all, but in the same room at home where they started. My grandsons are clones themselves, as it happens – identical twins – though the information that characterises them seems to be remarkably different. The unease that we feel about clones – that some people are said to feel about identical twins – is our version of the anxiety that some remote societies are said to feel about photographs and images in mirrors. The essence of the individual is his uniqueness, which is given expression by its embodiment in one particular piece of flesh, at one single physical location.

But even in the purely physical world, even in the world of sand and rocks and frozen pizzas, in spite of all our understanding of Heraclitus and indeterminacy, we are reluctant to dispense with the thinginess of things. Philosophers, worried by the ontological implications of all those nouns – 'sand', 'footprint' – once made great efforts to analyse propositions about them in a way that separated out the nature of the object from the question of whether it in fact existed, so that we could talk meaningfully (as we do) even about sand and footprints that *don't* exist. (So that 'I can't see a footprint' becomes something like 'There is no x such that x is a footprint and x is seen by me.') This reduces the solid thereness that seems to be attributed to the footprint (even the footprint that isn't there). It reduces its status as a noun to something more modestly adjectival – the quality of being footprintish that something might have, or that nothing in the world might have. But still that weasly x (the existential quantifier, as it is called) lurks in the background like the ghost of a noun, an uncharacterised something that somehow . . . yes, *exists* (even if the footprint doesn't) to serve as a repository for the sandishness or the footprintishness. Could we, as a more modern alternative, try analysing things not adjectivally but adverbially – into modalities of action: 'At this point in time and space things are going on sandishly; at that point footprintishly'? Or as a series of verbs: 'It sands, it footprints.' But even into these formulations nouns and pronouns have crept back, those faceless but substantive *things*, that shifty *it* that's also sometimes raining, or warm for the time of year, or going well, or Wednesday, or five o'clock.

Not surprising, though, that we should cling to a world of things. Instability and transience are relative, and what we see as things are indeed slower to change, and harder to analyse into movement, than the more short-lived phenomena we classify as events. They offer us a way of locating ourselves. It's hard to specify our situation in a world of shifting cloud – impossible, in fact, in the most literal sense, even for the most experienced pilots flying in cloud, to construct a stable framework of reference without the help of instruments. We feel a similar sense of vertigo when we see the world of things becoming fluid around us in a time-lapse film, or in some projection of ourselves in our imagination on to a more cosmic scale. We need a relatively stable framework against which the flux of what we experience as relatively changeable can be measured.

We need, for example, to find objects in the world stable enough for us to be able to enter into the relationship of possession with them. We need to set our mark on the world by owning pieces of it. Not clouds, not actions or events, but clothes and furniture, cars and crockery and books. Real estate, bricks and mortar, to be, as the contract assures us, enjoyed in perpetuity.

And yet the whole notion of possession begins to evaporate the closer we look at it. Think of the most mundane material objects – the food in your refrigerator, the car outside your house. The only way in which your possession of a loaf of bread or a pound of butter can be expressed in practice is to consume it – or to exchange it – after which you possess it no more. The only way in which your possession of your car can be expressed in practice is to drive it; or to look at it, or to wash it; or to let it out or to sell it; all activities which begin in time and end in time just as surely as any other experience.

Even these temporal events which in practice constitute ownership are highly restricted. The bread and butter can be eaten only until our appetite is sated. The events making up our ownership of the refrigerator in which the food is stored are limited to keeping the food cool. When we collect our new car from the showroom we seem to be taking possession of the freedom of the kingdom, of a world of summer days, of a lifetime of shared experience with the people who will ride with us. We discover slowly that we are buying only the right to move along certain narrow strips of land, in certain directions, at certain speeds, to stop only in certain designated places, to do either only in the increasingly limited spaces not occupied by other cars at the time. Yet when we look out of the window and see our new car sitting there at the kerb, how solid and continuous a thing it appears to be. And when we look again and find only a small pool of broken safety glass where it had been, how substantial and total seems the loss.

Not all possession is as hedged about by such narrow constraints. Think of ancient estates, possessed, and possessed absolutely, by proprietors unhindered by modern legal or moral inhibitions . . . Even so, what the great lord's possession comes down to in practice is hunting over the land when he feels like it, taking a share of the crop, humiliating and tormenting its inhabitants. However often he hunts, each hunt starts now, continues through the day, and finishes at dusk. His share of the harvest can be taken only when the harvest has been gathered to take it from. The satisfactory grovellings are made, and then the grovellers back out of the door. The floggings pall. The tenants are summarily evicted, and in being so pass beyond eviction's range.

Of course, the possession of your car and your house means more to you than picking up a renter and driving to your holiday villa. The loaf of bread and the packet of butter in your refrigerator have a different character from the loaf of bread and the dish of butter that your host puts in front of you, with instructions to help yourself. What they offer is the reassurance that you can, at any time you choose, drive, eat, occupy, enjoy. In practice, the essence of the relationship is almost always, almost entirely, this sense of possibility, and not the realisation of it. Indeed, one of the possibilities that possession offers us, and very much a part of its richness, is precisely the possibility of *not* realising it. So many other things we could do with our possessions! Sell them, donate them to charity, lend them to a friend, leave them to our children. Perhaps best of all, we know that we can do nothing with them. We can hoard them – leave the car sitting in front of the house undriven, shut up the house for six months of the year while we winter abroad. After all, if we do actually eat the bread and butter it's gone. If we do actually drive the car it becomes merely a dull and frustrating way of getting from one place where we don't want to be to another where we are soon going to discover, if we haven't already, that we don't want to be either.

We know all this, and yet still our passion to possess consumes us. We become collectors. We assemble unusual objects which locate us and no one else as their proprietors. We gather around us curiosities and freaks, fine wines and rare malts. We prize even more highly the unique – the autograph manuscript, the one known misprint of the stamp, the left testicle of the saint. The uniqueness of a painting we see as an integral element in its artistic identity. However many times the subject has been painted, however many times this particular artist painted it, he painted it exactly like this only the once. This is what painting *is* – the one single particular set of marks, by the one single particular author on the one single particular occasion. And when that uniqueness appears to be threatened by some process of reproduction that offers the possibility of copies which are for all practical purposes indistinguishable, and which would therefore surely embody all the qualities of the original except its uniqueness, we protect that uniqueness by declaration – by simply ruling that the original remains the one and only true example of itself.

We want to possess the patently unpossessable – our fellow possessors, other people. We want to possess them as slaves, and if not as slaves then as indentured labour, and if not as indentured labour then as our social and emotional dependants. Men want to possess women, women men. And here they have much more in mind than passive acquiescence. They want to command emotions and loyalties, to grasp some inmost essence of the other, to hold every inch of their lover's flesh in their hands at once, to surround that flesh, to be within it.

And of course we aspire to fix the flux of experience itself, to solidify the endlessly evanescent present into a graspable past, by recording it in our diaries, by shaping it into the stories we tell about it and the memories we run and re-run in our heads, by capturing it in snapshots, on film and videotape; to prolong the sudden flare of passion into an eternal flame. But the myths and memories take shape by acquiring an internal structure of their own, whose relationship to the events they are supposed to immortalise vanishes as surely as the events themselves. Those snapshots that we carry around in our wallets, and look at over and over again, notoriously come to replace the memories they are supposed to embody. We stop the video wherever we choose, and yes, now, with the help of this magical device, we hold time fixed in our hand . . . But we know that even as the frame freezes it becomes just another snapshot like all the others – that the life in it has died before our eyes. The eternal flame burns on inside the monument, noticed only on the appointed annual day of remembrance.

Look at the rows of books and records gathering dust along our shelves. Of course, it's useful to have them to hand when we need them, but our feelings about them go well beyond mere handiness. Art again, but no question of uniqueness here – these things have been published, issued in multiple copies. What we aspire to own here is not the object but its contents. A moment's reflection, of course, and we see that literature and music are as unpossessable as the light of heaven. Music exists as a performance in time. Possessing a recording of it means that we can command that performance at will; but however often we do so command it, the performance still begins, occurs, passes, and ceases. Even a book has to be read; and however many times we repeat the experience, on each occasion the reading begins, occurs, passes, and ceases. We can read extracts, and reread them, remind ourselves of paragraphs, sentences, particular words. But each extract exists in time as we read it and reread it – each paragraph and sentence, each word.

And yet, when we look at the rows of books and records on our shelves, they don't seem to need reading or playing. They seem almost to need *not* playing. We know that if we play the record, even of the music that means so much to us, it will pass before we can fully grasp it; just as it did the last time we played it, it will elude us once again. The mere possession of the record seems to be not just a time-saving alternative but a purer one, like the passion of the lovers on the Grecian vase, forever warm and still to be enjoyed, forever panting, and forever young.

We abandon the idea of possession and more modestly resign ourselves to contemplation. Even contemplation, though, is a process, a dynamic traffic between the contemplator and the contemplated. However calmly we gaze at the scene before us, or the picture that portrays it, the eye has to keep moving, ceaselessly moving, to bring the whole of the visual field before the fovea, the tiny pit no bigger than a pinhead at the centre of the retina where the packed receptors are closest to the surface – the only part that is capable of intense vision, the only part that is capable of actually seeing, in the sense of clearly distinguishing. And if we could somehow limit our ambitions yet further, to looking at a single fragment of the world that was small enough to accommodate in its entirety on the fovea, the eye would still have to keep moving fifty times or so a second, because if it stopped the mind would become almost instantly accommodated to the signal, and the image would cease to register on the consciousness. Even the stillest observation involves continuous change. We close our eyes and turn to inward contemplation. To keep even the simplest idea before our mind, however, requires a continuous effort, a continuous shift of perspective and emphasis, a continuous exploration of ramifications, consequences, and antecedents.

Do those who learn to meditate upon abstractions and religious entities escape from the endless shifts and changes of process? I wonder. Does the flying spot of light that forms a television image rest when the camera pans to darkness?

Our cities are ruined by traffic, we all agree. Not only in the specialised sense of vehicular road transport, but by traffickings of every sort. Our high streets are disfigured by supermarkets and fast-food outlets, by chain stores selling electronic goods and the current fashions in clothes. All this is something new, which has grown up since we were young, or since our parents were. If only we could get rid of all the cars and crowds, suppress the smell of frying onions, replace the supermarkets and electronics shops with honest self-employed saddlers and silversmiths, we could get back to the real city, the city as it once was.

So we set to work. We pedestrianise the streets. We ban fast food. We restore the old shopfronts and reserve them for small tradesmen and craftworkers.

Now we have quiet, salubrious streets lined with shops selling handmade jewellery, little pots of home-made honey, glass paperweights, and picture postcards. The people wandering where the cars used to be are not hurrying blindly from place to place. They have time on their hands. They are all from other cities and other countries, and have hours to kill before they can reasonably sit down again for the next meal of the day.

Is this the city as it once was? Look at the old photographs. The shopkeepers stand in front of row upon row of dead animals and cheap dresses. The unregulated chaos of carts and cabs and omnibuses, with the horses that pull them, is gridlocked at every major crossroads until the end of time. Read Pepys on the subject of pedestrians forced off the pavement by others who 'took the wall', of impassible traffic jams of twitching horseflesh. Think of Jo in *Bleak House*, and all the other crossing-sweepers who laboured to maintain corridors of dry land through the swamps of churned mud and horse-shit. You think they didn't fry onions at Bartholomew Fair, you think they didn't play loud pop music?

Whether the council's labours with the cobbles and wrought iron have really restored the past is in any case a somewhat academic question for you and me, who live in the place, because we don't go to these quiet streets any longer. We don't go there because, whatever we want to buy when we go shopping, it's not handmade jewellery, little pots of home-made honey, glass paperweights, or picture postcards. The life of the city, the life that we who live in it are actually a part of, has moved elsewhere, round the back of the set, to the approach roads where the lorries are unloading, to the suburban shopping malls where they now sell lavatory paper and thirteen-amp plugs and insect powder, to the car parks where the boot sales are held.

Am I against pedestrianisation, then? Am I against restoration and zoning and controls? Of course I'm not, any more than I'm against owning books and recordings. But it's difficult to have the quiet and also keep the city, because the city isn't a collection of buildings among which the traffic moves; the city *is* the traffic. That endless stream of cars and buses represents the movement of people to and from the encounters that constitute the fabric of their lives. The jams and grid-locks are the tangles that all movement and interchange involve, the choking exhaust gases the emanations from the consumption of energy necessary for all activity.

Some small part of that endless movement from place to place is no doubt being undertaken for its own sake, for the pleasure of movement, to soothe the baby to sleep, to try out the new car. Most of it, though, is for practical purposes, to enable people to stop being at the place they are leaving and to start being at the place they are going to, and what happens between those two points is merely the time that has to be lived through to achieve this. And yet that time, and the movement that occupies it, are also parts of the travellers' lives. The act of travel develops its own interest, its own hugely complex technology, its own scale of satisfaction and frustration.

This is true not only of motorists and cyclists and pedestrians, who are active participants; it goes for the most passive and numbed of commuters. For the best part of forty years a friend of mine left his home in Cambridge each morning and sat on a train for three quarters of an hour to get to his work in London, then left his work in London each evening and sat on a train for three quarters of an hour to get to his home in Cambridge. His conscious life was lived at home in Cambridge and at work in London. What happened on the train each way? He didn't work, he once told me, he didn't read, he didn't think – he sat there in a daze. That ninety-minute daze of transit, though, that dim background of the pressure of the seat on the buttocks, of acceleration and deceleration, of sway and clatter, of the passing flicker of fields and suburban houses, that ancient litany of unread station names, that drifting daydream of half-formed thoughts and half-felt feelings, were as much part of his life as all the other hours before, between, and after.

In any case, the movement of people from place to place is only the most visible aspect of the city's traffic. The bulk of it is out of sight – the business for which all those journeys are undertaken: the confrontations between buyers and sellers in the shops, the exhortations of the sellers, the alternating caution and recklessness of the buyers, the contracts, the phone calls, the money in the cash dispensers, the hooting, the rage, the tiredness, the crime. The streets and the buildings they lead to are merely the conduits of all this. They are the tidemarks left by the current – and the current that flowed here yesterday, because by the time a street or a building has been projected, discussed, drawn, financed, and built, the purpose it was intended to serve has already changed. The numbers of people it was designed for have grown, or the business has declined. The standards of comfort demanded have risen, the cost of the fuel to heat it has soared, income has fallen, the building materials have been found to be poisonous, the belief has grown faint, the rite has been reformed, the aesthetic has fallen out of favour. Buildings are like snail-shells – the residue of last year's growth, the record of last year's traffic.

Think of a tree in its essential form, and you think of a trunk and branches. The roots are out of sight and forgotten, the leaves dismissed as frivolous optional extras that come and go with the seasons. We fail to see the trees for the wood, because in truth it is precisely the leaves and the roots that are the essential tree. They are carrying out the process that constitutes it. The branches and trunk are merely the connections between the two – the streets that guide the traffic, the spoil heaps left behind as the leaves reach towards the sunlight. If there were no other trees around competing for light then the tree would function perfectly well with the leaves springing directly out of the roots.

This is one of the reasons that young people seem more interesting than older ones. They are more leaf than trunk, whereas older people are mostly the trunks of their lives. History, art, literature are also trunks – the no-longer-essential remains of the vital part of the process. So, too, of course, is philosophy. The Oxford philosopher Gilbert

Ryle memorably suggests the parallel between a finished theory and a farmer's path, and between the complex and untidy struggles involved in building each of them.[3]

Divert the traffic, certainly. God preserve us from the weather, for that matter, and keep us in the shelter of our homes. But if there were no rain and no sun out there we'd have no wood to build the home, nor fuel to warm it. If there were no traffic there'd be no street to divert it from nor frontages to reveal.

All life is process, traffic, trade. But some processes are slower than others, some trades less rough, some traffic less noisy. The building is built, is lived in; is repaired, is restructured; is abandoned, is pulled down. The flower becomes the fruit becomes the seed becomes the flower.

And these cast-off shells, these solidified traces of process, these *things*, have their place in our traffickings, just as cowrie shells and dead timber do. The cars crawling along the street have all been bought, and will all be sold – sold on until they are crushed and melted down and used as the raw material for further cars. What are we doing inside the buildings? Buying and selling, certainly – and among the things we are buying and selling are the buildings that we are doing it in. The cars and buildings are the currency we use to purchase our mobility, our safety, a survivable climate. We buy other evanescent qualities with them as well: a sense of physical well-being and mental equilibrium; self-esteem; the expression of our place in the social hierarchy.

Everything we touch is a token; it has a value in the commerce of our lives. Not all values have things in which to inhere. Values can be – often are – expressed in entirely disembodied symbols, as numbers written in books, as minute traces of magnetism. But what has made the objects of the natural world separable from their backgrounds and identifiable to us was in the first place the value that they carried. An apple is an apple because it has a certain value in the fuel market. We bring the apple into being because it is good to eat. Its utility to us charges the particles, makes them cohere thus and so, separates its sphericality from the shapes around it, its green from the surrounding colours, its blush from its green. We can exchange it for euros as well as for dollars – not only for nutrition but in the aesthetic market, for instance, or for its nostalgic references to simplicity, purity, childhood, summer days; or, conversely, for its sly allusion to sexual temptation, to rebellion against divine command.

Just as special relativity makes it possible to see the world not as things but as energy, and general relativity suggests a view of force as a part of the world of things (as a distortion of space-time), so you could dissolve out things in another way, and see the world as a sign system like language. This is the world as form, in which things are discriminable in the way that signs are, because of their differentiability from other things and signs.

Even when we recognise the transience and mutability of things we think of them as concrete and continuous entities during their trajectory through time. But their transience is more radical than this. They are like the particles in a cloud chamber, which leave what appears to be a continuous track. This track, though, we know to be composed of a series of discrete objects – separate droplets condensed out from the vapour through which the particle is passing. Each droplet has condensed around a separate discrete event – a collision between the particle and a molecule of vapour. In the same way even the most stable physical objects – the pyramids, the Parthenon – are given reality only through their involvement with us at different times and in different ways: through our seeing them, our thinking about them. Buying and selling them, for that matter. Building them and destroying them. They are series of interactions with the creatures who come into contact with them. Events, therefore. Or, at any rate, complexes of potential events. Even this, though, is to see them as indelibly marked with our fingerprints, because an event is an event only once it has been distinguished as such.

Empty space is the same. Every centimetre of it resonates electromagnetically on infinitely many different frequencies at the same time, waiting to be read as the radiation from each of the perhaps ten billion stars in each of the perhaps ten billion galaxies. You could see the universe as one single, complex, infinitely interconnected potential event.[4]

Waiting only for you and me to touch it here and there in passing.

Our most fundamental traffic with the world around us is our perception of it. In our understanding of perception we move between two extreme views. We think of it first as our passively receiving the world at large, like tireless hosts who are always At Home to anyone who calls. Then we become more sophisticated and believe we are the sole authors of the game, as if our great At Home were a novel where we made all the guests up inside our heads. We say, easily, that we come to see certain sorts of landscape because they have been seen for us by painters. This may be so, but it pushes the real question one stage further back: how do the painters 'see' this new landscape in the first place?

The truth is that the situation is a complicated one. We are neither giving a party nor writing a story. It's more as if we were casting a play, and trying to persuade actors to play the parts we have written for them. We are employers, trying to staff our great enterprise with the more or less suitable candidates sent round by the employment agency. We are in a kind of dialogue with the world.[5]

You might argue that a dialogue is precisely what it isn't. In a dialogue two parties speak. In our supposed dialogue with the world only we on our side are speaking (also writing the lines, announcing the stage directions, explaining how the machinery works and when the coffee breaks are due). But our potential employees are doing *something*. They are indicating acceptance or recalcitrance by their body language. They are somehow suggesting to us, by their very shape and and their manner of carrying themselves, the jobs they might do for us, the ways in which their role could be developed. The world is gesturing silently to us. Our mutual dialogue is like that familiar party game that the Victorians used to call dumb crambo; the world, forbidden by the rules to speak, desperately waves its arms and pulls frantic faces, while we attempt to guess the phrase that fits them. 'Wind, wind! Something to do with wind . . . *Wind in the Willows . . .!* No, because it's a film! *Gone with the Wind . . .!*' And what helps us is that most of the time, as in dumb crambo, the expression for which we are searching already exists – what we have to do is not to invent it from scratch, but to find it. In other words, we are attempting to identify something already given an identity by its maker's specification, already categorised by its origins.

In dumb crambo, of course, the performer is on our side, as desperate to communicate to the rest of us in his team the expression he has been set as we are to understand it. In our game with the world the raw material we are presented with has no interest in whether we understand it or not. It's going through its performance regardless of its audience. Forget dumb crambo – we're like fond adults watching a baby and interpreting its random actions for it. 'Oh, look at him! He's trying to get out of his cot and go off into the world to make his fortune! Now he's thinking, "I wish everyone would go away and leave me to get on with my thinking in peace!"' And even the most whimsical construction that we place upon an action might serve to identify it, might name it, if the action is repeated often enough to, well, take on a character. 'Oh, that's his I'm-going-to-be-a-big-film-star face! Now he's going to start banging with his gavel again to bring the meeting to order!'

The balance of this dialogue between the partners is delicate. The limiting case at one end of the scale is the exclamation. Someone jabs a pair of dividers into us and we scream. The jab sets the tone of the exchange; the exclamation is our helpless response. At the other end of the scale are hallucinations in the dark, where there is no external given at all, and everything has been supplied from our end. In between come a range of possible conversations, which seem as difficult to examine, as vanishingly elusive, as all the other conversations we have – as beyond examination or description as a row or a reconciliation with someone we love.[6]

A man's cock may rise to the touch as helplessly as his lower leg to the doctor's hammer, but most sexual arousal begins long before there is any physical contact. It may be beyond our conscious control, but it isn't merely mechanical; it involves the same complexities as any other sort of perception. Even if we learn to become sexually aroused as a conditioned reflex to certain set erotic images, like dogs salivating at the sound of the bell which they have learnt to associate with the serving of lunch, we still have to see that the pole-dancer is a pole-dancer, that the pole is a pole, and that the one is twined around the other. We have to *understand* what we are seeing!

In any case, a great deal of sexual arousal isn't like this at all. It involves much more subtle perception and much more complex imagination. You have to respond to physical characteristics that you have *never* come across before in precisely this form, to a situation and behaviour which allure precisely because they are *terra incognita*, and you have to project forward, through a chess game of intervening moves, to the possibilities they suggest. Even the apparently spontaneous reaction to danger demands the exercise of the same faculties. Most dangers don't arrive painted red, with the word DANGER stencilled on them. We have to perceive them and understand them, and project the possible scenarios that arise from them.

You might argue that there are some things so simple and self-declaratory as to need no interpretation at all. The blueness of a cloud-less blue sky, for example. You look at it and you see pure blue. You don't have to construct the blue, or use any powers of recognition or imagination. No other reading of it is possible.

Well, look at the sky again, and this time pay closer attention to the actual instant-by-instant visual experience you're having. Slowly, as you force yourself to observe and not to take for granted what seems so familiar, everything becomes much more complicated, and much less determinate. A black spot drifts across the blue, and then a small, faint, tangled haze. You know that these are floaters – events on the surface of the eye, and you have automatically discounted them. But now even the blue field itself beyond these local disturbances begins to look far from simple. It's not smooth at all. There's something patchy about it. The blue is uneven. It seems to evade you as you search for it – to be always just out of reach among a surface which is mottled with unblue. That simple blueness that you imagined yourself seeing turns out to have been interpreted, like everything else, from the shifting, uncertain material on offer. Even with pure blue you have taken in something far from pure, and by no means blue. The pure blue you have dreamed.

One of the difficulties in understanding these conversations we have with our surroundings is that the subjects of them are so various. Some of them are unfamiliar, some nebulous. An event we hadn't expected to see suddenly occurs in front of us – an accident, a crime – and we find it very difficult to offer any coherent account of it. Stare for as long as we like at some scenes – the shifting whites and greys that surround you when the mist comes down in the mountains, for instance – and you still can't respond. Even the most familiar things seen from an unfamiliar viewpoint – the back of your hand, if you put your eye close to it – can appear so complex, so irregular, so devoid of graspable and nameable features, that you wonder how we can ever establish communication with anything.

But then most of the conversations we have with the world are not like this at all, any more than most of the conversations we have with other people are about infinity or indeterminacy. Most of the time we talk about familiar, well-worn subjects; we see a world of objects we have seen before. And even when we are seeing them for the first time, they have for the most part a reassuring graspability and a reassuring similarity to objects that we *have* seen before. So much so that we take it for granted. Things have sharp edges that differentiate them from their backgrounds,[7] and we see these edges as part of the world's fixtures and fittings, as clear and definite as national frontiers – lines on maps dividing pink land from yellow, turnpikes across roads where passports have to be shown, barbed-wire fences and free-fire zones separating freedom from tyranny. A moment's thought, of course, and we realise that all these things are merely the markers of something else – agreements established by negotiation or usage, which have meaning only in so far as they are enforced by continuous guarding and patrolling. They have significance only in so far as there are differences between what lies on either side. They are the limits up to which certain laws and languages and practices run.

The physicist is in no better a position than the rest of us. He has no exact knowledge of the qualities of any object, or of the initial conditions on which any projection of its state or behaviour is based. There is always some imprecision in measurement,[8] and in any case the measurements are not of the individual particles that compose the object, but of broad averages of their behaviour.[9]

I said that the world has no interest in whether we understand it or not. This is a first approximation. A lot of the objects in the world have been designed, by man or nature, to signal to us. The redness of the apple has been selected by nature because it makes the fruit conspicuous and fruit-like to birds, and further selected by growers and supermarkets because it makes the fruit conspicuous and fruit-like to us. A principal function of the design of a can of beans or a pack of margarine is precisely to enable us to identify it from the surrounding noise on the supermarket shelf as a can of beans or a pack of margarine. Even when the market is not so clamorous, manufacturers know how much we value recognisability. Why do all cars look so car-like – so much like each other? Partly because they all have four wheels to balance on, an engine to house, and certain equipment required by the car's function and the law. Partly, though, because manufacturers know that we wish our car not only to be a car, but to announce itself as a car.

In our urban environment, almost every single thing we can see, except the sky, has been designed or adapted by man, has been launched into the world already organised, already specified and named. As we withdraw ever deeper into our cocoon of artefacts, so the world becomes more and more pre-accommodated to perception. Now we are beginning to move to a deeper layer still, the cyberworld, a world of meta-artefacts, where everything is a representation of an artefact. Just as human beings learned to vary raw food with cooked, so, in the old age of the race, we are moving on from the cooked to the pre-digested.

We look around at our fellow men, who built this environment. They all have heads and a similar selection of limbs, shaped by the inherent repetitiveness of genetic reproduction (and differentiated by the marginal randomness built into it).

We go out into the countryside, into the natural world. The fields have been cleared and shaped and planted by man. They are recognisably fields because they were made to be fields. The woods are woods because the underbrush has been cleared – because they have been left standing *as woods* in contradistinction to the fields. We move on to the uncultivated high moors. They are moors because they have been stripped by the flocks men graze upon them. They are moors because they are not fields or woods, because they are not covered by tarmac, not scored out by passing feet, like the road or the path from which we are admiring them.

We hack our way into the depths of the jungle, where no human being has ever ventured. Even before we look around, we have one conceptual hold upon our surroundings: we are in a place where, unlike almost all the other places we are acquainted with, no human being has set his mark. We gaze at the dense curtain of unfamiliar vegetation around us. The leaves on this tree – they're all alike! There is a pattern! In fact they are recognisably similar to the leaves on *that* tree. They are even not totally unlike the leaves on the trees in Hyde Park. There is a good reason for this. They are alike for the same reason that cars and electric toasters are – because they have been produced to serve specific purposes. And, like cars and electric toasters, they have been shaped by standardised processes of mass production which are more or less the same from Detroit to Singapore, from Hyde Park to the Matto Grosso. We can talk, and think, about the world in standardised, and standardising, formulae, because we live in a world – a natural world as well as a man-made one – of standardised objects.

We leave the teeming warehouse of the jungle, and trek on into the middle of the desert. Not only has no man ever set his mark here, but no plant grows. The great leaf factories of the forests, the great blade-of-grass factories of the plains, have found no foothold in this market, or else have been driven out of it. What do we see? What can we say? Quite a lot, even before we have begun to organise this unfamiliar new world to our purposes. For a start we can see the familiar sky, and its familiar division from the earth. And when we come to look at the earth, certain regularities begin to emerge. The dust consists of rather even grains. It has been smoothed into level surfaces. We have seen level surfaces before. The surfaces have been everywhere cut into channels and valleys. We have seen channels and valleys. There are regularities because the landscape has been organised by certain standard forces. The weather has broken down the irregular rocks; the rivers have ground them to powder; the wind has smoothed the powder and heaped it into waves and dunes; the rain has cut pathways through it. We recognise these regularities quite easily, because we have seen them before, and we have seen them before because, wherever we have been, the same standard forces of nature have been at work.

Even out of sight of land, on all the oceans of the world, the same shifting winds – no two alike, but all of them winds – raise the same familiar patterns of crest and trough: no two waves alike, but all of them following the same determinate general structure. Sea and sky are separated into the same universal disk and bowl by the same universal gravity. Even the chaotic individualism of the clouds is sorted into general classifiability by the constants of wind and height and temperature. We look closer, and see drops of water which are one much like another, because they have been formed by the same forces acting on them, the same surface tension. We see the endlessly different shapes produced when the droplets freeze – but which remain identifiable as snowflakes because their differences are within a range produced by a certain set of variables. On the outermost fringe of what we might want to call things, we find particles cloned by the innermost processes of the world.

And when we look up into the clear night sky we are saved by the apparent constancy that reassured Frost. We are looking into chaos, a vague cloud of solidified droplets, arranged in patterns that have no discernible regularity. But the relationships within this cloud, however irregular, however incomprehensible, seem to remain unchanged night after night, year after year, century after century, millennium after millennium; and gradually our skill at metaphor, our ability to see shapes where no shape is offered, enable us to map even this final wilderness. We slowly learn to read into them similarities to the regular objects of the terrestrial world. A saucepan – a W – a rectangle collapsing to one side just *so*. Even if we get no further than the Plough, Cassiopeia, and Orion, their gratifying discernibility gives us a hold upon the rest of the universe. Lurking in that sea of other stars, we know, however undifferentiated to our eyes now, are shapes which one day, when we have a little more time, we will learn to see with edges just as sharp, just as clearly differentiated from the ocean of stars around.

There is in any case a circularity here. The only evidence we have for the existence of these general laws and rules is the classifiability of the particular cases that they must have generated. But then the laws and rules transcend the particular cases we are familiar with. They take on a life of their own, just as the possibility of identifying constellations does; we see them as prescribing by analogy patterns and regularities lurking within the denser parts of the jungle, where we can perceive none.

If all the world were stars, with no saucepans or nebulous mythical hunters to serve as models, we should find things much harder. If the stars visibly shifted in relation to each other all the time, like the water droplets in a cloud, we should have no relationship with the world at all. Even if our visual field were entirely filled by the back of our hand we should be in severe difficulties.

Not completely at a loss, though. Internal relationships would begin to emerge. Some of the shifting stars would be larger, some smaller; some brighter, some less bright. This part of our hand would be pinker and rougher; that part whiter and smoother. Our attention would fix upon a largest and brightest star. We should locate ourselves by a particular prominence in the flesh over a bone or a vein. And we should impose another set of defining relationships on the scene by our own presence. We should locate up and down, left and right, near and far. These relationships, too, would have an orienting universality, because we are universally present in our world. We bring them with us, impose them upon the world in front of us by the fact of our presence. Soon we should begin to feel some grateful attraction to the identifiability of that brightest star, that highlight in the flesh. We should have begun to locate better and worse.

*

One last whiff of the arbitrary lingers about all these otherwise sharply distinguished products of genetic reproduction and unvarying natural laws, even about the artefacts that we have so carefully crafted to fit our hand and to announce themselves to our eye. One last set of boundaries often remains curiously fuzzy – their boundaries in time.

What, for example, could have more gratifyingly distinct spatial frontiers than a car? What could be more specifically designed to fit our purposes and fantasies, to sell itself to us? What could be simpler for us to engage in practical and imaginative dialogue with? But now follow it through time, from its beginnings in vague discussions between designers and sales directors; as it finds some changing symbolic shape in sketches, plans, models, and production drawings; as it acquires a physical existence part by part on the assembly line; stands motionless but gleamingly desirable in the saleroom; achieves fulfilment in the daily trips to work and the family picnics; suffers damage and repair, sale and resale; declines into old banger; is reborn as handcrafted streeter; subsides into inert and cannibalised hulk; undergoes compression, meltdown, absorption into the fabric of other cars, into tin cans and bicycles. When did it start being a car in any normal sense of the word? Somewhere this side of the preliminary discussions, certainly. When did it cease? Somewhere before its transmutation into cans of baked beans.

When something important turns upon it we often make great efforts to be definite about these temporal boundaries – and almost as often definiteness eludes us. When does a human being begin and end? Important questions, on which all manner of ethical and practical decisions depend. A great many people are perfectly clear about when it begins – at the moment of conception. They are as clear about it as if they were watching a light being switched on – so clear about it that they are prepared to use it as a criterion for condemning others to prison and damnation.

Well, this is an identifiable stage in the human process, certainly, just as putting the first part of the chassis on to the assembly line is an identifiable stage in the process of the car. But the process was going on for a long time before that, as the spermatozoa closed in beyond recall upon the ovum, and earlier still as the ovum ripened and the hotel room was booked, as phone numbers were exchanged, as eyes met. Even when fertilisation occurs it is a complex process, extended in time like every other process. Once it is complete, though, the new human being has unambiguously begun. Though perhaps only to cease to be a human being again, and to be flushed away with a tampon in its first weeks of life even before its presence has registered upon the world, unmourned and indeed totally disregarded even by those most definite about the absoluteness of change from non-existence to existence.

There is a strange contrast here with the boundary at the further end of life, once our human being has emerged into the world, and taken on all the familiar spatial characteristics of the breed. The breathing stops, the line on the electroencephalograph becomes flat. The cooling construction of flesh is no longer a human being – certainly not for those who believe that some defining spiritual entity entered it at conception and departed it with its last breath. And yet we treat it, all of us, religious and non-religious alike, with more reverence than we did in life – certainly more than we have ever shown to discarded tampons. It still *looks* like a human being. It hasn't *quite* ceased to be a human being for us, even as we screw the lid of the coffin down, and despatch it into the flames or the cold earth. There remains just a last touch of horror for the faint shadow of murder we are committing.

How reassuringly solid and autonomous the world once was! God saw all. Saw it all in one go, continuously and eternally. And since God was everywhere, he saw it not in perspective, not from some particular viewpoint, but from every possible viewpoint. From all sides of a cube simultaneously, for example. From an angle of ninety degrees to each of those sides – from an angle of one degree, eighty-nine degrees, seventeen degrees. From a millimetre off and a mile off. From every point inside the cube looking out. He could see up your trouser-leg and down your trouser-leg. See your vest through your shirt and your chest through your vest.

But then human beings, too, had a considerable capacity to see things from several places at once, to judge from the pictures they painted. We were standing as close to people in the background as we were to people in the foreground. We could see inside buildings at the same time as we saw the outside of them, the end of the story alongside the beginning of it. Was it a gain or was it a loss when Brunelleschi and Alberti invented perspective, and reduced the visible world to what I and I alone can see of it from the particular place I am standing, at the particular period I am standing there?[10] In any case it must have cut two ways. Just as the painters found a way to adapt their representation of the world to correspond more closely with their actual experience, so the sight of a world represented in perspective must have modified our experience of it. Humanism humanised all things, and made man the measure of them; it also humanised the man who was its measure.

Science held out for longer than art. It survived the Renaissance with its panoptic view of the universe intact. All things, in classical science, were seen from everywhere and nowhere, just as God had always seen them, so that the world became a great palace into which the public was admitted only on sufferance, only to gaze at in respectful wonder from certain restricted positions. It was not until the first quarter of the twentieth century that perspective invaded science. First relativity and then quantum mechanics reduced all those loftily impersonal measurements, those authorless statements of fact, to personal observations, actual or potential, made by me, made precisely now, precisely here, or made by you, made precisely there, precisely then. Philosophers at the same time, for rather different reasons, began to analyse statements about the external world into statements about our experiences, and to see the world through a foreground consisting of our own corporeal envelope and the internal sensations it generated.

It's difficult not to think that most of these changes were for the better – that they gave us a truer picture of the universe, and of our position in it. But if we follow their logic to its conclusion, and try to reduce the world still further, so as to correspond more precisely with the reality of our traffic with it, then problems arise.

When we see the arcade in a Renaissance Annunciation dwindling away into the distance, even though we find ourselves restricted to one particular spot at one particular time, we are, at that spot at that time, presented with a scene which seems to have a solid and continuous reality throughout the period of our observation. When we travel aboard our imaginary spacecraft at something close to the speed of light, and look at the clock to compare the passage of time with the passage of time experienced by the people we have left behind us, we do imagine ourselves presented with a clock which continues to be a clock throughout our experiment. Even when we look through the microscope to examine the light scattered by a particle moving without fully determinate position and velocity, we do seem to have an apparently determinate microscope.

33

You seem no less definite, for that matter, no less solid and continuous, as we sit across the table from each other talking. When I stop to think about it, though, in terms of what I actually experience, I realise what a very erratic and flickering creature you are, a mere will o' the wisp. For a start you're a half person, because whether you have any lower half behind the table-top to go with the top half I'm looking at I have no idea. Only quarter of a person, in fact, because the back of you is missing as well. I glance out of the window at the weather as you speak. The weather's fine, but *you* aren't – you've become a voice without a body. I look back – you reappear, and the world is weatherless. But as you drone on about phenomenalism I cease to listen because I'm thinking about the prospect of dinner. Your mouth opens and closes but you've become dumb, a silent movie. I politely keep my gaze fixed on you, but so vivid does the prospect of dinner become that now it's not you that I have in front of my mind's eye but a plate of risotto, each grain separate, swollen and gleaming with rich juices . . . So where are you? Dead and gone to heaven, for all I know.

I make a great effort, and refocus my attention on you. You're resurrected, as good as new – until I blink, and you've gone again!

Wherever we turn, all the partners in our great dialogue with the world sputter into life and gutter out again like this at every instant. And when you think about it more closely still you realise that the world around us has almost *no* continuous existence at all. What the human eye sees at any one moment, sees clearly enough to distinguish (as we noted earlier), is the narrow slice of the scene in front of us which sends rays to the fovea. Even when I'm looking at you most intently across the table only a few square inches of you ever come into existence. I look you in the eye, and you're a quadruple amputee, a trunkless head. You're mouthless, earless, noseless. You're one-eyed.

And yet you're not. However much I blink, however much my gaze or my mind wanders, the life in you goes on burning as strongly as ever. I've stored you away in my memory, accumulated you bit by bit. In fact I think I've gone further. I've formed some idea of you inside my imagination, and this idea of you has, in some way very difficult to describe, two eyes and even two feet, a back as well as a front, a past and future as well as a present, a character, an identity. An idea that will persist even as I see you to the door, and watch your figure dwindling down the lines of my perspective until it becomes a dot, and vanishes over the horizon. An idea that will remain even after you are in another hemisphere, even after you are dead. I shall be able to think about you and talk about you in your absence in much the same way as I could see you and think about you in your presence. Your availability to me is something continuous. It has extended seamlessly from the expectations I had before I first met you, through the period of your intermittent presence, and on to your absence again. I am not, in practice, affected by metaphysical anxiety that the world ceases to exist as soon as I look away. Perception and memory have imperceptibly merged. Like an accomplished hostess we have kept the conversation going with a roomful of the most uncommunicative guests.

Do I mean, then, by my awkward locution about your 'availability to me', what I suppose a phenomenalist would mean – that I *could* see you if I opened my eyes, or if I travelled to wherever you now are? Not at all. Things we can't see, people who are absent, are not merely conditional objects of our perception. They are independent of us; they have an asserted life of their own. They are like children who have grown up and passed outside our jurisdiction and control. If I imagine you, then yes, I remember you as I saw you when you were here, I visualise you as you would be if you were here now, I imagine you as you are now; I see you from a selection of the single viewpoints I have or should have. But I don't have to imagine you to talk about you, or to keep you in mind. The way I think about you when you're away is a bit like the way a physicist thinks about particles. He *can't* have in mind what he would see if he looked, and he can't imagine it, because the particle has no definite state – and in any case would change its state if it *were* looked at. That doesn't stop him talking and thinking about the particle, though.

There is something rather familiar about this version I have of you, which is not conceived from one single particular viewpoint at one single particular time, not even really from a combination of different viewpoints, but which is an idea allowing me to see you from close to and from far off, from the back as well as the front, from the inside, looking at me, as well as from the outside. It is not a portrait of you. It is an icon of you.

It's like the icon for a program on a computer, which can be maximised and worked on, then minimised and kept at the edge of the screen, hidden altogether, and recovered. In form an icon is like a common noun, which signifies any member of a whole class of objects, while a portrait is like a proper noun, which refers to only one specific thing or person. But a common noun may of course be used with a qualifier which limits it to naming only one member of the class that the noun itself designates, and which in effect turns it into a proper expression. There are many computers in the world, for example, but only one which is named by the phrase 'My Computer' in the lists of files at the corner of my screen. The phrase is attached to an image, a generalised picture of a computer which in itself is to be read as representing my computer, your computer, or any other computer in the world, but which in this context, with 'My' in front of it, in the corner of *my* screen, is used in a proper sense. It refers to the one machine in my possession just as surely as the phrase beneath it does. I should understand the icon, in this context, in exactly the same way even if I couldn't read English, just as the generalised picture of a man's face in a Byzantine church might well be understood to represent a particular saint – St Zenobius, say, if it said 'St Zenobius' underneath it, or if it hung in the church of St Zenobius, or if there was a tradition to that effect.

Exactly the same picture, hung in the church of St Zephaniah, would be understood to portray St Zephaniah. The same words, 'My Computer', and the same icon, when they appear on *your* screen, likewise name a different machine – yours, not mine. You might personalise the icon by replacing Microsoft's generic image with a photograph of your very own computer, showing the snaps of your family that you keep stuck round the screen.

And perhaps the representation of me that you work with when you talk or think about me, and the representation of you that I work with, are rather like what you've now just created on your computer – iconic functions, with personal portraits stuck on the front of them to limit them to a particular reference.

The purely iconic aspects of this symbol, like the purely iconic aspects of the symbol on your computer, could probably be replaced without much loss by words, and by common nouns at that. The hidden legs and the unseen back of your head, for instance, I probably don't particularise – they are a standard pair of legs off the shelf, a back of the head out of stock, unless you have most strikingly idiosyncratic fittings in these departments. Even the portrait of you on the front begins to look pretty small and generalised by the time you've been in Australia for a year, and I've first minimised it, then shifted it off the bottom of the screen to some menu I rarely open. (Though when I do maximise it, back it comes at once in all its particularity.)

If I try to portray to you someone you don't know (in a biography or a novel, for example) I can't rely on a proper noun, a name, to do the job. I have to resort to a series of common nouns. And once again I find myself assembling something from stock parts. I try to get beyond the baldly iconic outlines, of course – the two eyes, the nose and the mouth that differentiate my subject from a frozen pizza but not from anyone else on earth – by specifying particular sub-classes of feature. I order up not just a face from the face drawer, but a brooding, shrewd face, from the brooding, shrewd section of the drawer. In it I put not just a couple of standard eyes, but wistful brown eyes; not just any old nose and mouth, but a twisted nose and a mouth like a sour raspberry. The storeman might have found himself despatching the wistful brown eyes in another face altogether, a small heart-shaped one, and packaging the twisted nose with a sensuous mouth, in the middle of a face like a pickled walnut. The sour-raspberry mouth would have remained in stock for another occasion – perhaps never to be brought out at all before the world ends.

Not that any of this stops you reading the result in some way that I never intended. I describe to you the Chairman of the West Midlands Regional Hospital Board, and you at once identify him as Jesus Christ, as the Devil, as yourself. Or you see him as someone completely generalised – suffering humanity at large. But then if I copied the portrait of your computer with which you've replaced the 'My Computer' icon on your machine, and used it to serve the same purpose on *my* machine, I should see it now as signifying my computer, or all computers in general, however many snaps of your family were stuck around the screen.

So the world around us is irregular and confused. Its most enduring and solid features turn out to be transient and deliquescent to the touch. Its fabric is a series of events, fleshed out in our minds from an even sketchier set of events – the highly restricted and fleeting contacts that we have with it. Understanding this is where any inquiry into the nature of things has to begin.

But not to end. Because none of this confusion and mutability is arbitrary. We know that, underlying everything we see, the source of both its regularities and irregularities, are certain principles. These principles are expressed in intangible and invisible entities with a very different character from the objects of the visible world: scientific laws, together with the conceptual framework on which those laws are constructed – the procedures of mathematics and logic, and the forms of space and time. This is what we need to look at, in order to find things that we can be confident are the way they are quite independently of our presence – things with which our relationship is a little more modest, a little more adaptable to some kind of reasonable synthesis. Not at the fabric of the material world itself, but at the principles underlying it.

First, obviously, the scientific laws – the laws of nature – themselves.

2

The Laws of Nature . . .

. . . And the nature of laws

Scientific laws must be on a different logical level from the tangible world, and a logically superior one. They *govern* that world; and they govern it successfully – more successfully than any other laws have ever governed anything. They are *obeyed* by it – and obeyed absolutely. They are not part of the traffic – they are the codes that make the traffic possible. Unlike anything in the realm that they order so well, they are themselves changeless. And the more thoroughly scientists come to understand them (as many books now struggle to make clear in terms that the lay reader can understand), the more universal, elegant, and rational they seem. Scientific laws are the lurking spirit of the French garden that we might have expected the universe to resemble.

Or are even these transcendent entities not quite what they seem?

We take them so much for granted now that we forget how late in the history of the universe they first appeared – how late even in the history of mankind. It was only late in the seventeenth century (more than a hundred years after the discovery of the Americas) that the first of them were set forth: Newton's laws of motion and gravitation.[1] Like the Americas, they had surely been there long before they were found. Since the beginning of the universe, in fact, waiting only to be put into words and figures, like seeds waiting for the spring. It was the laws that had shaped the universe, long before the tiny disturbance of its fabric caused by the emergence of life and consciousness, and they would continue, immutable and eternal, until long after that disturbance had passed away.

And yet it was only life and consciousness that had revealed them.

Slowly but irresistibly, over the course of the next quarter of a millennium, Newton's laws and the laws that were derived from them, or modelled on them, replaced divine revelation as the foundations upon which our understanding of the universe was based. They assumed a picture of things that was too obvious ever to be remarked upon, of a world in which you could say precisely where everything was and when events happened, and precisely how each thing would be affected by each event.[2] The reassuring solidity, certainty, and universality of the religious universe were preserved under the new dispensation.

And then, just over two hundred years later, rather as doubts had first surfaced about divine revelation, complications began to appear, as complications sooner or later always do in even the most perfect schemes.

The first sign of trouble was an anomaly that seemed very abstract, and peripheral enough at first to be swept under the carpet (even by some of the physicists who first discovered it). The problem was this: Newton's laws require a smooth continuum, which is what our world seemed to be. In the early years of the twentieth century, though, it became apparent that heat energy, at any rate, was not like this. Heat, it turned out, is composed of indivisible minimum packets, rather in the way that every currency has a smallest coin. No cash transaction can be conducted in any units smaller than farthings or cents, or whatever, so every sale and purchase has to be in whole-number multiples of this minimum value. The problem spread from heat energy to light, since heat and light radiate in the same way, and from light to matter, because atoms change their state when energy is taken in or given out, and if the energy comes in farthings, then the atoms change in farthing jumps.[3] The physicists who thought that this was merely a passing procedural difficulty, and that a way would eventually be found to circumvent it, were soon to be disappointed. When the transactions were between very small entities – atoms and the particles that compose them – the implications of this limitation turned out to change everything.

There was another problem emerging at the same time. The failure of classical physics to describe the sub-atomic world was discovered by the classical method – observation. Now the act of observation itself came into question.

It had always been assumed to be external to the universe that was being observed. The particular standpoint from which the experimenter observed it was of no significance. The universe was seen not from here or from there, but panoptically, in the way that God, who is present everywhere simultaneously, had presumably always seen it; the scientists had simply inherited his magisterial indifference to time and place. Einstein realised that these observations and measurements are not external to the natural world. They are events within it. They have to be made by an observer – and they have to be made from that observer's particular viewpoint. Any observer, even a purely notional one in situations where no real observer could imaginably be, has to be situated at some particular place, and the measurements he makes depend entirely on his position and on the way in which he is moving (or not moving) in relation to what he is looking at. All time, Einstein realised, is local – and so, therefore, are all other measurements.

And then, as quantum theory developed, a rather similar situation appeared. It seemed that here, too, you needed a possible observer to make the mechanics interpretable. The behaviour of physical objects could never, even in theory, be completely specified. The limitation was insignificant in the case of everyday objects moving at everyday speeds, but became critical in the case of fast-moving particles. A choice always had to be made about the relative accuracy with which different aspects of an object's behaviour were to be established – and choices make themselves no more than measurements do.[4] So objectivity is gone, and also any possibility of precise prediction, because if we can't know both how fast something is moving and where it is there is no way of knowing when and where it's going to be.[5]

In fact the implications of quantum mechanics went even further: the very nature of the physical world, at the sub-atomic level, seemed to be dependent upon the act of observation. An unobserved particle turned out to be able to follow more than one different trajectory simultaneously, and its behaviour could be accommodated only by interpreting it not as a particle, a thing, but as a wave function, a mathematical abstraction representing the relative likelihood of its being in any particular place at any particular time, where its ambiguity is given expression in purely mathematical form, rather as some numbers (such as the square root of a negative number) can be understood to have two definite but different values simultaneously. As soon as an observation is made of the particle to resolve the ambiguity, however – which can be done only by intercepting it in some kind of way, and therefore disturbing it – the ambiguities allowed by its mathematical expression vanish, and it becomes an object in a world that is subject to the laws of classical mechanics.[6]

Classical objectivity continues to work in practice most of the time because most of the time we are concerned with a relatively stable, relatively slow-moving world.7 Throughout the sub-atomic world, though, a completely new mechanics had to be constructed to replace the Newtonian system. This mechanics was based not upon absolute connections between individual causes and individual effects, but upon averages and probabilities. The laws that had seemed self-evidently absolute had proved to be approximations, and what had seemed self-evidently universal didn't hold if you looked into matter closely enough.

The effects reached out far beyond the practical. They destroyed the whole picture of a clear, definite, and objective world. All relation of one object to another turns out to be from a specified viewpoint, and any expression of it is dependent upon the use we want to make of it. No description of the world is possible without the theoretical presence of an observer.[8]

If this is true then its implications become more and more paradoxical. It means that the huge events at the beginning and end of time which are described by cosmologists, and the tiny ambiguous events of quantum electrodynamics, all of them remote from any possible human eye, are events only in so far as they are nominated as events, only in so far as they lie at the end of chains of reasoning that lead back to a human mind. Not only did the supposedly immutable laws need amendment, but it seemed as if the presence and participation, even if only notionally, of organisms which are part of the world that those laws supposedly have brought into being, and which have arrived only in the last few moments of cosmological time, were essential to make them work.

So where does all this leave the status of scientific laws? Are they the independent entities that they were once assumed to be? When they are called into question, is it simply because of changes in our understanding and expression of them? Or are they human artefacts that have no real existence outside our statement of them?

A non-scientist (like myself) is a fool to trespass in this great palace of thought, which is surely among the most glorious of man's creations. Modern science, particularly physics, is a structure of the most amazing complexity and elusiveness. Venturing anywhere near it without mathematics, for a start, is like trying to fly without wings. The only way I can begin to approach it is through the supposedly 'accessible' books that some scientists write for laymen, and I can't honestly claim to understand more than a fraction even of these. There are many warnings from scientists of the futility of the enterprise. Even Richard Feynman, the American physicist who devoted so much time and ingenuity to explaining physics with amazing informality and clarity both to beginners in the subject and to those who know nothing at all about it, is very sceptical about how much can be achieved. 'All the intellectual arguments that you can make will not communicate to deaf ears what the experience of music really is. In the same way all the intellectual arguments in the world will not convey an understanding of nature to those of "the other culture".'[9] And he insists that, however simply he attempts to translate physical laws into plain language, he can't really make you understand the beauty and force of them if you can't understand their mathematical expression.[10]

Fools though we non-scientists may be to try to understand the thinking that has shaped the world we live in, we're even bigger fools if we don't. We can at any rate begin by looking at what scientists themselves say – at least some of those who have expressed their views (in so far as they can) for the general public. The result might give us a little encouragement, because it turns out that they are deeply divided about what science and its laws are. In the course of the last hundred years, since the beginnings of relativity and quantum theory, some of the scientists most closely involved, and some of the most observant philosophers of science, have taken the view that the laws of nature were:

> invented by man (Einstein, Bohr, Popper);
> not invented by man (Planck);
> expressions of a real underlying order in the world (Einstein);
> working models justified only by their utility
> (von Neumann, Feynman);
> potentially deterministic (Einstein);
> inherently probabilistic (Heisenberg, Prigogine);
> a dialogue between man and the world (Prigogine);
> a dialogue between the possible and the actual (Medawar);
> steps on the road towards complete understanding
> (Feynman, Deutsch);
> steps on a road that has no end (Born, Popper, Kuhn);
> forced upon us by the world (Planck);
> forced by us upon the world (Popper);

potentially all-embracing (Feynman, Deutsch);
inherently piecemeal (Cartwright);
likely in the end to be not only comprehensive but simple
   (Feynman);
accounting for less the simpler they are (Cartwright).

My categorisation of these differing views is very broad and gener-alised. (There is a fuller account of what these various authorities actually said, which is usually more nuanced, sometimes more self-contradictory, and always more interesting, in the notes for this chap-ter.[11]) It makes reasonably clear, though, that the one phenomenon for which scientists have most notably failed to provide any generally agreed account is science itself.

Most of them seem to concur, more or less, about one thing: that the laws are some kind of construct arrived at through the interaction of the physical world and the scientists who observe it. Curiously, though, one of the points that divides them most sharply is precisely the role of the observer in quantum theory. The majority of them are prepared to accept the probabilism of the Copenhagen Interpreta-tion,[12] which after all can be given well-established mathematical expression. What sticks in the throats of so many of them is the 'anthropocentricity' of the interpretation – its insistence that what we can observe is necessarily modified by the act of observation, and that the world cannot be understood without the idea of an observer.

Each of the dissenters in his turn kicks the importunate intruder unceremoniously down the steps. Feynman, for example, briskly attributes the ability of 'some people' to imagine that the centre of the universe is man to the unscientific limitation of their horizons.[13] Popper, who says he wishes to preserve 'the objective character of physics',[14] finds it likely 'that the world would be just as indeterminis-tic as it is even if there were no observing subjects to experiment with it, and to interfere with it'.[15] Murray Gell-Mann (the physicist who helped instigate another revolution in elementary particle theory as co-discoverer of the quark) is very definite:

When first formulated by its discoverers, quantum mechanics was often presented in a curiously restrictive and anthropocentric fash-ion, and it is frequently so presented to this day . . . This original interpretation . . . is not wrong, but it is applicable only to the situ-ations it was developed to describe . . . The universe presumably couldn't care less whether human beings have evolved on some obscure planet to study its history; it goes on obeying the quantum-mechanical laws of physics irrespective of observation by physicists.[16]

Prigogine, likewise, insists on 'the need to eliminate the subjective element associated with the observer'.[17] To his way of thinking, he says,

Through his measurements the observer no longer plays some extravagant role in the evolution of nature – at least no more so than in classical physics. We all transform information received from the outside world into actions on a human scale, but we are far from being the demiurge, as postulated by quantum physics, who would be responsible for the transition from nature's potentiality to actuality. In this sense, our approach restores sanity. It eliminates the anthropocentric features implicit in the traditional formulation of quantum theory.[18]

But, having thrown the wretched observer out of the front door, most of the dissenters seem obliged to let him slip in again round the back to do the lawmaking itself. Feynman explains that physicists are always having to change the laws because they have been 'guessed' or are 'extrapolations'.[19] Well, whatever nature does or doesn't do, it certainly doesn't guess at its own constitution or extrapolate itself. That's a job that can only be carried out by the guesser or extrapolator, the human observer who has been discreetly rehired for the purpose. 'Every theoretical physicist who is any good', says Feynman, 'knows six or seven different theoretical representations for exactly the same physics. He knows that they are all equivalent, and that nobody is ever going to be able to decide which one is right at that level, but he keeps them in his head, hoping that they will give him different ideas for guessing.'[20] If this is so then we can't do without the theoretical physicist in question, because nature doesn't have six or seven different theoretical representations of itself, or even one. Whatever 'the same physics' is, it remains unrepresented until someone invents a way or ways of representing it.

If scientific laws are 'dialogues with nature', as Prigogine holds, who or what must be maintaining the other side of the dialogue? It can surely only be some member of the human race – once again, presumably, our illegal immigrant from the local theoretical physics department. The laws and theories of Popper's objective physics have an entirely subjective basis. He will 'admit a system as empirical or scientific only if it is capable of being tested by experience . . . it must be possible for an empirical scientific system to be refuted by experience.'[21] What experience is this, if it is not that of our unacknowledged observer? Indeed, all of us are roped in to service. '*We*' are the arbiters of what constitutes a valid law, and our judgement is final:

> Every test of a theory, whether resulting in its corroboration or falsification, must stop at some basic statement or other [i.e., one about a particular observed state of affairs] which we *decide to accept*.[22]

> From a logical point of view, the testing of a theory depends upon basic statements whose acceptance or rejection, in its turn, depends upon our *decisions*. Thus it is *decisions* that settle the fate of theories.[23]

The result of this Popper expresses in a brilliant sustained metaphor that sorts oddly with his defence of 'objective physics':[24]

> The empirical basis of objective science has thus nothing 'absolute' about it. Science does not rest upon a solid bedrock. The bold structure of its theories rises, as it were, above a swamp. It is like a building erected on piles. The piles are driven down from above into the swamp, but not down to any natural or 'given' base; and if we stop driving the piles deeper, it is not because we have reached firm ground. We simply stop when we are satisfied that the piles are firm enough to carry the structure, at least for the time being.[25]

Even though the universe, in Gell-Mann's view, couldn't care less about the existence of us human beings, Gell-Mann himself sees no objection to treating quantum superpositions as alternative 'histories' or 'narratives':

> The quantum state of the universe is like a book that contains the answers to an infinite variety of questions . . . The questions always relate ultimately to alternative histories of the universe. (By 'history' we do not mean to emphasise the past at the expense of the future; nor do we refer mainly to written records as in human history. A history is merely a narrative of a time sequence of events – past, present, or future.) . . . Completely fine-grained histories of the universe are histories that give as complete a description as possible of the entire universe at every moment of time.[26]

So the universe is writing histories now? It's telling stories? The shameful truth about the universe, though, is that it's illiterate. It couldn't write so much as 'wish you were here' on the back of a picture postcard, let alone a history. It can't even open its mouth to speak. It couldn't tell a 'doctor, doctor' joke, never mind a narrative about all the bosons and fermions in the world. If any histories are getting written or stories told, we can be sure that the author must be the same overworked jack-of-all-trades who makes the measurements and maintains the dialogues and does all the representing and guessing and extrapolating and deciding – the human observer in whose existence the ungrateful universe takes so little interest.

You want to say, 'But there are histories out there waiting to be recorded! There are jokes waiting to be told!' No, there aren't. There is the *material* for jokes, once you've decided what's funny. There is the *evidence* from which histories can be written, once you've decided what's relevant to a particular interest. Until then there's just a great undifferentiated, overlapping tangle, without sense or even sequence, waiting for someone to discover a few loose ends and pull out a few usable threads, then to weave them together into a usable fabric.

Feynman's version of quantum theory also involves the notions of the 'summing over' of 'all histories' of a quantum event. The physicist John C. Taylor agrees that giving a proper definition of this is a difficult mathematical problem:

> What Feynman did was to replace continuous space and time by a fine mesh of a large but finite number of points at a large but finite number of times. Then the notion of 'all histories' is perfectly clear: it just means hopping from point to point at successive times in all possible ways. This procedure gives some approximation to what is required. It is assumed that this approximation can be made better and better by making the mesh of points and times finer and finer (that is, having more and more points and times).[27]

But there is no 'fine mesh of points' out there – certainly not a finite number of them – until somebody says there is. Elsewhere, talking about the problem of 'summing over' in the case of space-times, which can also be varied continuously, Taylor asks: 'How do we enumerate them all? What would one mean by enumerating all possible shapes of a cup. What do we mean by *sum* in such circumstances? In practice, we must approximate the sum by a discrete set of "representative" space-times.'[28]Once again, though, there is no approximation and no improving of it without an approximator and an improver, no selecting of a representative set of anything without a selector.

Schemes have been proposed for avoiding the staffing problem, and for getting the physics supported by its own bootstraps. Gell-Mann gives the credit for pioneering what he calls the modern approach to quantum mechanics to Hugh Everett III, who developed an alternative to the Copenhagen Interpretation that explains superposition in terms of a multiplicity of parallel worlds. Or at least, this is how Gell-Mann says that Everett's interpretation is often described. Gell-Mann himself, however, insists that there is no need to become 'queasy' trying to conceive of this, because he believes that what Everett really means is not 'many worlds' but 'many alternative histories of the universe'[29] – once again we have had to send for the historians. Gell-Mann's queasiness at the idea of many parallel worlds or universes, all equally real, is evidently not shared by David Deutsch, another follower of Everett, who has produced an extreme version of the theory, one that gets round the difficulties of superposition by postulating trillions and trillions of co-existent universes in which every possible version of events has actually occurred, is occurring now, and will continue to occur.[30]

Part of the difficulty in giving any consistent and generally agreed account of science is that, as the philosopher of science Thomas Kuhn makes clear, it is not a single kind of activity. Indeed it is surely more heterogeneous than even Kuhn implies – not just a single discipline alternating between stasis and revolution, but a whole family of loosely related disciplines, as various in their procedures and strategies as the activities of politics or art. Not all science is concerned with the establishment, confirmation, or rewriting of underlying laws, or with the construction of models and explanations. A great deal of the work in some – in geology and botany, for example (even in particle theory) – is taxonomic. Geologists go on looking at the rock strata in new sites, botanists go on looking for new species of orchid, not necessarily to test a theory, but simply to explore the possibilities of the world, to extend the range of their material. Some of the greatest psychologists and neuropsychologists (William James, Aleksandr Luria, Oliver Sacks) practise what Luria calls 'romantic science', directed not at the reduction of complexity to simpler underlying forms but at grasping the richness, idiosyncrasy, and subjectivity of experience. At least one of the recognised cornerstones of science, heliocentricity, doesn't relate to a class of phenomena at all (or didn't when it was first introduced, before other stars were observed more closely), but to one single phenomenon, our own Sun and its planetary system.

There are plainly quite different goals being pursued even within the same branch of the same discipline. Everyone knows the story of the gold-rush scramble to identify the three-dimensional shape that threw up the baffling patterns caused by the diffraction of X-rays in DNA crystals. This von Neumann would have recognised as fundamentally a search for a model (even though the goal seems to have been not the model itself but the glory that finding it would bring). But I once asked Max Perutz how he had solved a similar problem – the structure of haemoglobin, one of the first two proteins to yield – and the explanation he gave was the simplest, most down-to-earth, and (I suspect) the most universal of all. He had no theory that he was testing, he said, no model he was looking for; he simply wanted to find out what was the case.

The majestic simplicity suggested by the concept of 'the laws of science', like the majestic simplicity of a distant mountain range, breaks up into a more complex landscape when you approach more closely. Some laws are supposed (even if not by Popper) to be inductive generalisations from the observation of particular cases, and of these, some are supposed to determine individual cases, some to indicate only probabilities. Other laws are not inductive at all, but purely deductive. The inverse square law of the propagation of light, for instance, is a geometrical theorem; given that light travels in straight lines outward from its source in every direction, its intensity at every point follows diagramatically. (There is also an implicit contingent assertion that the diagram does represent the behaviour of light in the physical world, but if our practical experiments to test the law began to find contrary instances – because of the gravitational bending of light rays, for instance – the geometrical model would nonetheless remain intact.)

Lumped in with the laws are various 'theories' and 'principles', such as relativity, uncertainty, and natural selection, which are not really law-like in character at all, but conceptual clarifications of an almost philosophical generality, and difficult to categorise. Is the principle of natural selection (one of the most richly explanatory concepts ever introduced into human thought) deductive or inductive, necessary or contingent? Darwin derived it from observation, without any understanding of the genetic mechanism involved. But couldn't it have been derived by a modern geneticist without any voyage to the Galapagos Islands, purely from a consideration of the concepts of reproduction, random mutation, and competition for resources? Relativity was experimentally corroborated by the success of its predictions about the precession in the perihelion of Mercury and the gravitational bending of light. But it was derived by Einstein from first principles – and would have remained unassailable even if the predictions had not been fulfilled; what would have been open to challenge would have been the applicability of the principle to those particular situations. The uncertainty principle, conversely, was first put forward to help explain the anomalies in experimental results; but, like relativity, it derives from first principles, and could have been hit upon before it was necessary to explain anything, simply from an abstract analysis of wave/particle duality.

In amongst these heterogeneous entities is what seems to be a sub-class usually referred to by scientists and other writers on the subject as the '*fundamental* laws of physics', or the 'laws of nature'. The terms suggest a kind of scientific gold standard – a set of unchanging, unchallengeable commandments, graven on tablets of stone, from which all else is derived and against which all else is measured.

Which precisely, though, are the laws that qualify as 'fundamental'? The writers who so often refer to them rarely trouble to specify, and when they do their lists vary widely. If one tried to piece together a catalogue from the references it would seem to include the various conservation laws,[31] the laws of motion and gravitation, the laws of thermodynamics, and perhaps the principle of least action. A modern scientist would probably include the concepts of symmetry and invariance.

This is once again a very mixed bag. It's no clearer than with the general list quite what their logical status is: whether they are all generalisations of what has been observed, or whether some of them, at any rate, are more like rational principles around which our observations can be organised; or whether they are perhaps some combination of the two. Symmetry and invariance (never in my experience clearly differentiated in the literature) are really not laws but meta-laws – declarations that the laws to which they relate apply uniformly, without regard to any particular frame of reference, change of orientation, or other transformation.[32]

Are these 'laws of nature', first of all, really quite as fundamental as the term suggests? They can't *all* be fundamental, or at any rate not equally so, because some of them can be derived from each other. The principle of least action is said to be derivable from the laws of motion,[33] and two of the conservation laws (of energy and momentum) from Einstein's formulation of the law of gravitation.[34] The first law of thermodynamics *is* one of the conservation laws. Are the laws as a group to be regarded holistically, as supporting each other like the axioms of geometry, where more or less *any* theorem, once accepted, can be regarded as the source from which the rest of the system can be generated?

They have also changed (like the other laws), both in number and character, as physics has developed. Gravitation, presented by Newton as an unexplained remote force, was reinterpreted by Einstein as a geometrical concept. The law of conservation of mass became otiose when mass was subsumed under energy. Some forms of symmetry and invariance have turned out to be broken by the so-called 'weak' (short-range) force involved in radioactive decay.[35] The weak force, once an idiosyncratic outsider with its own code of laws, has been aligned with electromagnetism under the common umbrella of electroweak theory. One day, it is hoped, a Grand Unified Theory will be constructed which finds the common ground between electroweak forces and the so-called 'strong' forces that hold quarks together.

Changes like these to the codex are not superficial – they are themselves fundamental, because one of the goals to which science is directed is the discovery of the underlying patterns beneath apparently diverse phenomena. 'Few, if any, scientific theories are final,' warns Taylor; 'each one is destined to be subsumed into some more complete theory.' (This mutability, as he says, makes it dangerous to draw philosophical conclusions from science – 'grand deductions may be premature.') Does this suggest that the 'fundamental laws', even if not in a self-evidently fundamental form at the moment, are becoming ever more fundamental – perhaps approaching asymptotically ever closer to some irreducible perfection, even if never finally reaching it?

Scientific laws seem to obey a kind of conservation law of their own, however – the conservation of complexity. Reducing their multiplicity usually comes at the expense of making each of them more elaborate in itself. Einstein's reinterpretation of gravity as curvature of space-time may make it theoretically simpler than Newton's unexplained force, but it also makes it much harder to understand in non-technical terms. Energy has changed from a quality familiar to all of us in our everyday lives into an abstract one which is uninterpretable in any ordinary non-mathematical sense.[36] The catalogue of elementary particles, already as long and complex as an ironmonger's stocklist, will be extended still further in one of the proposed Grand Unified Theories by the introduction of a dozen extra varieties. String theory, which proposes a common underlying structure for all these multifarious entities, both established and conjectured, will do it only at the cost of an extra seven (or in some versions eight) spatial dimensions. Grand deductions may pass straight from being premature, as Taylor warns, to being too ponderous ever to be deliverable.

The status and origin of even the earliest of the fundamental laws, Newton's laws of motion, were cloudy from the very beginning. In the 'General Scholium', the remarks that conclude the *Principia*, Newton says that in 'experimental philosophy', as he calls science, 'propositions are deduced from the phenomena and are made general by induction' (precisely the account of scientific laws rejected by Popper). But is this how he arrived at the laws of motion? They are often said to be generalisations of the laws of planetary motion established by Kepler. For Newton's Third Law ('To any action there is always an opposite and equal reaction; in other words, the actions of two bodies upon each other are always equal and always opposite in direction'[37]) he gives some credit to Kepler, who certainly derived his laws from the 'phenomena' – the astronomical observations laboriously collected by his predecessor Tycho Brahe. For the first two, however ('Every body perseveres in its state of being at rest of moving uniformly straight forward, except in so far as it is compelled to change its state by forces impressed,' and 'A change in motion is proportional to the motive force impressed and takes place along the straight line in which that force is impressed') he offers no derivation. Each of the laws, as it is introduced in the *Principia*, is accompanied by a few homely instances of it – what happens when you throw a ball, spin a top, or press your thumb against a stone. Are these supposed to be the phenomena from which the laws are deduced, or are they simply practical illustrations to make the abstract principles more readily comprehensible? Much later in the *Principia* Newton describes various practical experiments he has conducted to test air and water resistance, but first he works out the consequences of his laws by pure geometry. As in geometry, everything is hypothetical. 'If . . . ', 'Suppose that . . . ', 'Let such-and-such be the case . . .' And in Book 3 he says that up to this point, 'I have presented principles of philosophy [i.e., natural philosophy – science] that are not, however, philosophical but strictly mathematical.'

So what is the basis for the first two laws? It is now accepted[38] that they are derived not from Kepler but from the first two of the three 'laws of nature' put forward forty years earlier by Descartes in his own *Principia* ('Every body, so far as it is altogether unaffected by extraneous causes, always perseveres in the same state of motion or of rest,' and 'Simple or elementary motion is always in a straight line.'[39]) What Newton's laws have in common with their Cartesian predecessors is not just their form and style but the novelty of their content. Before Descartes all motion, at any rate of heavenly bodies, had been regarded as inherently curvilinear. This was not unreasonable. All bodies freely moving in space travel in curves of one sort or another, because (as we now think of it, since Newton) they are necessarily deflected by the gravitational attraction of another body or bodies. Universal curvilinear motion became unreasonable in earlier astronomy only because it was carried to ideological extremes, so that, purely for aesthetic and metaphysical reasons, the orbits of planets had to be analysed into *perfect* circles, or combinations of them, which turned out to be impossible. Once Kepler had discovered that orbits were elliptical, the unreasonableness vanished.

What Descartes did, and what Newton seems to have adopted, was to analyse curvilinear movement into two (or more) separate rectilinear movements. Descartes, at any rate, certainly hit upon this scheme by the power of thought alone. The justification for it is the conceptual and pragmatic advantage it offers, because it makes it possible to construct an explanation for the tendency to move in curves, which had seemed to be simply an inherent primary quality of the world, and to make it an emergent feature with quantifiable origins. This is not, however, a discovery in the sense that the finding of Tristan da Cunha or the planet Neptune was a discovery. It is an invention. The straight line in which it is now said that a body would move if it were not acted on by other forces is not an actual constituent of the physical world. It is a fiction, just as Einstein said, an intellectual construct, the notional product of a definition, and it was borrowed from Euclid, not the world of phenomena. It is as artificial an idealisation as the perfect circles into which the earlier astronomers were trying to reduce the more complex curves of the real world. It just happens to be a suggestive and calculable one in a way that they were not.

So *how* did Descartes think his way to this model? *How* did Newton think his way to improving it (or reinventing it)? If they had never observed something in the real world – a ball being thrown, or a top spun – what could either of them have exercised their powers of pure conceptualisation *upon*?[40]

The conservation laws are also of mixed parentage, and also often difficult to characterise precisely. The first of them (of matter) was derived by Lavoisier from exact measurements of the ingredients going into chemical reactions and of the products emerging from them. This seems also to have been the source from which Carnot derived the conservation of energy, because he wrongly assumed that heat was a gas – a form of matter. And yet a prototype of the first (and perhaps also of the second) was formulated by Democritus some four hundred years before Christ without any experiment at all.[41]

The laws of thermodynamics are another conceptually cloudy area. They were effectively derived from the conservation laws. According to Clausius,[42] who made the original formulations of the first two laws, he came upon the Second Law purely by examining Carnot's earlier law that motive power cannot be created out of nothing.[43] But the Second Law can also be derived deductively from the concepts of order, disorder, and random change. (See, for instance, Gell-Mann: 'To the extent that chance is operating, it is likely that a closed system that has some order will move toward disorder, which offers so many more possibilities.'[44])

Why do we call all these strange entities 'laws'? The earliest use of the term in English in this sense dates only from the seventeenth century, when systematic science began to take off. The first two examples traced by the *Oxford English Dictionary* are dated 1665 – one from the Transactions of the Royal Society and one from Boyle – and they relate to a universe set and maintained in motion by the command of God. The 'laws of nature', notes the *Dictionary*, were viewed by those who first used the term in this sense as 'commands imposed by the Deity upon matter',[45] and 'even writers who do not accept this view

54

often speak of them as "obeyed" by the phenomena, or as agents by which the phenomena are produced.'

Edgar Zilsel[46] identifies Descartes as the author of the usage, in *A Discourse on Method*, published some thirty years earlier. Descartes says at the beginning of his account of natural philosophy that he has 'observed certain laws established in nature by God in such a manner, and of which he has impressed on our minds such notions, that after we have reflected sufficiently upon these, we cannot doubt that they are accurately observed in all that exists or takes place in the world.'[47] And the three laws of motion that Descartes presents in the *Principia Philosophiae*, published seven years later in 1644, he calls '*Regulae quaedam sive Leges Naturae*' – certain rules or laws of nature.

Zilsel claims that the concept of physical law was virtually unknown to classical philosophy (which saw the universe as determined by reason rather than by law), scarcely hinted at in the Bible, and unconsidered in the Middle Ages, which 'perceived the reign of God much more in miracles than in the ordinary course of nature', so that comets and monsters were more worthy of examination than the daily sunrise and normal offspring. There is no mention of laws in Copernicus, or even in Galileo,[48] while Kepler makes no use of the term himself in introducing what are often described as the first truly scientific laws – his three laws of planetary motion. (The first two he expounds without characterising, the third he calls a theorem.[49])

Zilsel suggests that Descartes synthesised the concept from two sources: on the one hand the working practices of craftsmen, which were also the inspiration for Galileo, and which involved acquaintance with physical regularities and quantitative rules of operation; and on the other the biblical notion of God's eternal law, as developed by St Augustine and Aquinas. Descartes derived his laws not from experiment, but 'with no other principle upon which to found my reasonings except the infinite perfection of God'.[50] He held that the necessity of logical and mathematical truth was the product of God's will, but he somewhat diluted the explanatory function of the divine origin by suggesting that even God might himself be constrained by his own perfection. He has endeavoured, he says, 'to prove that . . . even if God had created more worlds, there could have been none in which these laws were not observed'[51] – i.e., that lawfulness reflected divine perfection in a way that the arbitrary exercise of will would not have done.

55

From Descartes the expression was taken up by Hooke and others in their contributions to the Royal Society.[52] It first became general in the scientific vocabulary, however, when it was used by Newton in the *Principia*, where his three laws of motion were set forth under the general heading '*Axiomata, sive Leges Motus*', and where he also refers to 'the law of gravity'. It is difficult to believe, say Newton's editors, 'that he was not (even if unconsciously) making a direct improvement on the laws announced by Descartes'.

Newton, according to Zilsel, never says that the laws he has formulated have a divine origin. God remains in the background, however, even though his relationship to the laws is never fully stated. 'This most elegant system of the sun, planets, and comets', says Newton in the General Scholium, 'could not have arisen without the design and dominion of an intelligent and powerful being.' God is also responsible for the 'diversity of created things', which 'could only have arisen from the ideas and the will of a necessarily existing being'. This appears to be an answer to the puzzle still facing cosmologists of how the vast irregularities of the observable universe arose, if it developed in accordance with uniform laws. As Zilsel comments, 'Altogether in the *Principia* theology has retreated from the laws to (as the modern physicist would put it) the initial conditions.' The corollary of this, however, once those conditions have been established, seems to be that the laws take over God's sovereignty. 'No variation in things', says Newton, 'arises from blind metaphysical necessity, which must be the same always and everywhere.' A blind necessity, uniform in space and time . . . This may be the point, in spite of Newton's intense piety, where religion and science finally begin to part company.

So the notion of 'laws of nature' was in the first place adapted from the notion of laws laid down by some kind of central authority. I'm uncertain about the theology here. God's laws relating to human behaviour are always being 'flouted' by the wicked creatures who have been given free will (as I understand it) precisely to allow them this option. But if God also wrote laws specifying the form and functioning of the material world, which does not have the benefit of free will, it presumably has no choice but to conform. In both cases, the idea that the laws 'govern' their subject matter, which in turn 'obeys', seems appropriate (even if, in the case of human beings, they can choose not to be so governed, and to disobey). The monarch who imposed the laws has now been deposed by scientists, but the legal usages remain. Phenomena still 'obey' the laws; the laws still 'govern' the phenomena. So now a much odder analogy lingers: with the laws written by human beings themselves – the ordinary citizens to whom the laws apply.

Whether the analogy is with laws made by an absolute ruler or by a democratic assembly, though, is a secondary question. What exactly does the legal metaphor imply in either case? Juridical laws, whoever writes them, may be constitutive or regulatory. They may bring the institutions of state and government into being and establish their functions, or they may attempt to regulate existing institutions, and to control conduct that is already seen to be occurring.[53] So which sort are the laws of nature – constitutive or regulatory? Is the suggestion that there would be no material world – no rocks, no stars, no dust beneath our feet – if there had been no laws first to call them into being, just as there would be no National Radiological Protection Board or right to maternity leave? Or is the suggestion that electric charge and angular momentum, for example, would get up to all kinds of strange tricks – would start getting themselves lost or even created, perhaps – if there were not laws to stop them?

The idea that the laws of nature are regulatory seems on the face of it to be the more ridiculous interpretation. The law forbidding housebreaking has a function because people do in fact break into houses. What would be the point of a law requiring entropy to increase if it always increased anyway?[54] Popper suggests something a bit like this, however, when he talks about scientific laws 'prohibiting' certain possibilities: 'Not for nothing do we call the laws of nature "laws": the more they prohibit the more they say.'[55] This is a dramatically back-to-front way of expressing his idea of falsifiability. What he means is that if we discover behaviour not allowed for in the law – and the more precise the law the greater its significance if it remains unfalsified – then the law has to be abandoned. It would be a curious sort of criminal law, though, that was abandoned as soon as it failed to prohibit what it was enacted to prohibit. And the implications of falsifiability makes the analogy with juridical law even odder. If a law can in theory always be falsified, then it can *never* be stated in any definitive form – or can never be known to have been finally stated; as if the current Finance Act had to remain a bill forever in draft in case some unforeseen loophole relating to offshore investment cast doubt on the enforceability of all its other provisions.

So are we to think of the laws as being constitutive? If so, then they must surely have been in existence before the phenomena that they brought into being (or at the very least have come into existence in the same moment as the phenomena, and in such a way as instantaneously to determine their nature). What could this mean? What form could they have taken without the material to manifest themselves in?[56]

What kind of existence do they have, for that matter, except in so far as they are *stated*? The laws of God and man are expressed in language. The laws of God are written in part on slabs of stone and in ancient manuscripts, presumably extracted from a complete codex stored inside his head – no doubt including as yet unknown ones in the regulatory section waiting only to be revealed by the evil ingenuity of man in contravening them – or guessed at by latter-day exegetes. The laws of man, likewise, are written in statute books, in the decisions of judges and the commentaries on them, or at least (like God's) in someone's memory. The laws exist only in so far as they are expressed. Their expression is an enactment. They become laws by virtue of being laid down.

When scientists state a scientific law, however, we don't see this as an enactment, as the instrument of its author's will. We regard it as an attempt at embodying a pre-existent truth. The expressions of the same law are often very different one from another – think of the varying accounts you have read of the Second Law of Thermodynamics.[57] Feynman (who believes, as we have seen, that 'every theoretical physicist who is any good knows six or seven different theoretical representations for exactly the same physics') offers three radically different ways of stating the Law of Gravitation: Newton's causal formulation; through changes of potential within a field; and as a minimum principle, according to which an object follows the curve for which the difference between kinetic and potential energy is least.[58]

So what is it that these different 'representations' are representing? What is this pre-existent truth? If Bohr is right in insisting that our physics (and our chemistry and history, too) is what we can say about the world, the question arises as to *why* we can say it. For Einstein (even though the laws themselves are 'fictions'), the objective correlative (if his biographer Abraham Pais is right) is some kind of underlying harmony. For Popper it is certain 'genuine regularities in nature'.[59] Popper sees that this is a metaphysical problem, and it is surely the same one that recurs over and over again in relation to the interactions between human beings and the world. What is the ineffable 'fact' expressed by a proposition, without whose existence the proposition itself is held to be false or meaningless? What, for that matter, is the 'proposition' that lurks behind all the different sentences (in English, French, sign, etc.) that embody it? What, again, is the physical object that I see as golden, blazing, and high in the heavens, and that you, from where you are standing, see as red, pale, and low above the horizon? How can we come at the *fact* except through the proposition that it validates? At the *proposition* except through the sentence that it gives meaning to? At the *sun* except through what you see and what I see? And at *the laws of nature* except through the particular forms we find to express them?

Is this a 'a dialogue between mankind and nature', as Prigogine says? Well, yes, in the same weird kind of way as in all our representations of the world. A dialogue between the silent performer in dumb crambo and his team, as I suggested before. Or between a ventriloquist and his doll, perhaps, where the doll has a mind of its own and feels free to mock the ventriloquist – but can do it only in so far as the ventriloquist puts the words in its mouth. Between Hamlet and the Ghost, where the incorporeal figment of Hamlet's uneasy imagination tells him things he didn't know – but can do it only in the words that Hamlet imagines for him.[60]

And in the case of dialogues that result in the 'laws of nature', what existence can this elusive partner have apart from the particular cases that instantiate it? Feynman's philosophical casualness is revealing, when he refers to the law of the conservation of energy as 'a fact, or if you wish, a law'.[61] So, if we *don't* wish, it's not a law at all? But in any ordinary usage laws and facts are very different things! There is a law forbidding housebreaking – and there is a fact that houses are broken into. The law and the fact are not in any way contradictory. Laws don't attempt to state what happens to be the case; facts don't attempt to 'govern' anything. But Feynman has surely let slip the truth. Facts are what 'the laws of nature' actually are – broad generalisations of the way things do happen to behave (or would, if we could separate out this particular aspect of their behaviour from all the other aspects, and from the behaviour of all the other things around them).

You are stirred to protest. The laws of nature aren't just inert records of past and present practice – they're *predictive*! Well, the predictiveness is something that *we* supply. We have confident expectations that what has proved to be generally true in the past will turn out to be true in similar cases in the future; and this confidence is buttressed (psychologically, not deductively) by our having in the past had similar expectations that have proved (also now in the past) to be in general fulfilled. There's nothing specially scientific about this. Isn't it much what happens in any kind of factual generalisation? When we say that the natives are friendly, or that daffodils come before the swallow dares, doesn't this imply that we expect the natives to be friendly on our next visit as well, and the daffodils to bloom early next spring? Wouldn't these statements be falsified, just as a scientific 'law' would be, if next time the natives threw bottles at us and daffodils didn't flower until July? Do we need to see these statements as laws, of a rather specialised sort, to explain this?

For that matter, don't even statements not about general classes but about individual cases arouse similar expectations? You tell me that – I don't know – your father is senile, or that your house is too small for you. Isn't there an unstated implication that your father is likely (though not of course certain) to go on forgetting who you are? That the problem with the house will persist until you extend it or move? Does this say anything very profound about the world? Only that the mental deterioration of old people tends not to go into reverse (though sometimes it does), that houses don't suddenly sprout loft extensions and conservatories of their own accord.

But the laws of nature are *universal*! – so is the reluctance of houses to throw out playrooms and granny flats spontaneously. This is as true of my house as it is of yours. It is true of all the houses in the world. Why is its truth so wide-ranging? Because a house is a house is a house! We might generalise it further, and say that this kind of reluctance is true of inanimate objects as a whole. We could write even a law to this effect, along the lines of Newton's First Law of Motion: 'Every body perseveres in its state of extension or of dilapidation, except in so far as it is compelled to change its state by the efforts of some home handyman or builder.'

This universal universality – this propensity of a house to be like a house, and of a hydrogen atom to be like a hydrogen atom, with all that this implies – we shall return to later in this part of the book. In the mean time, as a sobering reminder that universality has a downside as well as the strange metaphysical glory that has dazzled so many observers, we should bear in mind the judgement of Nancy Cartwright, a practical physicist as well as a philosopher of science: the more fundamental and general the law, the less true it is of any actual situation in the real world.

What goes for 'the laws of nature' goes also for the general scientific principles (entropy, relativity, etc.) which are not exactly statements of fact, but clarifications of our concepts. They have no existence independent of the concepts to which they relate. Our taste for ontology, which is yet another expression of our yearning to locate enduring landmarks in the restless flow of the traffic, raises our hopes of the laws and principles of nature; but they, too, are emergent. They, too, are aspects of the traffic.

Even von Neumann's down-to-earth account of science as the construction of models isn't quite as simple as it appears.

You think of a model and what comes to mind first of all is perhaps something like a model train. What intrigues us here is the precise mapping of the original in miniature. The shape, the colour, every little detail – they are all reproduced, only smaller. And it moves! It moves out of a miniature station hauling miniature carriages full of miniature travellers carrying miniature suitcases. We scarcely notice the real trains that we travel on each day. We put our eye down to the level of the miniature platform as our miniature locomotive pulls in, however, and we have an impression of size, weight, and power that we rarely have, dulled as we are by familiarity, from the vastly greater size, weight, and power of an actual train. We have a sense of reality more real than reality itself. We feel that we are looking at something which is at last graspable. Literally so. We could grasp it in the palm of our hand and for once actually take hold of the world that so persistently eludes us.

The kind of model that von Neumann has in mind – a mathematical construct with verbal interpretations attached – seems on the face of it to be very different. It doesn't have the same shape and colour as what it represents, nor the authentic logo on the side, nor the little driver in the cab. It doesn't map the world in that literal one-to-one manner. What it does have in common with the model train, though, is surely that same graspability. It renders down the grimy confusion of the world into something clean and neat. It turns its gross physicality into a series of symbols, and of operations on these symbols that can be done in the head, or written out on paper, with clarity and elegance. It makes the ineffable effable.

Graspability is achieved not only by a change of scale or the use of symbolism. Anything, smaller or larger, simpler or more complicated, natural or confected, might serve as a model if it brings out some salient feature in the original. A chocolate Easter egg can serve as a model of a hen's egg because it echoes its shape, in spite of its failure to share the most important functional feature of the hen's egg – the capacity to protect and nurture the chick that will burst out of it. A jack-in-the-box, on the other hand, might serve as the *functional* model of an egg, even though it has a completely different shape. Indeed, you might think that the chocolate emptiness of the Easter egg on the one hand and the pregnant cubicality of the jack-in-the-box on the other contribute to their effectiveness as models by throwing the differing common features of each into a more striking contrast.

There is no limit to what might serve as a model of something, for one purpose or another. You might model a hen's egg in – I don't know – a flash of lightning, say, which would dramatise the brevity of its existence, and the elegance of its solution of the equations expressing various physical forces. Even with the apparently simple one-to-one scale model of the train we have to choose which features it is to display. The gleaming paintwork of the original, yes – but probably not the griminess of every surface to the touch. The numbers on the dials in the cab, certainly – but not the graffiti sprayed on the carriages. The curtains at the windows of the Pullman coaches – but perhaps not the curtain of dried vomit outside some of the other windows.

Suppose the model was *really* one-to-one – a hen's egg as the model of a hen's egg, a jack-in-the-box of a jack-in-the-box, a full-size train of a full-size train . . . What would be the point, except to establish the limiting case where meaning is extinguished in tautology? Not so fast, though. An egg *is* the model of an egg, and a very useful one, in the sense that what enables us to see an egg as an egg in the first place is its similarity in appearance and function to another egg, or eggs; and this primitive modelling is no more vitiated by differences in colour and speckling, or by the fact that some eggs have chicks in them and some don't, than the more usual sort of modelling is by even the most radical variances. In the case of the egg's function a simple description might serve as a kind of model, in something like von Neumann's sense. (And of course you could model its shape mathematically, in precisely von Neumann's sense, which might bring out more clearly than another egg its kinship not just with fellow eggs but also with drops of water and the orbits of the planets.)

With the idea of a language model we come close to the alternative description of science as *explanation*. In explanation, as in modelling, we seek to bring out salient features by expressing them in a different mode. This seems entirely straightforward – explanation isn't a concept that itself needs explanation. Or is it? The more you think about it, the more different kinds of procedure you can think of that count as explanations, and the more different ways there are in which explanations function in different contexts.

We explain the unfamiliar in terms of the familiar. The unfamiliar to you in terms of the familiar to me: 'How do you tie a clove hitch? Like this – just watch me.' The unfamiliar to you in one context in terms of the familiar to you in another: how could such a nice woman beat her husband to death with his own niblick? Well, remember when you were so angry with your own husband . . .

We explain the less tangible, the less graphic, by comparison with the visible, the concrete: you know about the world going round the sun? – well, that's what the inside of an atom is like. No, hold on. You've seen waves on the sea? It's more like that. (Now try to see it both ways at once!)

We explain the particular in terms of the general. The failure of this particular firm is just one case of a trade depression affecting the whole region. The depression affecting this region is an example of the depression affecting similar regions in other parts of the world. This worldwide pattern of depression is part of a more general trade cycle . . . But we also explain the general in terms of the particular. If you want to understand the world trade cycle, look at what's happened to this particular firm.

We explain departures from the regular and expected – why the train is late, why the summer is colder than usual. But there is another sort of explanation – and this is what scientific explanations are mostly like – not of the unfamiliar or the unusual but of the familiar and the normal. Not of why things sometimes fail to function as they should but why they ever function right. Explanations of this sort locate the unfamiliarity lurking beneath the surface of the familiar. This is the real insight in the discovery of gravity; it makes noticeable for the first time what people had always had in front of their eyes but never seen. Lifting a sack of apples plainly required an effort – it was a reversal of the natural order of things. The fall of the ripe apple, however, seemed natural, effortless, not in need of explanation.[62] What the notion of gravity did was to turn the second picture upside down, and hold it alongside the first. Now we see the parallel; we feel the strain in the earth's biceps.

The first difficulty in this sort of explanation is to isolate the one feature of the scene that needs explanation. The apple falls in the orchard, and if anything needs explanation then everything does: the gradual increase in the apple's weight as it grows, the movement of the breeze among the leaves, the weakening of the stem as the apple swings; and for that matter the displacement of air as it falls, the thud as it hits the ground; then again, the planting of the tree, the ownership of the orchard, the development of fruit-growing; and, yet again, the transmission of the image to our eyes and the sound to our ears; the movement of the signal to the brain; its processing in the brain; the question of consciousness . . .

Open your eyes at random and you are looking at more than could be described in a thousand years, and more than could be explained in a million. Lift your gaze to the complex, shifting irregularities of the clouds above your head. What sort of description could you ever give that would exhaust their possibilities? And what sort of explanation? Well, you certainly feel that it's possible to give a general explanation, just as it is to give a general description. A meteorologist could identify all the forces involved – winds, updrafts, temperature, moisture content, etc. A physicist could name all the optical effects governing the appearance that this particular distribution of water droplets has for us – the angle of the incident light, its reflection or absorption by the droplets, etc.[63] And very useful this might be for all kinds of purposes. But you mean more than this. You want an explanation of how these general forces have produced, not this general effect, but precisely this particular array of phenomena that you see in front of your eyes now.

Well, we can ask the meteorologist to quantify all the variables. He'll have to quantify them not in the way that he does in a weather report, as average values over broad areas and extended periods of time, but at each of a series of arbitrarily close points in the visible heavens, at each of a series of arbitrarily close moments in an arbitrarily long approach to the present. It's difficult to imagine what any complete statement of these quantities – or any model that represents them – would be like. The only thing we can be certain of, though, is that it will have to be at least as complicated as the cloudscape it is explaining. What point could such an explanation serve? What would it tell you that you can't see more graphically and immediately just by lifting your eyes and looking at the sky?

Close your eyes, and even what you're seeing now defies your powers of description or explanation just as comprehensively. Limit your consideration to one small, isolated, and relatively defined element of the scene behind your closed eyelid – to that floater which is drifting slowly upwards, now turning and drifting down again. What would constitute a complete explanation of it? Of its provenance and position, its shape and colour, its speed and trajectory? The more you think about it the longer your explanation goes on – and the longer your explanation goes on the more it merges into the universal sea of explanation, in which all things are lost.

Answers can be offered, if at all, only to particular questions. Why does the floater move upwards at that particular speed? Why is it here and not there in the visual field? And we pull back from the specific to the general, from the behaviour of this floater to the behaviour of floaters in general, to the force of gravity balanced against the viscosity of the fluid in which the floater is moving. Why does the earth draw apples towards itself? And we pull out to the behaviour of other planets, of stars and atoms – and of apples themselves. You feel that if you can pull back this far you can pull back for ever, and ask: why is everything the way it is? But the further you've pulled back already, the less room there is to pull back further; the more general the question, the less there is to take in as explanation. You can't pull back to show more of a scene that includes everything already.

This is why Popper, like Einstein, rejects the traditional view of science as proceeding from the observation of phenomena to the construction of theories that explain the phenomena. How do we select what we observe without first having a theory that determines what's relevant and what isn't? As illustration of the absurdity of the approach he is attacking ('still so widely and so firmly held that my denial of it is often met with incredulity') he tells the story of 'the man who dedicated his life to natural science, wrote down everything he could observe, and bequeathed his priceless collection of observations to the Royal Society to be used as inductive evidence'.[64] He says he made the same point to a group of physics students in Vienna by instructing them:

'Take pencil and paper; carefully observe, and write down what you have observed!' They asked, of course, *what* I wanted them to observe. Clearly the instruction, 'Observe!' is absurd . . . Observation is always selective. It needs a chosen object, a definite task, an interest, a point of view, a problem. And its description presupposes a descriptive language, with property words; it presupposes similarity and classification, which in their turn presuppose interest, points of view, and problems. 'A hungry animal', writes Katz, 'divides the environment into edible and inedible things. An animal in flight sees roads to escape and hiding places. Generally speaking, objects change . . . according to the needs of the animal.' We may add that objects can be classified, and can become similar or dissimilar, *only* in this way – by being related to needs and interests.

This is wonderful; but perhaps a little too sweeping. We can't help but see patterns and possibilities in the world around us, even without any purpose or theory. If I had been in the lecture hall in Vienna when Popper issued his instruction I should no doubt have found it eccentric. But if I'd been sufficiently in awe of him to feel obliged to attempt the task I should certainly have been able to write down *something*. Features of the scene would have caught my eye – most notably, I should think, our distinguished lecturer and his curious demands. Does this mean that I must have had a theory, even if I had never consciously formulated it, of what a lecturer should be like, and what he ought to say, a theory that Popper was either confirming or falsifying? Well, possibly. Some general conception or preconception of lecturers and their ways, at any rate, that Popper seemed a striking exemplar of, or contrast with. But equally I might be noticing for the first time a similarity between Popper and other lecturers, which could be not the result but the beginnings of a theory about lecturers in general. And, even though this theory might have to be abandoned in the event of some future experience of a different sort of lecturer, seeing that he was different would itself depend upon my having noticed the similarity between the earlier cases.

Isn't it also perfectly possible that if I wrote down my observations of Popper it was because I was simply – well – struck by Popper as

Popper? That I might be struck enough to observe and describe him even if I'd never been to a lecture before, or heard any stories about lecturers and their supposed dullness, or their propensity on the other hand to issue attention-grabbing challenges? And even if I *was* measuring him against a conception formed from my experience of other lecturers, I must have begun to form *that* conception by observing some first lecturer in the series . . .

Or was I comparing and contrasting even that first lecturer with my experience of non-lecturers? Well, to some extent, no doubt. But we can't go on like this indefinitely! I must have been struck by some uncovenanted aspects of the world at some point or I should never have got started on seeing anything! I must have been struck by the sight of my thumb waving about in front of my face above the cradle long before I had any theory of thumbs, or any purpose in seeking them out.

Here's a new and more plausible version of Popper's challenge: program a computer to notice what it has not been programmed to notice!

We model an amorphous group of stars in the night sky by holding up a swan against it; but before we can do that we have to be familiar with what a swan is, which we do by holding up a swan against a swan, and that swan against another swan, and that other swan against another thousand swans . . . We understand the propagation of light from Alpha Centauri because we understand the propagation of light from the reading-lamp on the desk, and we understand *that* because we have set the reading-lamp beside the Wolf Rock lighthouse and a candle flame and a match and the sun . . .

What makes it possible to do this – and this is a point to which I shall return later, because it's what makes the whole system of human perception and language possible – is that swans are so much alike, and so is the light from this match and the light from that candle. Why are swans so much alike? Because of the iterative nature of biological reproduction. Why is this candle so much like that candle? Because of the iterative nature of candle manufacture. Why is the light that the candles produce so recognisably like the light from the sun? Because candles are designed to imitate the function of the sun. All right, then: why is this shapeless, functionless clod of earth so much like that shapeless, functionless clod of earth? Because they are formed from the same common material, which breaks up in much the same way in

Mandalay as it does in Middlesex. Why is this molecule of calcium carbonate in the clod of earth so much like that molecule of calcium carbonate? Because they are both composed of atoms of the same elements, which were formed by the same transmutative processes out of the same common material.

You protest that you want more from an explanation than mere analogy. When you peep inside the conjuror's hat after the performance, or reach the last chapter of the detective story, you don't want to be fobbed off with a list of tricks, or a catalogue of stories in which corpses are found in locked rooms. You want some kind of revelation – a false bottom in the hat, a sliding panel in the study with a secret passage behind it. (Or you half want it, and half want to find the mystery still unsolved.) But even a revelation is a kind of analogy. It explains by showing the parallels between this situation and all the other fictions we have come across where puzzling phenomena have turned out to be explained by hidden strings, mirrors, secret codicils in wills, concealed paternity, etc.

An explanation in terms of scientific laws and principles appears to be an opening up of the surface of the universe, and the revelation of some genuinely other world beneath, made of a stuff that is different in kind from the physical stuff that confronts us; a world of abstractions whose nature is incorporeal. But scientific laws and principles are not separate from the physical world. They are generalisations of the way in which the physical world behaves. They have no existence outside that world. The move is once again from the less general to the more general. We are not opening the lid on to a world within – we are holding up a wider selection of this world.

The only material we have out of which to construct either models or explanations is the world that we are confronted with. When we achieve some understanding of it we do so through our commerce with it. As we search for what we need we become familiar with the shape and colour of its parts. As we handle it we feel the weight of it. As we bend it to serve our purposes we learn the stiffness or weakness of it, the fragility or pliability.

The different shapes and colours we fix and identify to ourselves by analogy with each other. The sizes and weights we establish by comparing this object with that. The breaking-strain and the melting-point we identify by breaking and melting.

*

What we are seeking with all our laws and models and explanations is understanding; and understanding, at any rate, is surely something we understand well enough, without any need of the explanation that explanation itself seemed to require.

Or do we? The more we look at it, once again, the more elusive the idea seems, the more clouded by the vagueness which is characteristic of so many familiar concepts. I understand French. Yes – but not in quite the same way, sadly, as I understand English. If you speak too fast, or you use too many vernacular shortcuts, then I won't understand at all. Though there also comes a point where my English will give way as well under the same stresses. I understand your feelings, and your position on the question we are discussing. Of course I do – though they don't seem to me quite as understandable as *my* feelings, I regret to say, or quite as comprehensible as *my* position. I understand quadratic equations, in the sense that I could probably still solve one, and give some kind of explanation of them to my granddaughter. I understand a cash dispenser. Well, I can usually get money out of it, though I've only the haziest general notion of what's going on behind the scenes. Do I understand my computer? Yes, I do, as a matter of fact, more or less, when it's working right, for most of the modest everyday uses I want to make of it. But when something goes wrong, and I have to look a little more deeply into the workings of the program I'm using, let alone the hardware under the lid, then I don't understand it at all.

But then even the most familiar faces seem suddenly strange, and even the friends we understand the best spring huge surprises upon us, which make us wonder with hindsight whether we ever had any grasp of their reality at all. From 1687 until 1916 people understood gravity, understood everything there was to be understood about it. And then they discovered that they'd misunderstood the whole basis of it, and that it couldn't be fitted in among the other fundamental forces of the universe at all.

We *did* understand all these things, though, we *did*! We got money out of the machine, we read *Eugénie Grandet*, we grasped why the tides followed the moon around the earth, we knew enough about our old friend's thoughts and feelings to get on perfectly well with him for years. Donald Winnicott introduced into child psychiatry the idea of the 'good enough' parent – a parent very possibly remote from a mythical ideal whose failure to exist in the real world seemed by implication to lurk accusingly in the background of earlier psychopathology, but who does actually succeed in bringing up reasonably well-adjusted children.[65] What we seek, likewise, is the good-enough understanding of things that will enable us to deal with them for the purposes in hand.

Our traffic with the world becomes ever more sophisticated. With ever longer and ever finer tongs we learn to handle objects more and more remote from us in time and space, on a scale ever larger and smaller than our own. But if in the end we come to understand something about the quarks that compose the protons and neutrons that compose the nuclei at the heart of the atoms that compose the molecules that compose the cells that compose the bacteria that keep us alive, or about the quasars that blaze in conditions entirely removed from any we have ever come across even in the depths of our own galaxy, and upon principles which seem to subvert the physics of the world we know, at the very edge of the perceptible universe and the very beginning of reconstructable time, then that understanding comes at the end of a long chain of analogy which has its beginnings in our most humble dealings with the everyday world around us. It comes out of our recognising the food we need to put in our mouths, and in our interactions with our mothers and our enemies and the checkout clerk in the local supermarket.

None of these questions, however, about the existence of specific objective correlatives to the specific scientific laws we formulate, casts any doubt upon the objective existence of the general principle underlying them: causality, the principle that every state of affairs must be the result of some earlier state of affairs. This is the bedrock on which all scientific (and historical) investigation rests – and the next stratum to be surveyed.

3

Events and their Ancestry

An almanac of causal lineage

The principle of causality could be defined as the law of the conservation of cause (and might perhaps be seen as a more general way of stating the law of the conservation of energy). The causality of a system (like its energy) is invariant. States of affairs can no more come into being out of nowhere, or pass out of existence without consequences, than energy can. There must in theory always be an earlier state of affairs to account for them, and a later state of affairs for which they in their turn account.[1]

The objective existence of causality was famously challenged by Hume, who argued that the only connection between cause and effect was in our own minds – it was simply an association of ideas. We expect it to kill someone if you stop them breathing, because we know that it always has in the past, but there is no logical connection between the fingers on the windpipe and the death, and no logical reason for believing that it will happen in the future.

Not that this was ever thought to be a *practical* problem. The predictions we have made in the past on the basis of our experience have worked perfectly well, and not even Hume actually expected them to stop working today or tomorrow. And it wasn't just our particular predictions that worked – it was predictivity in general. Kant attempted to refute Hume by arguing that if we can make sense of the world at all – and we can, we do, we're doing it now – then it tells us something about the nature of the world. It implies that it has a certain stability, a graspable continuity in space and time, and an orderliness in the articulation of its parts, including a regular and comprehensible connection between cause and effect. I suppose a really determined sceptic would argue that, if everything else about the world might suddenly curdle tomorrow, the contingent truth upon which this argument is pinned might also fail – we might suddenly cease to be able to make sense of the world, so that the fact that predictivity in general has worked up to now offers no better a reassurance about whether it will in the future than any of our specific experiences.[2]

The trouble with arguments of this sort is that they don't pursue their own logic far enough. If all bets about the future continuation of predictivity are off, then so are the bets about everything else. If stopping someone breathing may tomorrow not cause death, then 'stopping someone breathing' may tomorrow not mean stopping someone breathing. Why should semantic connections be any more secure than causal ones? 'Death' may not be death. 'Tomorrow' may not be tomorrow. In which case it's surely impossible to make any meaningful statements *at all* about the future, even predictions that predictivity may cease. 'Predictivity,' when tomorrow dawns (whatever *that* may turn out to mean), may not be predictivity.

This traditional anxiety of philosophers about 'justifying' induction – finding an *a priori* guarantee for our predictive generalisations from past events – is a kind of philosophical neurosis. It's like some people's fear that, however many times they wash their hands or check that they've turned the gas off, their hands are still dirty, and the gas may still be on. Patients with an obsessive-compulsive disorder of this sort have unfeasibly absolute standards of cleanliness and certainty. Doubts about induction, likewise, arise because deduction has been set up as the gold standard against which induction has to be measured. Philosophers find it difficult to forget the faith they once had that the syllogism could offer not only certainty but new knowledge, rather in the way that cold fusion seemed to offer the promise of free energy. But of course there is no new knowledge generated inside a syllogism. The really cautious philosopher might in any case remain a little wary about the certainty of the syllogism. If I can't be sure that hydrogen and oxygen will tomorrow combine to form water, then I may be a little rash in my confidence that A and (if A then B) will tomorrow combine to imply B. There have been many more arguments among logicians about entailment than there ever have among chemists about the composition of water. If I were forced to choose, I think I might put my money on the immutability of $2H + O = H_2O$.

Causality has been much more seriously challenged by two developments in modern science. The first was quantum mechanics. Even though Heisenberg took a little time to grasp all the other devastating implications of indeterminacy, about this one he was clear from the very beginning. Already, before the publication of the paper in which he introduced the concept, he told his future colleague Weizsäcker that he thought he had 'refuted the law of causality'.[3] In what he called 'the sharp formulation of the causality law' (if we know the present, then we can predict the future), he wrote later that same year, 'it is not the consequence but the premise that is false. As a matter of principle we cannot know all determining elements of the present.'[4]

Indeterminacy, he wrote later, renders traditional causality meaningless – not invalid, but 'empty of content', because (as he had shown) it is impossible, as a matter of principle, ever to know everything about the whereabouts and behaviour of a particle. Since we should need to know (among other things) exactly where it is now, plus how fast it is travelling in what direction, to know exactly where it will be and what it will be doing in the future, all predictions about the future situation of an individual particle are necessarily probabilistic.[5]

The second challenge comes from cosmology. If the theory of the Big Bang is correct, and the universe had a beginning, then there does seem to be a rather extensive state of affairs that has come into being without any precedents to account for it at all. The theory also, of course, undercuts all the traditional conservation laws relating to energy, charge, and so on. The jump from nothing to something is a very difficult moment in every way for cosmologists,[6] who believe they can pursue developments back to the end of the first second after the event, but no earlier. There are some who sidestep the difficulty about causality because they see the beginning of the universe as being part of a cycle of expansion and collapse, so that events in this phase of the cycle are determined by events in previous ones; and there are others who propose that this universe is a local development in some larger meta-universe, perhaps causally related, which is inaccessible to us.[7]

Even if you leave aside quantum mechanics, the Big Bang, and Hume, causality isn't quite as simple as it seems. The 'sharp formulation' of causality, as stated by Heisenberg, I take to be deterministic,[8] and what it suggests is something like a game of cosmic snooker. The cue-ball hits another ball. This propels the second ball into motion, which as a result hits a third ball, and so on. So the impact of the cue-ball produces a chain of consequences, and however complicated the ensuing sequence of collisions, each of them is entirely determined by that first impact. Because the cue-ball is travelling on precisely this course at precisely this speed it hits black at precisely such an angle and displaces it precisely so far in precisely such a direction; with the result that black in its turn hits red at precisely *that* speed and *that* angle; until eventually all the energy transmitted by the cue-ball has been distributed, and the precisely calculable friction of the table-top, air resistance, and the resistance of the net at the bottom of the pockets has brought all the balls to rest. The complete chain of immediate consequences has now come to an end, but the position of the balls which has resulted from it will modify the results produced by the next stroke. And so on, until the game ends.

But when you stand back from the snooker table for a moment you can see that this narrative seems so clear and definite only because (as with any other narrative) we have left so much out. We have discounted the pockmarks worn in the ancient cloth of the table in Lord Bagshot's billiard-room, and the vibration from the rock group playing in the bar next door to the table in the North Balham Sporting and Social Club. The only events of which we have taken note are those that are defined as relevant by the rules of snooker.

Well, we could, now that we've thought of them, factor the irregularities and the vibrations into our causal chain. But are we now sure that we have thought of all possible contributions to the process? What about the gravitational effect of the moon . . . ? The planets . . . ? How far afield do we have to look? Where, in any case, does cause stop and effect start? It seems plain enough with the snooker balls. Cause: white hits red. Effect: red goes into the middle pocket. Hold on, though. What about the front and back of the red ball? The energy transmitted by white to the front surface of red has to travel to the rest of the ball before it moves. So is the compression of the front of red the cause of the effect on the back? What about the sections between front and back? Do we have to come right down to the individual molecules that make up the ball? Is the absorption of kinetic energy by the first molecule in the chain the cause of its transmission to the second? What are we going to say about the atoms that compose the molecules . . . ? And of course in some circumstances (if we are doing physics) we may want to consider any of these things as part of the causal chain. In others (if we are writing a sports report on the game) we may not. But the satisfyingly solid and distinct click-click-click with which causality seemed to work is beginning to sound more like a muddled and continuous buzz. Which out of all this overlapping confusion of clicks do you want to hear, and which do you want to suppress?

In any case, we have left one huge causal factor out of our account of events on the snooker table: precisely the one that makes the game interesting to play, and not just a philosophical metaphor – the stroke that first set the cue-ball in motion. The contribution of even the most expert snooker player introduces an indeterminacy that far outweighs all the other indeterminacies put together. The behaviour of the human musculature, of eye-hand coordination, are notoriously imprecise, and remain so, however hard we try to render them as mechanical as the events on the table itself. This is what makes snooker a playable game and not a foregone conclusion.

You gesture impatiently: this is the given, the subject of a separate commission of inquiry. But, even if we start from the point where the cue-ball is rolling, the events it sets in motion are going to have consequences for future events on the table which are not fully describable by the laws of mechanics, because the identity of the player who will make the next stroke is dependent upon whether the sequence has culminated in success (a ball being potted or a cannon made) or failure. You gesture impatiently once again. But you can't make even the broadest predictions about future events without taking this into consideration!

The sequence of events on the table that we were considering may have other consequences, too, not governed by the rules of snooker, but still part of the game and again crucial to the events upon the table at the most simple mechanical level. The player making a successful break may be encouraged, and play better; his opponent discouraged, and play worse. How much better? How much worse? Impossible to quantify, even in theory. Difficult to characterise even in the vaguest terms, even as between positive and negative. The successful player may become over-confident, and play worse; his trailing opponent more determined, and play better.

Now the causal chain in which those simple spheres are moving is beginning to seem remarkably complex, and remarkably indefinite. Even to begin to trace it through we need to know something, heaven knows what exactly, about the early experiences of the players, the confidence or lack of it instilled in them by the attitudes of their parents, etc.; then the biographies of the parents, the grandparents, and so on. This is what the concept of causality involves: not a chain but an inverted cone, in which any single event can be traced back to ever wider circles of causes, until in the end it has spread back to the total state of the universe at some point in the past; and then presumably back still further, through ever narrower circles as the total state of the universe shrinks to the moment of the Big Bang – where everything that was ever going to happen, even your small triumph against the reigning champion of the North Balham Sporting and Social Club, was set in motion by the structure of that first dense object.

And if you trace the trail of effects forward into the future the same vague spread occurs. The kinetic energy transmitted by the player to the cue-ball has been transformed into a rise in temperature of first red, then black, then the surfaces of the table and the air above it; the shock wave of vibrations from the impact of ball upon ball travels outwards through the building. Textbook examples of classical causality, but when the vibrations show up on the seismograph measuring earthquake activity in the laboratory on the floor below, they're dismissed in the same way that we dismissed the irregularities in Lord Bagshot's billiard table, as irrelevant to the system under consideration. There is no action we can take which has only the effect desired. In war the effects of 'collateral' damage can notoriously outweigh the damage to the target. In medicine, as Oliver Sacks discovered from his experiences in treating patients with L-dopa, there is no such thing as a silver bullet, a drug that treats one single symptom.[9] Now molecular biologists hold out the hope of one again, by altering just the single element of the gene that controls the symptom we wish to treat. Yet again, though, we note with unease the appearance of other unforeseen results in the creatures on which these modifications are tested. It turns out that this one cell also had other roles to play – in the immune system, perhaps, or in the take-up of apparently unconnected chemicals. (Some biologists have in any case now come round to the view that many genetic effects are produced not by particular individual genes but by genes working together as an indeterminate and adaptable system, rather in the way that the brain does, as evidenced by the capacity it often displays to redistribute function after neurological damage, or as society does with so many major communal undertakings.)

The effects of any political intiative, even the best-intentioned, are even more haphazard. This is the exhausting truth that reformers never learn (and perhaps *must* never learn, if they are ever to do anything at all). The bigger and more drastic the dose of reform that is administered, the vaster the area of devastation – and the smaller the proportion of it that overlaps with the target. Revolutionaries discover that it is easier to destroy the whole of society than to cure snobbery, or anomalies in the distribution of sealing-wax, just as the British discovered in the Second World War that it was easier to bomb entire cities than to hit particular factories. The totalitarian state slaughters tens of thousands in the camps, and it's true that it also improves the provision of nursery education. Important, of course. But across the frontier they also improved their nursery education, and at the cost of nothing more drastic than a few late-night sittings in the national assembly.

And yet we trace the trails of cause and effect at every moment of every day. We shape our actions on the basis of the effects we forecast for them, and do it sufficiently well most of the time to survive and prosper.

How is this possible? For a lot of the time, of course, when we talk about causes and effects, we are not using the terms in the sense of this particular debate. We are referring to broad concepts and classes of event: 'Religious belief is one of the causes of terrorism'; 'Violence begets violence.' Although these generalisations must be logically derived from particular individual cases, they don't seem to raise difficulties as acute as the cases themselves. Not even in scientific contexts, whether they are used to express connections which are supposed to be universal and exceptionless – 'Tides are caused by the gravitational pull of the moon and the sun'; 'Moving an electrical conductor through a magnetic field causes a current to flow in it' – or plainly probabilistic (and often disputed) ones: 'Smoking causes cancer.' We understand the basis for statements like these: the connections that have been observed in similar cases in the past. And, even if we have no logical basis for expecting the connections to recur in the future, we know how to test the assertions by seeing whether they in fact do.

Individual causes can't be sourced and tested in quite the same way. 'Her outburst was caused by stress'; 'My fear of dogs was caused by an unfortunate incident with a wire-haired terrier when I was a child.' Probably we are deriving them by applying the previous sort of generalisation to the particular circumstances. Other people's outbursts have seemed to result from stress, other people's fears from childhood experiences. We can't test them in the same way, though, because the chances of observing similar connections of stress and outburst in this particular subject are more limited, and will come too late to verify this particular case – and there is no chance at all of my repeating my childhood and its experiences. But in everyday cases like this we tend to be speaking in a loosely explanatory way that none of us would expect to represent any very precise, or precisely testable, aspect of the world. And with large-scale single instances the frontiers and the truth-conditions are all so vague. We know that we are talking about things that feel comfortable. As with a woolly cardigan, an explanation that feels comfortable is a good enough fit. There's no need to call in a tailor to make it sit to every contour of the body.

The problems really begin when we focus our attention upon specific individual micro-situations, as we have to if we are to make sense of causality in the snooker-ball way. 'Tides, etc.' – yes, fine. But when we come to the effect of the moon on this particular piece of seawater . . . How much seawater exactly? Where are its natural boundaries in space and time? How do we exclude the effects of wind and current, of passing ships and shoals of herring? 'Stress, etc . . . .' It seems reasonable enough. But precisely which demands made, hours worked, consequences threatened . . . constitute the stress? Precisely which words spoken, nerves tensed, facial expressions forming . . . constitute the outburst? Now it begins to seem difficult not only to know quite which generalisations to refer these particular instances back to, but how to get our fingers on any isolable situations that constitute before and after.

Philosophers who examine causality in this sense expend a great deal of ingenuity in trying to discover some statable objective basis for causal relations – what the 'essence' of a cause is, how different bits of the causal chain can be identified and related to each other – and to constructing logically justifiable formats that express the situation. What *kind* of thing, first of all, constitutes a cause? Is it an event or is it the fact that the event occurs? This is a subject in which Jonathan Bennett (who at the very beginning of his career was my tutor in philosophy) has specialised. He has woven logical nets of amazing complexity and subtlety to catch the elusive prey, and accepts that the idea of a single cause for anything is a chimera:

> We cannot easily formulate a precise, detailed truth about *the* cause at time T of a given fact's obtaining, because that would be too big a mouthful: it would involve antecedent conditions of such richness that nothing could possibly stop the consequent from arising from

them. So we have no practical use of 'the cause' in its strict and literal sense. In ordinary speech and writing, 'the cause of x' always means 'a salient cause of x' or 'the most salient cause of x' . . . [10]

But even 'the most salient cause' is a concept that depends entirely upon the point of view of the proposer, and the use to which it is being put. 'A collision with an iceberg caused the *Titanic* to sink.' What cause of what effect could be more salient than this? Bennett has ways that could be used to analyse the event into its causal constituents, as it develops through the shearing of the plates, the ingress of the water, the change in the ship's specific gravity, etc. Even leaving all this aside, though, the salience of the cause is still relative. For the commission of inquiry charged with establishing responsibility for the disaster, the connection between impact and sinking might seem to be a given, and the salient cause might appear to be the captain's decision to set the particular course that he did. For a glaciologist it might be the separation of the iceberg from its source, because this is something to which his studies are relevant. For the ship designer it might be the failure of the supposedly iceberg-proof hull, because this is the kind of thing that he is responsible for, and that he can affect in future. For the business analyst it would be the competitive pressures of the North Atlantic passenger trade that were crucial. For the political historian the logic of capitalism. And so on. You might argue that you could agglomerate all these causes into one single complex, rather as the first simple statement that we began with telescoped the developments in the situation between impact and sinking from one moment to the next. But what purpose would it serve? What use would it be to the ship designer, to distract him from hulls and plates by burdening him with considerations about glaciers, or to the glaciologist to wave balance-sheets in his face?

We are able to talk about causes and effects precisely because we *don't* in ordinary usage search for essences of isolable atoms of event – because we enter into our usual dialogue with the world around us, and select out of all the possibilities it offers us which events are relevant: how much of which events are to count as causes for the practical purposes in hand, how much of which as effects; where the causes are to begin, and where the effects are to end.

God, as you might say, proposes; man disposes. Causality – even classical causality – involves human participation, like everything else.

*

So the boundaries of the determinate were always indeterminate, even before Heisenberg.[11] But now it turns out that there are some events that we don't have the option of locating in determinate causal chains at all. The behaviour of individual particles is random. It cannot, even in theory, be precisely predicted.

The concept of the random, of course, exists in classical systems. There, however, it has two applications which are perfectly compatible with determinism. The first is as a synonym for noise – phenomena (like the vibrations caused by the snooker upstairs when they show up on the seismograph which is being used to study earthquakes) which, even if they are fully determined within their own context, are to be dismissed as irrelevant to the one under consideration. Or it is introduced into experiments and observations as an artificial construction, by way of random number tables and generators, again as a way of excluding the unwanted – in this case the appearance of misleading patterns and regularities in results because of uncovenanted patterns and regularities in the choice of data.[12]

In quantum mechanics, however, the random is neither a construct nor a form of exclusion. It arises naturally, and it is inherent to the phenomena under consideration. The only predictions you can make about quantum phenomena, therefore, are probabilistic ones – derivations from general averages, as if the only predictions you could make about your own activities at 11.00 a.m. tomorrow had to be derived from statistical generalisations about what people of your sex, age, nationality, class, and educational background are usually doing at that time of day. (Generalisations which in the everyday world are derived precisely from particular instances of what you and other particular individuals do on particular occasions.)

You might feel that indeterminacy is significant only in the world of sub-atomic particles, and that this represents some local anomaly which could perhaps be explained without prejudice to the general principles of causality and predictivity. But indeterminacy is *not* confined to the world of sub-atomic particles. No full description can ever be given of a moving bus any more than of a moving electron. The only difference is that, in the case of the bus, the anomalies are so tiny as to be beyond practical consideration. But some of the *practical* implications of indeterminacy in the macroscopic world have also turned out to be vast, most obviously in biology. It is the staff of life – the source of the random mutations which provide the differential biological forms for natural selection to operate upon.[13] Quantum randomness, though, played its part in the structuring of the universe long before the biosphere evolved, and on a cosmic scale. Victor Weisskopf has shown[14] that the development of the galaxies was a series of indeterminate events. You might also say that, in the absence of any possible determining cause, the supposed origin of the entire universe in the Big Bang can be understood (if it can be understood at all) only as a random event.

Many physicists have baulked at the loss of deterministic causality apparently implied by quantum theory. One of the aims of multiple universe theory, at any rate as propounded by David Deutsch, is to preserve determinism, by declaring that the events in each possible universe are fully determined by the history of that particular universe; though another gaping hole in the structure has appeared, because now the relations between the different universes are not deterministic – seem, indeed, not to be causal at all – perhaps even not to exist.

Prigogine, the prophet of complexity and instability, tries to develop what he calls 'an "intermediate" description that lies somewhere between the two alienating images of a deterministic world and an arbitrary world of pure chance',[15] by showing that probability is itself an objective quality of the universe – that 'the quantum theory of unstable dynamical systems with persistent interactions leads, as in classical systems, to a description that is both statistical and realistic . . . As in classical physics, probability emerges from quantum mechanics as a fundamental concept.'[16] In fact he believes that, even at the macroscopic level, the laws of nature deal not with certitudes but with possibilities, and 'describe a world of irregular, chaotic motions more akin to the image of the ancient atomist than to the world of regular Newtonian orbits'.[17] He sees science as being 'on the eve of the triumph of the "probabilistic revolution", which has been going on for centuries. Probability is no longer a state of mind due to our ignorance, but the result of the laws of nature.'[18]

The suggestion that probability has ever been regarded as 'a state of mind', and that it was 'due to our ignorance', seems extraordinarily confused. It's obviously untrue in the case of mathematical or statistical probability, but how could even the probability that it will rain tomorrow in Des Moines possibly be a state of someone's mind? It is an assessment of a situation entirely external to the person making the assessment, and if it's the state of that person's mind then so is any judgement of an objective state of affairs – or any expression of it whatsoever, no matter how definite. The assessment might be limited or distorted by that person's ignorance, whether of the current weather situation in Iowa or of meteorology in general; but it might equally well be derived from a knowledge of everything that could, today before the event, be known. Whether it actually *will* rain tomorrow in Des Moines is not something that can at present be known, and our failure to know it is no more the product of ignorance than our failure to know whether the square root of minus four is minus two or plus two.

But then the view that Prigogine sees as supplanting this straw dummy isn't any more plausible. There is certainly a quantifiable probability that it will rain tomorrow in Des Moines (though I can also make statements of probability without any quantifiable basis at all). This probability is based upon either the past record of weather in Des Moines at this time of year, and/or a projection from the developing weather pattern in Iowa as it has been observed today. So is this probability for tomorrow an objective quality of the weather today? There is, after all, also a quantifiable probability, derived from exactly the same sort of information, that it was raining in Des Moines yesterday, and another one that it is raining there right now. The idea that these probabilities might be objective qualities of the weather is not appealing, though, because we know that a fully realised situation does already exist. It either did rain yesterday or it didn't, it either is raining now or it isn't, and we can find out by phoning someone in Des Moines; at which point the alternative probabilities are revealed as ephemeral assessments which have now become obsolete.

But then the same fate awaits statements of future probability. The probability can be stated more and more precisely as the moment for its realisation or non-realisation approaches – until, at the moment itself, the probability is extinguished in certainty. (Some probabilities can be extinguished earlier. For instance, the probability that it *won't* rain in Des Moines during the next twenty-four hours, if at any point during that time it does.)

Is the situation any different with (for example) the behaviour of a particle? Our failure to have complete knowledge of the present state of a particle is nothing to do with our state of mind or our ignorance – it's part of the logic of the situation. The probability of its being in this or that particular state is objectively quantifiable, and although it can't be extinguished by events, like the probability of rain in Des Moines, it still arises only in the circumstances of its being expressed.[19]

Of course, it's possible to generalise about the behaviour of more than one particle, and the individual unpredictabilities will tend to cancel each other out, just as the movements and moods of individuals in a crowd become subsumed in the general movements and moods of the crowd as a whole, so that the higher the numbers of particles about which we are generalising – as when we talk about the behaviour of an object that the particles make up in the macroscopic world – the more broadly established the generalisations become, so that they are effectively deterministic. A macroscopic object is a population of quantum particles. There is no contradiction between the applicability of deterministic laws at the macroscopic level and probabilistic ones at the quantum level. The two are equivalent, just as the series of probabilities about our individual holiday choices, for example ('I, on the basis of what I did in years past, am forty per cent likely to go to Italy . . . you are likewise twenty-nine per cent likely . . . your brother-in-law is zero per cent likely . . . '), is generalised to the definite prediction that 'seven per cent of the British population (or whatever) will spend their holidays in Italy', which in due course will be verified or falsified in a way that the individual probabilities cannot be.

If this constitutes determinism, though, then it is a determinism remarkably unlike the classical sort, which undertook to explain not just the general behaviour of classes of cases, but the particularities of every individual case. A professional snooker player would occasion surprise if he shrugged charmingly and said that there was no telling what effect his next stroke would have on the particular ball he was playing. It would be fortunate for him if it happened to hit black and made a cannon with blue; but for all he knew it might make the cannon with yellow, or go into the middle pocket, or split in two, or rise vertically into the air. The only thing he could say was that if he made the same stroke often enough, in a large enough number of identical situations, then on the whole he could be sure that the general trend would be to make the cannon with black and blue. The goals of the 'broad masses', as they used to be called in the Soviet world, are not the same as the rights of the individual.

Various confusions arise about this. Even a scientist as perceptive as Max Perutz claims that a genetic mutation is deterministic in the sense of being 'an event whose happening is determined purely by the laws of chance'.[20] But the case with 'the laws of chance' is the same as with 'the laws of nature'. Laws *determine* nothing, even if they express aspects of the universe which *are* themselves determined. And the laws of chance, unlike physical laws, don't even summate and codify our observations. They are simply ways of expressing the mathematical implications of randomness, of saying what it is for something to be *undetermined*. They merely state mathematically what the probability of an event is, in circumstances when it is impossible to predict it specifically. To say that the probability of getting a head on the next spin of a fair coin is fifty per cent, or that the more times you spin it the closer the total of heads will be to fifty per cent, is not to express a law, or a theory, or a generalisation. It is subject to no experimental verification or falsification – even if you spun the coin a million times and got precisely five hundred thousand heads and five hundred thousand tails it wouldn't verify the statement, any more than it would falsify it if you got a million of one or the other (though the latter case would cast some practical doubt on the fairness of the coin, or of the spin, or of the spinner). The statement is a hypothetical conditional like any other mathematical statement. '$5 - 1 = 4$' makes no assertion about the way things are. In so far as it relates to the world of things it says merely that *if* you have five of something and you take one of the things away you will have four. It determines nothing; it merely makes clear the ways in which the concepts of five and one and four are related to each other. The laws of chance, likewise, far from being the expression of any kind of determinism, make clear the mathematics precisely of undetermined situations (they don't even imply that such situations exist).[21]

Some of the confusion arises because the concept of probability is used with at least four distinct meanings:

1. Mathematical, following deductively from the enumeration of the choices. A die has six sides, therefore there is by definition a one-in-six probability of any single specified result – and if this particular die doesn't deliver these probabilities then its physical construction fails to reflect the mathematical formula that it is supposed to reflect.

2. Also mathematically deduced, but from observed contingent data. If twelve per cent of the population is blue-eyed, the prob-

ability that the next person to enter the room is blue-eyed is twelve in one hundred. This kind of probability, however, obviously depends upon how you specify the situation. Where is the room? Who is likely to have access to it? If it's in a district with a high proportion of black or Asian residents, or if the only people who ever come to call are members of your family, who are all brown-eyed, then the national generalisation won't yield a reliable probability.

The probabilities of dice *could* be calculated in the same way, from the records of previous throws. The correlation with the mathematical version might be interesting. You would get a more accurate prediction for any particular actual die, and you could use it to measure the trueness of the die.

3. In reference to present or past events such as the rain yesterday or today in Des Moines, where, even though the state of affairs may be definite, doubt exists as to what it is. Your aunt is probably at home now, and is probably not answering the phone because she can't hear it in the garden. Well, either she is at home or she isn't, and either she's not answering the phone for this reason or there's some other explanation. If you knew more you could state the situation quite definitely. Your aunt herself could (or probably could!). In the meantime your estimate of the probability can depend only upon your knowledge of your aunt's past behaviour and projected plans.

A confusion arises here because of the habit of physicists of assigning a whole-number probability value, 1 or 0, to past events whose outcome is known (1 – it happened; 0 – it didn't). This is misleading if it is intended to accommodate uncertainty to certainty, and in any case metaphysical. The concept of probability becomes extinguished at the limiting case.

4. In reference to predictions where rational judgement can be exercised, but where there is no way in which the grounds for this judgement can be properly quantified, as in predicting the outcome of a horse race or a literary prize. The resulting predictions in both cases, of course, can be quantified in the form of odds; but these don't reflect any objective facts about the situation, only the balance of judgements made by people who are betting on the outcome (or the predictions of bookmakers as to what

those judgements will be – a second level of the same sort of probability).

In fact this fourth category could be divided again, because there is a difference between horse races and literary prizes. In the odds on Derby runners the predictions of the punters are at any rate guided by *some* statable objective factors: the recorded past form of the runners, their present condition, estimates of the softness or hardness of the turf, etc. The state of the turf and the muscle tone of the horses will have a rather deterministic effect upon the result of the race, and you can imagine quantifying them on various scales. There are other relevant determining factors which are not quantifiable even in theory: the state of mind of the various jockeys, for instance, and the state of mind of the various horses. (Even horses have their good days and their bad days!)

But when we turn from the Derby to the Booker all objective correlatives to the odds seem to have vanished. There is no past form in the contenders which is relevant, because their previous record of wins and losses in literary prizes is unlikely to have any bearing upon the decisions of the judges in this case. The past record of the judges' tastes and prejudices may be some kind of guide. It may be the case statistically that they have in the past tended to favour works by white authors, or by black authors; short books or long books.[22] There is no way, though, in which this could be quantified – no way in which it could be given definite form of any sort. And even the judge who has taken a known position on particular categories of fiction may decline to categorise these particular entries in a way that makes them fit his position. He may in any case change his position – may even seize the opportunity to demonstrate his open-mindedness, or unpredictability. Moods and vagaries on the part of the judges, just as unquantifiable as the states of mind of the jockeys and horses, also come into it. In any case, even if the initial states of judgement, prejudice, and mood were precisely statable, the complex dynamics of a group decision often have a chaotic effect upon the outcome. Another radical difference: there is no clear decision procedure for settling the outcome – no fixed winning-post which one of the runners must eventually pass first.

A literary prize, in fact, is a paradigm case of a class of uncertain situations which will result in a definite outcome, and which are not random (or no more than partially random), but in which the outcome can be assigned no real objective probability. Assessments of probability are made, as in all human affairs, but this probability derives from no actual state of affairs in the world – not even to some unquantifiable one inside the minds of the assessors.

I suppose you might say that, as with quantum states, it derives from a series of different and mutually contradictory states of affairs: that A is going to win, and not B, C, or D; that B is going to win, and not A, C, or D . . . etc. With the passage of time these states then *decohere* (as physicists say when the multiple simultaneous possibilities of the quantum state are resolved into the single actuality of the macroscopic world). If this analogy has any substance at all it is only in situations where (as in a competition) the different outcomes are discrete (as with the different slits in the diffraction grating through which the particle might pass). Most of the Booker-type probabilities in the world, though, are not as clear-cut as the Booker itself. There is a continuum of possible outcomes. What will the state of the parties be after the next election? What will the company's profits be in the current financial year?

You might extend the types of probability to include the class of cases where the range, not only of possible inputs but of possible outcomes, is indefinite in range and form. This may be the most numerous class of all. What will the sky look like at six o'clock this evening? What will her mood be when she arrives? What will I be doing this time next week? The supposedly neatly branching pathways postulated in modern quantum mechanics become infinite in number, and shade into vague smears of possibility across the future landscape.

There is, however, one fundamental likeness between probabilities of all these different sorts: in none of them does the probability represent any wholly present state of affairs. There is no possible single snapshot of the present from which the probability could be derived. Probabilities for the future, even where they have a statable objective basis, necessarily involve not only the present but the past.

The existence of a past depends upon a standpoint in time, and therefore upon a possible observer. This is equally true in the case where the record of the past is a natural one (fossils, rings in tree-trunks, etc.), because this is interpretable as a record only in the understanding of someone who is perceiving it now. The only existential status a probability has is as an assessment made by a possible present observer who is in a position to relate past to future.

Even if the classical universe might be imagined to work without an observer (though I have argued that it can't), the probabilistic universe of quantum mechanics makes no sense at all unless it is inhabited; and nor does the probabilistic macro-world proposed by Prigogine and others.

How is it that we can, in the determinate world of classical mechanics, use things like the spin of a coin to generate random outcomes? Which is to say, outcomes undetermined by either previous states of affairs or by human choice. It's because, even in the classical world, there are different degrees of determinacy. The spin of a coin gives determinate outcome to a relatively indeterminate process. Human actions, however skilled, are not precisely controllable, so that the amount of kinetic energy transferred to the coin, its exact point of application on the surface of the coin, and the distance of the coin from the surface that will eventually bring it to a standstill in either this position or that, are fairly indeterminate. Once the coin has left the spinner's thumb, of course, it becomes an exemplar of physical laws which solve the complex equation relating momentum to gravity and air resistance, and the diameter of the coin to the speed of the spin and the distance to the floor, with absolute precision. Since the result of all this has to be expressed with digital unambiguity (heads/tails), the relatively indeterminate input has an entirely determinate outcome – but one that remains in practice indeterminable in advance.

Exactly the same process could be seen at work, but with an even sharper crossover from indeterminate to determinate, in a random-choice machine that I saw in action when I was in Moscow as a young man. It had been built out of household bits and pieces by a friend of mine called Valya, who was doing research in logic. We gambled on it for kopecks, and for the sake of argument in some debate we were having I played according to the Monte Carlo Fallacy – I pretended to believe that the laws of chance implied that a run of results against the odds increased the probability of an opposite result, and, entirely unjustly, won all the kopecks that Valya, his wife, and his mother had in their flat.

But what really interested me about his machine was how it worked, which was probably based on a familiar idea but was a novelty to me. It ran off the electric mains, which, in Moscow as in London, carry a fifty-cycle alternating current, so that the polarity of the circuit reverses fifty times a second. You pressed your finger on a button and then took it off. If you took it off when the polarity was in the positive phase the left-hand lamp lit up. If you took it off when the polarity was in the negative phase the right-hand lamp lit up. Which lamp lit depended entirely upon the moment you chose to take your finger off.

If human choice can be random then this would of itself introduce randomness into the machine. But, even if you regard human choice as entirely predetermined, it can't function deterministically here, for three separate reasons. Firstly, you have no way of knowing which phase the current is in, so that you're choosing blindfold. Secondly, even if there were an indicator on the machine displaying the phase, you could never perceive and discriminate an event lasting only a fiftieth of a second. And thirdly, even if you had been able to, you could never react fast enough to make use of the knowledge. So that, whether you take the view that your choice of when to press the button is genuinely free and sovereign, or whether you think it could be entirely explained by reference to the history of the world up to that point, the outcome of your choice, or 'choice', varied in a way that was entirely undetermined by it.

There was something vertiginous about the sheer speed, authority, and precision with which Valya's machine both delivered the product of your (apparently) unfettered choice and simultaneously expunged its significance. You might argue that the indeterminability of such an outcome is trivial. Left lamp, right lamp – a kopeck to me or a kopeck to Valya. But now imagine the game being played with one of the circuits connected not to a lamp but to a revolver, as a form of Russian roulette – or even, a few years later, when technological possibilities had improved, by a drunken Air Force general, with the machine wired to an intercontinental ballistic missile . . .

The element of choice (or 'choice') is in fact an irrelevance. The operating principle is, as with the spin of a coin, the interference pattern set up between two systems of different determinacy. (Though 'pattern' suggests a regularity which is precisely what is missing, except in the statistical evennesses emerging probabilistically over a long run of results.) It would still work if the input were provided not by a person but by another machine, provided only that its functioning was less precise than the cycle of the current on the Moscow grid. You could get the same results by setting a clockwork motor, say, to operate the button (whether at predetermined intervals, or even tripped by a device which was able to recognise the phase of the current), because the machine would show up the motor's imprecision in a way that a less precise marker (the hands of a clock, say) does not.

What would happen if you connected Valya's machine up to a system of equal determinacy – if, for instance, its output were used as the input to another machine of the same sort? Nothing. The output of the second machine (provided that it was connected to the same source of current) would be the same as the output of the first. The two machines would remain in phase (or with a regular drift in the phase caused by the travel time of the signals in the circuitry) and the output of the second machine could be entirely predicted from the output of the first. What, on the other hand, would happen if the machine were operated not by a human being or a clockwork motor but by a device even more accurate than itself – an atomic clock, say, together with an appropriately swift and precise switching device? Then it would function as a randomiser for the converse reason – by exposing the slight fluctuations in the mechanical functioning of the generators feeding the Moscow grid.

Mapping macroscopic phenomena on to the same phenomena considered at the molecular level would produce a similar random disparity to the one introduced by a human being operating Valya's machine; so would mapping the molecular on to the sub-atomic; and probably so on downwards through the hierarchy of elementary particles. The random behaviour of a particle is another example of the same phenomenon. The randomness becomes apparent at a similar crossover point between incompatible systems – the point of decoherence, when its indeterminate behaviour is resolved in the determinate environment of the macroscopic world.

And what are we to say about the twilight zone between the determinism of classical mechanics and the indeterminism of the quantum world – the kind of systems that are studied in so-called chaos theory? These are very common – the weather, for example, or turbulence in fluids. (Prigogine believes them to be characteristic of the real physical world in general.) They are systems that appear to be unstable or disordered because the events that constitute them are so complexly interconnected – and so often feed back into themselves – that very small differences in initial conditions produce very large differences in outcome.

The systems have determinate origins. Warm air rising locally, in an entirely law-abiding way, becomes involved with other currents of air, each of which is no less classically determined by local conditions; or a placidly flowing river, peacefully adhering to the tenets of classical fluid mechanics, tumbles into a gorge – and abruptly the warmed air and the falling water start to behave like a bank manager of unimpeachable probity who one fine day absconds with a lap-dancer and large sums extracted from his customers' accounts. Generalised descriptions can still be given, and generalised probabilistic predictions can be made, of the sort that the psychologist called in by head office can offer about bank managers having mid-life crises. But it becomes impossible to give precise specific descriptions, or to make precise specific predictions. In the past everyone knew exactly what the bank manager was doing and where he would be at every moment of the day. Now not even the psychologist can tell the bank exactly what he's up to with the dancer, or whether he will flee to the Bahamas or Australia or Budleigh Salterton.

Now think of the interplay between two unrelated chaotic systems – the cloud patterns in the sky above and their reflections in the rapids below . . .

'Chaos theory' is a paradoxical name, since what it refers to is a general understanding that the apparent chaos of these systems has a kind of order in it. They display patterns, though remarkably complex ones. These patterns arise from a logic that is often common to apparently unlike systems, and which transcends particular physical idiosyncrasies (the turbulence of the water in the rapids, for example, shares a logic with turbulence in the markets). The magnification of small differences, and the sudden transitions into complexity, often arise from iteration – the repeated feeding back of results into the same operation. The theory extends mathematics into realms where it previously seemed inapplicable, and it is in the spirit of the various other ways that have been found to use mathematical theory to describe elusive aspects of the physical world (such as the discovery that imaginary numbers were a good model for the paradoxical behaviour of particles in quantum conditions).

One chaotic system which is not, I think, studied by the proponents of chaos theory is the confusion of their views on its philosophical implications. Does it render the phenomena it studies 'predictable',[23] 'not utterly unpredictable',[24] 'unpredictable',[25] or even 'freed at last from the shackles of order and predictability'[26]? Most of them insist that chaotic systems are deterministic, by which they seem to mean that they exclude the random; though some believe that they 'randomly explore' their possibilities.[27] One of them even denies that the systems are probabilistic,[28] but in any ordinary usage the statements that these theories allow one to make are surely probabilistic when applied to individual cases, and approach the point where probability becomes so high that it merges with the determinate only with a greater and greater degree of generalisation, very much like statements about particles.[29]

We have plenty of familiar parallels in human behaviour. You observe a man (another misbehaving bank manager, perhaps) moving haphazardly about a field, and his behaviour appears entirely random, until you discover that he is searching for a wallet he has dropped, while at the same time taking a call from his mistress on a mobile phone which is almost out of receiving range, and trying to avoid being seen doing so by his jealous wife. You understand now that his moves are not arbitrary, that they are complexly purposive. You have no mathematical equation to model them, and no way of predicting his next move or any other particular move – even of producing any very useful predictive generalisation. His behaviour is highly indeterminate, but not random. Even if you can't begin to predict it you now feel that you can understand it. It is explainable; it relates to ideas you are familiar with.

Now suppose that neither the wandering man in the field himself nor anyone else can offer any account of his purposes. You might try to preserve the principle of explicability, even if not the practice, by insisting that there *was* some purpose in it, even if the man himself was entirely unconscious of it. But now we are back into metaphysics: there must in principle be an explanation because . . . because the principle requires there must in principle be an explanation.

And isn't there an element of the chaotic in even the simplest and least turbulent natural phenomena? I look up from my work and see the leaves on the trees outside the window stirring gently in the breeze. Are these part of a fully determinate system? It's difficult to know what to say. Not, presumably, if you include the movement of the air, which is part of the weather, acknowledged to be chaotic. But set this aside for the moment. There must in theory be ways of representing

the elasticity of the leaves and branches of this particular sort of tree
. . . though each individual leaf is going to incorporate slight variations
from the average shape . . . so that once again we are talking about
probabilities and generalisations. Now think of the light reflected into
my eye from the leaves as one branch moves gently in front of another – and both of them together in front of the branches of a different
sort of tree . . .

It is difficult to imagine the order of complexity that would be
involved in any mathematical representation of this simple, familiar,
peaceful scene. Now consider what it would be like to trace back the
causal history that accounts for it. You think of general laws relating
to gas pressure, and explanations of leaf development. But these, as
related to the behaviour of individual molecules and cells, are merely
probabilistic. To be used to present a fully determinate causal history
they have to be seen as the templates of astronomical quantities of particular events, tracing on the one hand the growth of this particular
leaf from this particular cell . . . to the seed excreted by this particular
bird . . . to the laying of the egg from which the bird emerged. Then on
the other hand the growth of *this* particular leaf . . . and the movement
of these particular molecules of air in response to the pressure exerted
by certain other molecules . . . molecule from atom binding with atom
. . . quark with quark . . .

Wider and wider, as with the simple game of snooker, grows the
cone of previous circumstances needed to explain each fragment of
the scene, just as your genealogy does as you trace it back. Your family tree, however, has regular mathematical outlines, because each
generation of effect has precisely two causal forebears, so that its ret-
rospective growth would go doubling backwards until at some point
it exactly matched the forward growth of the world's population.
(At which point everyone then alive would have been identified as
one of your ancestors, and your personal genealogy would be extinguished in generality. If you traced it back further it would taper
away again, through smaller and smaller complete populations of
homo sapiens and hominids, to the first single beginnings of bacter-
ial life.) The causal genealogy of that swaying tree, however, even if it
eventually found its way back to being coterminous with the entire
contents of the universe, and from there to the first beginnings of
everything in the inconceivable density of the supposed primal
object, would follow an indefinite path. Its shape would depend upon

what, from one antecedent to the next, you decided was a cause. Leaf from tree and tree from seed? Seed from tree + the bird that broadcast it? Tree from tree from tree . . . from bird from egg . . . from egg from bird m + bird f . . . ?

And perhaps the separate causal chains that we locate for overlapping phenomena can never be systematically related to each other, however far we follow them back. The trees live in symbiosis with the birds that nest in them and broadcast their seed, but each of the two families pursues its own customs and traditions, and meets for causal trade only at certain points – in nests and berries and seeds. Even the most classically determinate systems pursue their separate stories: the snooker balls on the table, the seismograph that the cannon disturbs, the earthquake that deflects the cannon and overwhelms the trace of it on the seismograph. You might say that these different families of events could be connected back historically to a common ancestor, just as two different individuals, or two different species, can always be, if you go back far enough. But by then so many other forces have been brought to bear upon each of the individuals, so many other genes have been introduced into the pool, that their cousinship has become too remote and dilute to have any meaning for our practical purposes.

Practical purposes . . . Once again we are driven up against this simple notion. How otherwise can we make sense of causality? Causes and effects have no clear natural frontiers, no status as causes and effects outside our finding them to be such. We judge what are causes and effects by selecting out of all the range of possible circumstances the ones that we think are relevant. Why do some circumstances seem more relevant than others? Because we see the analogy between these circumstances and other circumstances we have come across before. Often because we have already generalised and systematised a whole set of such analogies into a principle. Because, in other words, it's useful to us. Einstein said that the theory determines the observation. (So, he might have added, does the mathematics.) When we don't know the causes of something, but insist that we could always *in theory*, or *in principle*, establish them, what exactly *is* this theory or principle? It is (once again, as with the randomly wandering bank manager) the theory or principle that we could always *in theory*, or *in principle*, establish the causes of everything.

This is what determinism in the end comes down to – a circularity, a principle held up by its own bootstraps. We have a feeling in our bones that an explanation of some sort can always be offered, and this feeling is as self-suspended as the feeling that the world is governed by some sort of benevolent force. Perhaps this snug constriction offers the same kind of security to some that tight swaddling is said to give for Russian babies.

The clock-spring is wound, the regulator releases its energy according to a certain inbuilt periodicity, brass wheels mesh finely with other brass wheels, and all the future states of the mechanism should be exactly specifiable. This is what makes it function as a timekeeper. We are not greatly surprised, though, to find that it loses or gains a minute or two a week, which makes it plain that the specifiability is not as exact as all that. Then one day it stops altogether. There was evidently a speck of dirt in the mechanism somewhere. One of the bearings has become a little worn. Well, we could have found the piece of dirt, and worked out what it would do; we could have calculated the wear on the bearing. But we know that our predictivity is never going to be complete. We are never going to be able to isolate our clockwork completely from other forces in the universe, or completely calculate the possibilities for wear and breakdown that lurk within it.

Looking back from the present, the determinist says: 'I could in principle give you a causal account of *everything* about the present. All I should need to know is everything about the past.' What possible meaning can be attached to this? In what sort of way could *everything* about the past ever be known? How big would this gentleman's head have to be? The answer is: bigger than you think, because one of the things that it would have to contain, among so much else, is his head. Then again, in what sort of words would the knowledge be expressed? In what sort of equations? Would these words, once uttered, these equations, once written, then become part of the world they were supposed to contain? In which case . . .

Looking forward from the present, the determinist assures us: 'I could in principle give you a causal account of *everything* about the future. All I should need to know is everything about the present.' All the same problems arise, of course; the only complete model of the present state of anything would be that present state itself. And what an incomprehensible mess that always is! Its unsatisfactoriness is well expressed by the yokel in the ancient but profound Irish joke who is asked by a motorist how to get to Kilkenny, and who replies: 'Well, I shouldn't start from *here*.' Excellent advice, that applies to so many aspects of life; think of all the political reforms, all the scams and wheezes, all the human relationships, all the battle plans, all the schemes for doing anything, that might just possibly work if only you could start from scratch, with a clean sheet of paper, and not from the crumpled, scribbled-over heap of torn dog-ends which is all that ever comes to hand!

There is another problem, too, about wholly specifying the present state of even the most limited system, never mind of the entire universe: the present keeps dissolving into the past. In any case, there is always an element of arbitrariness, or at least of choice, in specifying the present state of affairs – or any other starting point in a causal system. This arbitrariness carries through the system, even in one as apparently closed and classical as a snooker table, so that our knowledge of the future position of the balls on the table at the end of the quadruple cannon, even if we exclude the indeterminacy introduced by the uncertainties of the human performance in which the shot originated, must always remain more or less indeterminate. And in the highly volatile systems studied in chaos theory, where the smallest variations in initial conditions notoriously produce large differences in outcome, the significance of this inevitable arbitrariness of the starting point will be likewise magnified. The only entirely definite and non-arbitrary starting point one can conjecture in the universe is the supposed starting point of the universe itself (which is always just a moment or two out of the cosmologists' range).

In which case the only entirely definite and non-arbitrary model of the universe is the universe itself.

The notion of universal causality is overwhelmed by its own universality. All the delicate crystal lattice we glimpsed for a moment has melted into mush. We are drowning in the great Sea of Generality!

Universal causality is a uboama.

The uboama is like the rainbow: the closer you approach it the less substantial it becomes, until, as your hands finally close upon it to capture it, it vanishes altogether. I discovered its existence when I was first writing a newspaper column. Regular contributors to a newspaper attract regular correspondents. One of mine was a man who lived in Barking, and who had written a private scripture under some form of divine guidance. I forget exactly how God had conveyed the revelation to him, or quite what he was reported to have said. I was more struck by the material upon which my correspondent had chosen to write the revelation down, which was brown paper and the margins of old newspapers. He sent me regular and substantial samples of the work, in bulky parcels which had to be returned after I had digested the contents. He explained that there was a much larger archive that he couldn't send me, because it had been sequestered by Barking Council in the attic of a house he had once occupied. He wanted me to use my position and prestige as a columnist to persuade the Council to return it.

He also wanted me to perform one other small service. Under divine guidance, he explained, he had founded UBOAMA, the Universal Brotherhood of All Mankind Association. In consultation with God he had worked out the Association's aims and constitution, and done all the rest of the intellectual and inspirational spadework. What he wanted me to do was to look after the clerical side: to use the secretarial facilities to which I presumably had access to get application forms printed and distributed to the projected membership, which was of course all mankind. Then I simply had to make sure that all the forms were returned and filed, and to issue the membership cards, after which universal peace would reign on earth.

I declined, I'm afraid. I had only a part-time secretary, and in any case it seemed to me that there was a flaw in the Association's constitution. I could imagine an association that included *almost* all mankind, but not absolutely the whole of it. It seemed to me that there had to be some group left over – if only the staff of Barking Council, who had proved so unreceptive to the Association's teachings in the past – so that there was someone for the members of the Association to revile, patronise, missionise, mock, enslave or murder as they, or God, thought fit. Experience suggested that the more members of mankind there were assembled in some grouping, the more likely that grouping was to encompass within itself all the differences that divide mankind at large; until, at the point where it included *all* mankind, those differences would become indistinguishable from the differences that divide mankind as it is.

I'm now beginning to doubt my recollection that the location as well as the nature of this enterprise was Barking. But the uboama has remained with me as the embodiment of what the positivising philosophers who taught me used to call metaphysical: a proposition which no state of affairs can falsify, or a concept without antithesis. If everything is blue, or false, or evidence of God's benevolence, then nothing is, because the words no longer discriminate anything from anything else.

If everything is causally linked to everything, likewise, then causality is trivial; it is a uboama. For causality to mean anything there must be some possible alternative. To establish any system of cause and effect we have to treat the world at large as noise, from the confusion of which we struggle to distinguish particular signals, particular patterns that might be isolated sufficiently to be modelled by some particular procedure we are offering. The possibility of order requires the possibility of disorder. The converse is also true: disorder implies order. The increase in disorder required by the Second Law of Thermodynamics can occur only if there are states of relative order to start from. (And, if entropy is the redistribution of energy so that it can no longer form part of a causal chain, then this also surely implies a distinction between the causal and the non-causal.[30]) The concept of predictability, likewise, requires the possibility of unpredictability. And if you should ever be tempted to think that the realm of science, at any rate, is concerned only with the predictable, remember that in information theory the information which it studies is defined precisely as the unpredictable.

The ways in which human beings mess up the supposedly elegant determinism of the universe's great game of snooker are legion. One of our untidiest habits is littering the place with counterfactual conditionals. 'If I'd known you were coming I'd have baked a cake,' we say, flinging it down as if it was the most natural thing in the world. 'If I'd remembered to check my messages I'd have known you were coming. If I hadn't been too distracted by thinking about counterfactuals I might have remembered to check my messages.'

Philosophers have always had a problem about counterfactuals, but the difficulty that has mostly exercised them is knowing whether statements like these are true, and if so why. The only way we could tell whether I really would have baked a cake if I'd known you were coming would be for me to have known you were coming – and, the assertion plainly implies, I didn't. The truth of the whole depends upon the truth of the antecedent – which *isn't* true, if the assertion means what it appears to mean. Nor are there any circumstances in which it could be made true, because it relates either to a past state of affairs that cannot be retrospectively altered, or to a present ('If you were the only girl in the world . . . ') which is also beyond being other than what it is.

The problem in reconciling counterfactuals with determinism, though, if we're being literal-minded, and taking them at their face value as assertions about how things actually are, is that they run right against the grain of causality. My claim about the cake implies that things could have been otherwise than what they were. It could have been the case that I'd known you were coming (though in fact it wasn't). And if it *had* been then it would have opened up a completely new causal pathway. It would have caused a cake where no cake was. After which – who knows? – to your choking to death on it, and then to your widow or widower remarrying and having another child, whose granddaughter would on 12 September 2063, at a solemn ceremony in Disneyland, become Empress of Europe . . . But if things could *not* have been otherwise, as determinism claims, and I necessarily could *not* have known, given the previous causal history of the universe, that you were coming, then all statements suggesting that things *could* have been different, and that I *could* have known, are surely either meaningless or perhaps simply false. They have no place in the history of the snooker table. 'If red had not kissed black, it would have gone into the middle pocket.' What do you *mean*? It *did* kiss black! It *had* to kiss black! It could not *but* kiss black, because it had been struck by the cue-ball in exactly the way it *was* struck! There is no toehold for statements about what would have happened if what happened hadn't happened![31]

And yet we go on saying these things, even about the behaviour of snooker balls, and we persist in thinking we mean something by them, and understand what we mean.

Now it has to be admitted that there is indeed a great cloudiness about a lot of counterfactuals which makes them very difficult to deal with. 'If Hitler had not invaded Russia in 1941, and if Japan had not attacked the United States in 1942, then there would have been no Labour government in Britain in 1945.' But even counterfactuals as hazy as these we do manage to discuss. Indeed, this is surely their function –

not to assert any particular state of affairs, like factual statements – but to stimulate debate and the re-examination of what *did* actually happen. (Compare the rhetorical function of one of those challenging assertions that are offered in exam questions: '"Democracy is death to the freedom of the individual." Discuss.') They are judgements, and like most judgements, not true or false, but interesting or uninteresting, helpful or unhelpful. They open up fictitious worlds in which we see our own reflected as in a distorting mirror.

But let's forget grand claims about cakes and wars, and look at a very simple counterfactual, not cloudy at all, where everything is here in front of us, right under our noses. 'If I had not written the previous sentence, this sentence would be the first in the current paragraph.' This is surely both clearly meaningful and quite simply true. You might reasonably reply that if it is true then it is true because it is like 5 + 3 = 8: because it follows not from its expression of any contingent state of affairs, but simply from the meaning of 'not' and 'previous' and 'first'. Of course – but it does also propose that the world might have been contingently different – that something which did in fact occur, my writing of that first sentence, might *not* have occurred. So even this trivial example of the genre remains problematical to a determinist.

I suppose a determinist might deal with counterfactuals in the same way as with the snooker and the tree swaying in the breeze. He might say that yes, the second sentence *could* have been the first – if the state of affairs which preceded the writing of that paragraph, and which determined the course of it, had been different, and so on – back, once again, to the beginning of the universe.[32] In other words, for my piffling little counterfactual to be meaningful the whole course of the universe up to now would have had to be different; my counterfactual would have to be part of one single giant overarching counterfactual rooted in some event that occurred before causality began, and reverberating on until causality is extinguished. What I am implying is not just a marginal modification to an insignificant corner of recent history, but a totally fictitious universe from first to last.

This might seem to be a rather ponderous solution to the problem, somewhat akin to tidying the living-room by rebuilding the house. Nevertheless, a somewhat similar theory of counterfactuals (only vastly *more* ponderous) has been put forward by David Lewis[33] and others in terms of modal logic. Modal logic is the logic of propositions involving possibilities, probabilities, and counterfactuality, and Lewis is a proponent of modal realism – the view that all possible worlds do actually exist. He seems to derive this from the unstated premise that for a proposition to have meaning it must express an actual state of affairs, so that possible states of affairs which are not in fact the case are given an ontological status by being treated as 'possibilia', and he believes that all possible worlds (which make up the 'pluriverse') have the same metaphysical standing as the actual world. Since the actual world contains rocks, electromagnetic waves, etc., so must all the alternative possible worlds.[34]

Counterfactuals, then, are simply true indicative assertions about actual states of affairs in particular alternative worlds. '"If you had walked on the ice it would have broken" means that *The ice broke* at every member of a certain class of worlds . . . '[35] Modal realists expend an enormous amount of labour and skill on finding technical elaborations of the theory that are intended to make its translations more palatable. I can only admire their determination and ingenuity; but I can't help feeling that the whole enterprise is intended as a tease, a thought experiment to see what happens if you extend the rule of logic to notoriously rebellious tribes such as counterfactuals. Apart from all the metaphysical difficulties of the theory, as a way of analysing counterfactuals it misses two features that lie at their very heart.

The first is that we surely intend counterfactuals to say something about *this* world. When I tell you that I would have baked a cake, I mean that there would have been a cake for you to see, the sight of which would have touched your heart, not that there was a cake which unfortunately didn't touch your heart because it was invisible from where you were standing because it happened to be in another world. You know, too, that there is a real difference in what I should be telling you about my feelings and our relationship (in *this* world!) if I said the opposite: 'Even if I'd known you were coming I still wouldn't have baked a cake.'

The second essential feature that the modal realist analysis misses is precisely the fact that counterfactuals are talking about what is *not* the case. This is the joy of them. They are a way of slipping the normal bonds of causality and actuality, of getting an outside perspective on the world as it is. They are comments cast in the form of fictions, dramatised particular projections from general truths (or untruths), or from parallel cases. They may be highly plausible ones based upon the most careful research and considered judgement, like novels where a fictitious character has been threaded into an otherwise historical record. They may even be open-and-shut, like the one about the first sentence in the paragraph, or the one about the ice if it *did* break when someone else *did* step on it. But for that matter they may be reckless libels, or purely fanciful, and still be meaningful.

Nor is this an abstract point. What gives the actual world in front of our eyes its significance and savour is that it is shot through and through with the iridescent gleam of all the other possibilities that we conjecture it might have embodied. Our understanding of what it is for the world to be the way it is is intimately bound up with our understanding that it might have been otherwise, and may be yet.

Another problem: if counterfactual truths have to be found a home in alternative worlds, then not only must there be infinitely many of these worlds but each of these worlds must automatically beget infinitely more. If, for example, the friend for whom I failed to bake the cake had been a modal realist (and he *was*, in some world or worlds!) he might well have replied (and *did*, in one or more of the worlds where he lived): 'So this cake that you didn't bake in the actual world, but must have done in another – what sort of cake was it?'

> Me: A fruit cake, of course, since I know this is the sort you like.
> My modal realist friend: Just so long as it didn't have nuts in.
> Me: Of course not. You're allergic to nuts. If I *had* baked a cake with nuts in it would have choked you.
> Modal realist friend: Another counterfactual! And one that implies the most murderously aggressive feelings towards me! Because in some world or other you *did* bake a cake with nuts in! When you knew perfectly well it would choke me!

Of course, he *couldn't* have eaten the cake with nuts in, or the fruit cake either, since they exist in worlds which have no causal connections with this one. But then I couldn't have baked them. Not this I. It must have been another I, perhaps known to me as 'he', but possessing (I assume) all my characteristics, memories, thoughts, etc., author of all my works, father of all my children . . . [36]

It's difficult in any case to see how modal realists reconcile counter-factuals with determinism, as Lewis and others claim to do. Lewis offers only two ways in which an alternative world might diverge from the existing one:

(1) Through an indeterministic causal transaction that starts out identically but ends up differently at the two worlds, each outcome being legal [i.e. in accordance with causal laws].

(2) Through a causal transaction – deterministic or not – that starts out identically at the two worlds and has one outcome at α [the actu-al world] and a different one at world w, the outcome at w being in conflict with α's causal laws.

Which is to say that, either determinism *doesn't* hold, or, if the causal laws of the alternative world are different from the ones in this world, things in that world have been different *ab initio*, just as Leibniz (see note 32) and my caricature determinist were reduced to proposing.[37]

If we consider counterfactuals as fictions, though, or as quasi-fictions, causality is irrelevant to our acceptance of the false antecedent. How it would have come about that I knew you were arriving, or that you stepped on the ice instead of not stepping on it, is outside the purview of the respective assertions, just as it is how Fabrice del Dongo man-aged to participate in the battle of Waterloo, when no such person as Fabrice del Dongo ever existed. Even if determinism is true we must be able to read *La Chartreuse de Parme* and understand it without too much metaphysical anguish. Never mind literature, never mind story-telling – we must be able to use our imaginations, and envisage things that are not the case, if we are ever to find solutions to even the most practical problems. And if we had to find a reason for my knowing you were coming, or your stepping on the ice, neither a random quan-tum event nor the complete suspension of causality as we know it would seem as likely as my getting an email from you, say, or your simply fancying a slide.

The untidiness of human mental processes has an even more pro-foundly unsettling effect in another way upon the clarity of the events on the great green baize if, as quantum mechanics seems to at least some physicists to imply, those events are specifiable only within the limitations of human observation. These limitations are severe. It's not just that we are required to be in one place at one time, or that we cannot measure conjugate variables with equal accuracy. Our senses impose limitations of their own, just like the instruments with which we extend them. The human eye will not discriminate below a certain threshold, so that when we place our ruler on the table to measure the distance travelled by red, or our protractor to measure the angle at which it departed from the track of white, there is always a slight haze of indeterminacy about the fit between the two, however many times we magnify it through the microscope to inspect it more closely.[38] And since, even within these limits, the eye sees only a tiny patch of the scene in front of it at any one time, and has to dance like a pea on a drum to take in any significant area, every observation is made not at a point in time, but over a period, during which the universe does not stand still. Then again the signals from the eye take time to reach the brain. The brain takes time to make sense of them, and to resolve them into figures and statements. Our vision of the world is always slightly blurred, and always slightly out of date. The lines in which the balls travel are not quite straight, the angles somewhat indefinite, the click-click-click not quite clean. And the indeterminacy in the behaviour of the particles making up the world we are observing is paralleled and reinforced by a similar indeterminacy of the particles in our nerves and brains whose behaviour constitutes the act of observation.

You might say that this introduces an element of counterfactuality into even the most indicative of our statements and the most definite of our measurements. We say: 'Red moved off at an angle of 10.3 degrees and travelled 15.92 cm before it hit black.' Translate: '*If it had been possible (which it wasn't)* to make an exact correlation between the protractor and the track of red, or between the ruler and the extent of red's travel, then 10.3 degrees and 15.92 cm is our best estimate of what the readings *would have been*.'

Fractal geometry (a development of chaos theory) is an admirable attempt to come to terms with the irregularity of the world – the irregularity in this case being a function not of the subjective indeterminacy imposed by human observation, but of an objective feature of the universe: the failure of so many phenomena to conform in any useful way to the regular lines and curves of traditional geometry. But here, once again, the observer, and the limitations of his viewpoint, have to be taken into consideration. What Mandelbrot (one of the founding fathers of chaos theory) realised is that measurement – and in fractal theory, more fundamentally, the number of dimensions involved in the measurement – depends upon the distance of the observer from what he is observing. Mandelbrot was well aware of the parallel with relativity and indeterminacy. 'The notion that a numerical result should depend on the relation of object to observer,' he said, 'is in the spirit of physics in this century and is even an exemplary illustration of it.'[39]

The 'Mandelbrot set' that results from this – the equations that locate certain regularities within the irregularity of the world – is an example of another category of representation that sits uneasily with determinism. The successive shapes thrown up by the set – the repeating structures contained inside each other like Russian dolls – are not probabilistic or indeterminate but *approximative*. Each figure of the set pursues the ever-receding horizon of exactitude in the way used in calculus, or in the computation of the numerical value of π and $\sqrt{2}$ – by a series of closer and closer approximations. Once again our understanding of the world emerges not passively, but through our active traffic with it, through the very process of approach and withdrawal.

The approximations can be pursued layer by layer, without ever reaching any final or absolute statement about the world. They are presumably halted at the level of the quantum, as the unpacking of a nest of Russian dolls is halted when you reach the solid doll in the centre. But a solid doll is exactly what the quantum is not, and the final and absolute elude us once again as approximativity merges with indeterminacy.

There is a destiny that shapes our ends, rough-hew them how we will . . . Yes, and just as there is a centrifugal force to balance the centripetal force of gravity, so there is another destiny that roughhews the ends we've tried to shape.

We have touched upon the indeterminacies that the vagaries of human performance introduce into snooker, and the way in which they make it a playable game, rather than a set of foregone conclusions. What makes it an *interesting* game, though, and not just another random result generator like Valya's machine or the spinning of a coin, is that, in spite of all our mechanical inaccuracy, we *do* in fact manage to exercise some control over proceedings. Snooker players can, by the policies they adopt at the tactical level in playing a stroke, together with the level of effort they elect to make in carrying those policies out, and by the policies and efforts they employ at the strategic level in developing their playing skills, reduce the indeterminacy of their input. Their choices and decisions are central to the game. Part of the causality of the game is generated by the players from inside themselves.

A determinist[40] would insist that this is an illusion; although we seem to make choices and decisions they are not autonomous – they are not fresh input into the causal chain. They are the results of anterior causes, like everything else – our genetic inheritance and the events which have influenced us. I detect the outlines of that same familiar uboama lurking in the bushes here. To prepare myself for my game of snooker with you I drove myself on to practise every day even when I was exhausted. I racked my brains to find the right shots. I bent over the table until my back was breaking, I squinted until my eye watered, I held my breath until everything went black in front of me. Now, counter-suggestible as always, you become a determinist, and tell me that I had no choice in any of this. Are you trying to persuade me, I scream, that really I was putting no more effort into all this than you were into that shot of yours when you casually jabbed at the ball, thinking about determinism instead of the stroke, and missed completely? Of course not, you tell me. All you're saying is that my squinting and grunting, my sensations of strain and self-sacrifice, my feeling of ingenuity in choosing to play the stroke in that particular way, were themselves determined, were part of the causal process. To which I reply that if I have to go through the same full agony even though I had no choice as I should if I really did have a choice then I don't see from where I'm standing what the distinction is. You beg me not to wave my cue in front of your face in that threatening manner. You are obliged to utter these words, of course, and to feel the urgency behind them, even though you know that they are really being uttered for you by forces outside your control, and that, if they affect my choice as to whether to jab the cue into you instead of the ball, they do so only in some weird way which, while it may be one of the absolute external determinants of that choice, doesn't imply that I actually have a choice to be determined.

In any case, the role that human choice plays in the proceedings of the universe is much more complex, and much harder to pin down, than the terms of the traditional debate about determinism vs. freewill allows. Even if our decisions were not decisions at all, but mechanical responses to stimuli, the mechanics would be vastly more complex than this simple dichotomy suggests. In the first place, what bears upon our actions is not the external events themselves, but our perceptions of them. What we perceive, and how we perceive it, open up fresh possibilities and demands for a subjective input, and even if we agreed that this input was determined in its turn, there is plainly more than one causal system involved here, with all kinds of strange new interference patterns (or non-patterns) forming between them.

The most strongly marked interference of all, though, is the one that arises between *your* intentions and interests on the one hand, and *mine* on the other. This is what makes the outcome of the game so difficult to predict, even with the most accomplished players – the fact that the game is not, ultimately, between player and balls, but between player and player. Not only among human beings, but throughout the biological world, predictivity is scuppered by competition. With hindsight, of course, everything can be explained, and determinism preserved: why Chelsea beat Arsenal, why the dodo became extinct, why *homo sapiens* prevailed over Neanderthal man, why Austin Motors was eaten by Morris Motors, and Morris by the British Motor Corporation, and the British Motor Corporation by British Leyland, and British Leyland by BMW, and BMW by Phoenix, and then ceased to exist altogether. But until each of these battles was joined, and each of these questions put to the crucial test, determinism held its breath.

The vagaries of human performance leave plenty of room for the concept of error, both errors of choice and decision (cue-ball hits black exactly as intended, but black doesn't go into the pocket because we'd failed to take into consideration that blue is in the way), and errors in the execution of those choices and decisions (cue-ball comes off the cue at an unexpected angle). Generalised explanations may be offered (tiredness, lack of concentration, demoralisation, etc.) but error in human performance is so unsurprising that often no explanation seems to be called for. To err is human; getting things even relatively right deserves prizes.

But can machines make errors? Of course – they plainly *do*. To err is also machine-like. Machines frequently fail to behave in the way that we expect them to, and the way they are designed to. They frequently fail to behave at all. But here we feel, as with causality in general, that there must always be an explanation. And even where the explanation offered is a generalised one – the machine was overheating, was being misused, was old – we are certain that it could in theory always be made absolutely specific. 'Overheating', yes – but what actually stopped the engine turning was that the components of this particular bearing expanded by seven micrometres, when the free play available was six micrometres. 'Misused' – loading of 107 kg applied to a part with a breaking-strain of 100. 'Old' – metal-fatigue after 27,756 flying hours, at accelerations of 2.3 G, in ambient temperature down to -37 degrees Celsius. All this may prove to be the case when we investigate, but the basis for our certainty that it must be so is . . . our certainty that it must be so, our picture of mechanical systems as entirely deterministic.

But mechanical systems aren't as deterministic as all *that*. There is room for negotiation in all these apparently precise formulations. A cable with a breaking-strain of 100 kg won't necessarily fail at 100.001 kg – or for that matter at 100.1 kg, or 101 kg, or even 110 kg. There is a possibility of shearing when the wing has flown twenty-five thousand hours, and a probability after thirty thousand hours; but it may stagger on for many more hours even after that. We can naturally claim that we could be more precise if we had more information – if we knew exactly how the fibres of this particular cable had been manufactured, or exactly what weather conditions the wing had been exposed to in the past. But even then some area of uncertainty would remain, which might I suppose eventually be traced down to indeterminacy at particle level.

In some mechanical systems error is not just possible – it is required. The perfection of living forms that so astonishes us arises from the imperfection that they embody – from random error in the transmission of genetic information upon which the precise and brutal mechanisms of natural selection operate. Quantum perturbations of particles, says Monod,[41] accumulate in cells until the aggregate is large enough to produce a macroscopic effect – an error in the form which may then be magnified by tripping a cascade of further errors.[42] If genetic material always reproduced itself without error there would be – well, no genetic material, for a start. Reproduction itself emerged only through selection, and selection only through the existence of variants from which to select. Fallible material is improvable material; infallible material remains unchanged.

So error in biological reproduction is not a random phenomenon. It is selected, determined by the process of which it is itself a determinant. But the particular errors themselves are not selected. Monod identifies two principal sources of their randomness.[43] The first is the interface between two different causal chains, as with the crossovers we examined earlier, but in this case between the one in which the mutational event itself is a link, and the one that connects the mutated genetic structure with its functional effects. (This is also, as in the case of Valya's machine, an interface between a non-deterministic system, or at any rate one in which non-deterministic events occur, and a deterministic one.) The second, and more radical, source is that mutations begin at the molecular level (where a small initial variation may trip cascades of further errors), and here quantum rules apply, including the uncertainty principle.

There are parallels with this logic in other areas of biology. Certain plants, in the Australian bush and elsewhere, can propagate only when the plant itself is destroyed by fire. This system of propagation has been determined by selection. It has been selected because it works; plants that propagate in this manner do manage to propagate, and thereby produce offspring that propagate in the same way. Bush fires, however, occur in haphazard relationship to the rest of the natural cycle, ignited by lightning strikes, or by the action of sunlight on complex concatenations of flammable materials.[44] And while the generation of static electricity in clouds, and the distribution of sunlight when the clouds clear, may be deterministic (in so far as weather systems are fully determinate; see above), and while the exact places where the lightning and the sunshine happens to strike may be fully determined by the interaction of the weather system and the geography and biology of the terrain, that strike, that sunshine, is part (as Monod would say) of another causal chain, and remains random within the life-cycle of the plant whose propagation it enables.

Once error has got into the system its consequences may be entirely deterministic. This is the stuff of classic tragedy. A fatal flaw in the hero's character, or a crossover between two moral systems like the crossovers we have noted between causal systems, and it sets in motion the relentless and unstoppable machinery of doom. It is also the stuff of classic farce. An initial misunderstanding occurs (a misheard name, a conclusion wrongly jumped to); or there is a crossover between the causal systems set up by different characters' intentions, whereby the lecherous compulsions of Monsieur Béchamel become entangled with the spiritual dispensations of Monseigneur Roux.

From this mutation in the fabric of everyday reality events follow with rigorous causality. The door slammed by Monsieur Béchamel in his haste to conceal Madame Soubise from the unexpected arrival of Monsieur Soubise strikes Monseigneur Roux on the behind just as he leans forward to implant a chaste ecclesiastical kiss upon the upturned forehead of the kneeling Mademoiselle Mornay, and propels him, with the strictest regard for classical mechanics, head over heels into the concealed bed where Madame Béchamel is waiting for Monsieur Soubise . . . As in the turbulent systems studied by chaos theory, a small anomaly in the initial conditions leads to complex and unpredictable consequences. This discrepancy in scale between cause and effects, according to David Deutsch, occurs only in classical systems and not in quantal ones, which suggests that quantal farce remains a challenge for the ambitious playwright.

Even some apparently random error turns out to be systematic when you look into it. You believe you are touch-typing the words 'I am a determinist', but when you look at it you find you've typed 'O s, s fryrt,omody'. An outcome only too characteristic of the indeterminacies of human performance, you might think – until it occurs to you that you were typing perfectly correctly, but with your hands one key to the left of where they should have been. So the errors were systematic, not random at all; though now the question arises as to *why* your hands were one place to the left, to which no clearly determinate answer presents itself.

So the supposedly universal causality on which the laws of nature depend has no more existence than the laws themselves, outside the man-made expression of it. It has substance, like the laws, only in the context of human thought and human purposes.

Something is out there, though, which is independent of us! That something can only be the universe itself, the great theatre of space and time in which the events given dramatic form by the concepts of laws and causality are played out. This must be a structure that endures, like any theatre, whether there is an audience in the house or not.

A tour of the building is next on the schedule.

4

Grand Theatre

The structure of space and time

What could be more familiar to us than space and time? It's often hard to trace the causes of things, and the appalling complexities of modern physics may be entirely beyond our grasp. But space and time are immediately, ubiquitously, and quite simply present to every one of us. There it is all around us – space. Here it is going by – time.

It's true that we're confused and dismayed, or possibly excited and even more confused, by some of the notions of time and space introduced by scientists, which seem to have little relation to that familiar pair of old boots we trudge around in every day. In relativity space and time are treated as a single entity, and one that can be subject to local distortion. Cosmologists talk about the *creation* of time and space in the first moments of the universe, as if they were not modes of the material universe, but parts of it, no less constructable and destructible than matter itself. Modern superstring theory proposes a unified explanation of the various forces and fundamental particles by introducing ten spatial dimensions (or in some versions eleven). The direction of time in any equation of classical physics is said to be arbitrarily reversible without upsetting the mathematics, so that the universe, having expanded from an infinitely dense point to some maximum size, may then go into reverse, and contract again to the beginning of time where it began.

Well, there are many aspects of the universe at large that are at odds with life lived on the middling scale of creatures like ourselves. Let us not be overawed, though. We need to keep our heads about us, now as always, because these middling-sized heads of ours are where all our great schemes for understanding the universe must in the end be accommodated.

We might note first of all that, even in our modest everyday experience, the concepts of time and space are intimately bound up with each other. We can separate them notionally, for the purposes of calculation, just as we can separate mass and energy, or voltage and amperage. But they have no independent physical reality. We can elect to give an account of some object as if captured whole at a mathematical point in time, but the concept is as artificial a construction as the point in time itself. In terms of empirical procedure we have to scan the object in order to describe it, and this is plainly not instantaneous. In any case, for our account to have any practical significance it must be assumed to hold over some period of time, specified or unspecified, so that a complete account of it would have to include any changes in its state during that period. Such changes may be small – may be imperceptible – may be beneath the threshold of consideration for our particular purposes. In theory, I suppose, no changes may occur at all. But if they don't, if the object we are describing is really immutable, then this would be part of the story. (And a quite astonishing part.) In any case, a full account of the object would have to specify these changes, or our disregard of them, or their absence, together with the length of time over which they are being considered.

This length of time, conversely, is defined in spatial terms. Firstly, because any event – not only a war or a sigh but also an observation – must occur in the flesh of spatially extended matter. Secondly, because the only possible measure of time is a comparison with other events (the birth of Christ, the progress of the earth around the sun, a vibration in quartz, etc.). These events generate signals that have to be brought into relationship with each other by an observer (or by an instrument standing in for an observer), and for the origins of the signals to be located in relationship to each other the distance they have travelled must in theory be specified. In other words, the spatial position of the observer is a term in the equation. In the case of imprecisely located and leisurely comparisons such as one between the birth of Christ and the repeal of the Corn Laws (both legs of which involve a chain of observations – eye-witnesses whose reports are heard or read by others, whose reports are heard or read in their turn, etc.) this is purely notional, but in the case of fast-moving objects like particles (as Einstein showed) it is crucial.

Time and space, in other words, are functions of each other. And in both, once again, the observer is central.

*

At once this introduces difficulties when we extend our speculations beyond the human scale. We begin to feel dizzy at the conception of a universe that goes on for ever in every direction – east and west, north and south, up and down, backwards and forwards. Our conceptions of space and time, after all, are practical; distances and durations are there to be specified – have meaning *only* in so far as they are specified. The difficulties of infinite size and duration, however, are overtaken by the difficulties introduced by the now common conception in cosmology that the universe began (and may end), and that it has in some sense a theoretically statable extent. We may feel not easier but queasier at the idea suggested by various cosmologists of a universe which is infinite but has a *shape* – a flattened sphere, a saddle, a doughnut, a set of dodecahedrons, the horn of an old-fashioned gramophone.

We can't help mapping back upon the universe not only the experience of our own life, with its beginning in birth and its end in death, but our general experience of all the things we are familiar with, each with its physical limits, and of all the processes and events we have ever come across, each of which begins, runs its course, and ends. Even the mountains, we know, were once flat and will one day return to level dust; even the sun began, and will one day grow cold. The beginnings and ends of which we have experience are all relative. The day began – but there was another day before it. The day will end – but another day will follow. The universe, though, the totality of things, outruns all possible subjective observation – outruns even the kind of mental projection from our everyday experience by which we conceive the birth and death of mountains.

The only possible analogy that our insignificant but ubiquitous observer can make is with the experience of his own life, before which, for him, there was nothing, and after which, for him, there will be nothing again. In the sense in which I give the world being, I bring it into being with my birth, and extinguish it with my death.

And yet, from inside my life, by projection, I also bring into being a continuing world that preceded me and will survive me. Without this projection I cannot make sense of my beginning and end. I don't in practice see my life, and my experience of the world, as emerging out of nothing. I accept, just as totally and naturally as I accept my own experience, that there were things happening before my life began (some of which explain my coming into being), and that there will be things continuing after it ends (some of which will be affected by my ceasing to be). What gives its beginning and end significance is surely this contrast between the short period when things of which I could have experience were, are, and will be happening, and the periods on either side of this when things were and will be happening of which I cannot have experience. So the analogy is a limited one.

What does it mean to say that fourteen billion years ago there was only a very small and dense object on the point of exploding? How can anything be small if there is nothing else that is larger, or dense if there is nothing else that is less dense? Well, with hindsight, of course, we mean that it was smaller and denser than it subsequently became. But at the moment of its beginning it had no qualities at all.

How can qualities that don't exist be meaningfully compared with qualities that do? What relation can something without qualities have with any possible observer, however notional?

And what does it mean to say that before this qualityless something there was nothing? How can there be a before for there to be nothing in, if there was nothing? How can there be a nothing to be before, if there was no before for it to be in? And are we saying something even more puzzling than this – that not only the fabric of the universe, but also the space and time in which that fabric exists, had a beginning? How could time be passing, how could there be space in which things could exist, if nothing *was* happening, if nothing *did* exist? But then how could things have suddenly started happening and being if there was no time and space for them to do it in?

These are familiar puzzles in which we have all lost ourselves. The two familiar companions of our daylight hours, when you wake and think about them in the small hours, assume gigantic proportions that paralyse the mind. When day returns, though, what does our actual acqaintance with these two suddenly mysterious and intimidating old friends consist in? It boils down to a series of practical considerations – of where things are located and how large they are, of when events occur and how long they last. Space, in our everyday life, is a matter of the position and size of particular things; time is a matter of the dating and duration of particular events; and none of these qualities causes us any great conceptual difficulty.

Can we agree on one simple thing for a start: that these four homely qualities are always – and can only be – attributes of particular things and events? That there is no stuff called size at large in the world, only particular sizes, and that for there to be a particular size there must be a something that has it? Likewise with position, and with dating and duration. To say anything about a size, a position, a dating, or a duration is to say something about a thing or an event – and not only about one thing or event, but about its relationship with other things or events. So that if there is nothing that has a size – nothing stretching from somewhere to somewhere else, together with something else to compare it with – then there are no sizes. To say there are no sizes is to say that there is no such thing as size. If there is nothing happening, likewise, and nothing else happening to measure it against, then there are no durations – and to say that there are no durations is to say that there is no such thing as duration.

You might reply that, even if there were nothing, the *possibility* would still exist of there being something, and that if so then there would be the *possibility* of a size and a duration, of sizes and durations, of space and time. It's a strange world you are committing yourself to here. If possibilities have some real existence, as you suggest, then in the long millennia before pomegranates had evolved there must already have existed the possibility of pomegranates. You wave your hand tolerantly. Why not? And the possibilities of bubble gum and the abdication of Edward VIII? Take a seat, you urge them both, and make yourselves at home! How about the possibility of elves, then? Still unrealised, this one, but who knows? A place at the conceptual table is being kept for them. What about the possibility of *shrdlus*? You don't know what a shrdlu is like, since I've only just invented it. *I* don't know what a shrdlu is like, since, in all the rush of inventing it, I haven't yet had a chance to think up any characteristics for it. But I could, if you wanted; the possibility of my doing so exists. And if you're prepared to accept the existence of a possibility of shrdlus, let me try you on qwertys . . . and so on, ad infinitum.

So am I saying that sizes and durations came into being along with the physical universe? Not at all, because it makes it sound as if there are the things of the physical universe *and* there are the sizes and durations they have. But surely what I've been arguing is that once there weren't any sizes or durations and now there are? So when *did* they come into being?

This is a weird way of putting the question, but it has a simple answer. Or rather, a multiplicity of simple answers, each relating to one particular size, position, or duration. The size of the Empire State Building and the duration of the Second Punic War came into being when someone turned their mind to these particular things. When they measured them, estimated them, planned them; thought about them; dreamed about them; rode up them in the elevator, lived through them. Just as the possibility of the Abdication came into being when someone first raised it in the particular circumstances of the time, and the chemical formula for benzene when someone first wrote it.

What might 'space' and 'time' mean over and above all the particular sizes, distances, etc., in the universe? Space and time come into being, just like sizes and durations, if you want to put it this way, when the question arises. It might be less misleading to see them as *adverbial*. Not as things, or abstractions, but simply the ways in which things happen. You can whistle cheerfully, and you can suffer cheerfully, but unless you do *something* there is no cheerfulness. (You tell me that in your case you can even more cheerfully do nothing – but this is only in the sense that doing nothing is a palpable alternative activity to whistling, suffering, etc.) There was no cheerfulness on earth before life began – nor at the time of the dinosaurs. Cheerfulness evolved as human beings evolved, and will cease when all human life comes to an end. Was cheerfulness a conceptual possibility, waiting to be realised when human beings came along? What could this possibly mean except that, once human beings came along, some of them were sometimes cheerful? What purpose other than mystification could be served by finding such a clumsy locution for such a simple fact?

Forget for a moment the difficult idea of everything (with or without time and space) beginning and ending. Imagine something more contained: a hiatus *within* the spatial and temporal structure of the present universe – a hole in which nothing exists, an episode during which nothing happens.

So I produce the perfect vacuum of the philosopher's dreams. It has to be enclosed, of course, and I do this with a sphere three metres in diameter. What is the diameter of the vacuum? Not 3 metres, evidently, since this is the distance between the surfaces that enclose the vacuum, and which are not themselves part of it. About the distances *within* those parameters there is nothing to be said without the introduction of rulers and rays of light, and thereby the destruction of the vacuum.

Now I manage to halt all the processes of the universe for a second. There is some difficulty, I realise, in imagining that the planets cease to revolve around the sun, the galaxies cease their headlong flight to the periphery of things, and the clouds of probability that mark the passage of electrons around the nucleus freeze into a kind of conceptual cirrus, even for a second. There is an even harder thing to imagine, though, which is how I am able to measure the length of time for which this hiatus lasts if there are no processes running to compare it with. My statement that it lasts for a second is entirely vacuous. It lasts for a second no more than it lasts for a million years. Time is suspended as the hiatus begins, and resumes only as it ends. Its ending follows seamlessly from its beginning. There is, in other words, no hiatus. The question of whether space and time had a beginning is not additional to the question of whether the universe had a beginning, it is synonymous with it.

The Cheshire Cat vanishes and its smile remains. The notion of a smile that exists independently of lips and a face is a charming one. So is the idea of a space and time that exist independently of things and events. At once it becomes possible to play all kinds of tricks. We can imagine invisible men, who somehow have physical powers of perception and ratiocination, even of speaking and doing, without a physical presence in the world. We can imagine people who go back in time to visit or revisit the past. All these, of course, are merely fictional devices, but now physicists explore mathematical models that suggest the possibility of twists and folds in the temporal fabric, of 'time's arrow' being reversed. David Deutsch, for instance, believes that multiple universe theory seriously implies the theoretical possibility of travelling back in time.

Well, I'll put down my money with everybody else and make a booking on the first time machine to be granted a certificate of time-worthiness. Where shall I go? Before I risk some major expedition back to eighteenth-century Weimar or classical Athens, or even a nostalgic trip to my childhood, I'll try the service out with a short hop back to yesterday.

So, it's 25 April today, the sun is shining, and I'm fastening my seat-belt for take-off. Scarcely has the stewardess had time to serve the complimentary cocktails before I'm looking at the date on my watch and – yes, it's 24 April. I disembark in the land of yesterday. The sky is overcast, and there is an odd rumble of distant thunder. This is right! I remember quite clearly – the sky *was* overcast yesterday! There *was* just such an odd rumble of distant thunder! This is amazing!

Hold on, though. What I mean is that there *will have been* a rumble of thunder yesterday.

I rapidly forget the grammatical niceties, though, because I suddenly realise with a shock why that old man approaching looks so familiar. It's me! And wearing the trousers I was wearing yesterday! Will have been wearing yesterday. This is as fascinating as watching the playback of a security video. Only of course there is sound as well as vision. I can hear the phone ringing. 'Hello,' says the old man (me, yesterday). 'Oh, hello! Yes . . . No, I did, but . . . Oh, I see . . . Oh, *no* . . .!' Not, on the face of it, a very interesting soundtrack. But it is to me, because I remember perfectly well who he was talking to – who it will have been – and what she was – will have been – saying . . .

But if I *remember* the weather and the trousers, and what my daughter said on the phone, then I *am* watching an old security video. I haven't gone *back* in time at all. I'm watching some kind of rerun of the past.

No, because by the time a public service starts we shall have got beyond the first crude prototype of the time machine, which will indeed be not much of an advance upon a security video. The new model is an interactive programme, and as soon as yesterday's Frayn notices me standing there we can have a conversation with each other. He will be astonished to see me, I imagine, particularly when I tell him what a delightful surprise he's going to have when he gets on the time machine tomorrow . . . Except, of course, that this can no longer be yesterday's Frayn as he really was, because yesterday's Frayn was *not* astonished by the appearance of his future self, and was *not* told that he would be coming back to see himself tomorrow. Just as well, because if he knew he would be coming back tomorrow to see himself today he would realise on reflection that he would then be destined to repeat the experience when today once more became tomorrow . . . and then repeat it again . . . and again . . . and he might have found the prospect a little less entrancing than it seemed at first glance.

But they will go on improving the machine. I shall wait, I think, until it enables me not simply to watch or even to meet myself as I was yesterday, but to *become* myself as I was yesterday. The right trousers, certainly – but with me inside them, and having no knowledge of whether they're the right trousers or not. The same phone call from my daughter, but this time entirely audible because the receiver is in *my* hand, pressed to *my* ear – and just as surprising as it was the first time round because it *is* the first time, and I have no idea what she is going to say . . .

So how shall I know that I have gone back in time? What will differentiate 24 April as experienced by means of the time machine from 24 April as experienced when it happened? If I can live through exactly the same 24 April for free, and never know that 25 April has intervened, why pay the fare?

Well, one can imagine reasons. If 25 April turns out to have depressing prospects – if on that day you suddenly see the long-dreaded mushroom cloud rising above the housetops, and realise that you're inexorably entering the halls of sickness and death – I suppose plain, ordinary 24 April, with its slight overcast, its friendly phone call from your beloved daughter, and its happy ignorance of what's to come the following day, might suddenly seem to have quite remarkable charms.

Of course, you'd soon find yourself back in 25 April and its miseries again. Well, you know the answer now – you can take another trip back to the 24th. Your daughter comes with you, presumably, if she's going to make that phone call. *Everyone* you meet on the 24th gets a reprieve. Don't they? The boy who delivers the newspaper that starts the day off – the journalists who write the newspaper – the lumberjacks who fell the trees that make the newsprint . . . You've saved the world!

But have they all opted for trips in the machine, too? Supposing the paper-boy has a death wish, and when *he* sees the mushroom cloud he simply stands there in the street and waits to enjoy Armageddon . . . ?

If my trip is on the more advanced David Deutsch model of the time machine, of course, then I shall avoid these difficulties by travelling to 24 April in another universe, a universe in which the choice of going back to 24 April has not yet been put to the paper-boy on 25 April because 25 April has not yet occurred. And when, tomorrow, it *does* occur, and the paper-boy still persists in his pathological obstinacy, no problem – I simply hop on another David Deutsch Timeways flight, and take off for 24 April in yet another universe where either the paper-boy has a more reasonable attitude or there is no paper-boy . . . Only in this one, when 25 April arrives, there turns out to be some sect in Texas who want to enjoy the end of the world . . . So on I go again, until I find a universe where everyone has exactly the same views as I do about how things should be organised . . .

One or two slight teething troubles to be sorted out here, I think, before time machines go into full commercial service.

Cosmologists have suggested something much more wholesale than a few day-trips to the past. They have proposed that the direction of time itself may change.

The idea of reversible time is not new. The fundamental laws of physics, based upon Newtonian dynamics, treat all processes as reversible. This assumption has been carried forward into relativistic and quantum physics because Einstein accepted Minkowski's treatment of time as simply a fourth dimension, logically equivalent to the three geometrical dimensions, so that there is nothing odder about going backwards as well as forwards in time than there is about going left to right as well as right to left, or down as well as up. The four-dimensional universe has become the common coinage of physics.

One of the attractions of reversible time to cosmologists is that it seems to make more digestible the notion of a universe that has no boundary in space or time. If the universe had a beginning, in the Big Bang, and will have an end, in the Big Crunch, it raises all the questions of before and afterwards that we have been looking at. Here, it seemed, was the answer. Something *was* there, something *was* happening, before the beginning of the universe as we know it. The universe was there, and what was happening was that it was approaching its beginning backwards. And when it reaches the end it will start to run backwards again, until it reaches the beginning, when it will run forwards . . . And so on.

There is a theoretical problem, however: the 'arrow of time'. The phrase (introduced by Arthur Eddington in 1958[1]) is a reference to entropy. The celebrated Second Law of Thermodynamics states that the entropy of the universe (or of any other closed system) is increasing; which is to say that, while the total amount of energy in the universe remains constant, it is more and more evenly distributed, and therefore less and less available as an agent of change. In other words, things can only be going forwards and not backwards. As Prigogine points out,[2] 'we have inherited two conflicting views of nature from the nineteenth century: the time-reversible view based on the laws of dynamics and the evolutionary view based on entropy. How can these conflicting views be reconciled? After so many years, this problem is still with us.'

Prigogine rejects the 'spatialisation of time' and the reversibility that a geometrical interpretation implies. He finds reversibility 'almost inconceivable',[3] which seems to be an appeal to the common sense we all share, and he also believes that it is inconsistent, not only with biology and psychology, but also with modern conceptions even of the inanimate.[4] He argues that the formulation of the Second Law by Clausius in 1865[5] was 'the first formulation of an evolutionary view of the universe'. The traditional laws of physics don't describe the evolving universe in which we live,[6] and he holds that we must 'discard the banalization of irreversibility' chiefly because 'we can no longer associate the arrow of time only with an increase in disorder. Recent developments in nonequilibrium physics and chemistry point in the opposite direction. They show unambiguously that the arrow of time is a source of *order*.'[7] He claims to have encountered a lifetime of hostility to the idea of irreversible time. Nonetheless, cosmologists now seem to have moved on from the notion of an eventual or intermittently contracting universe to one that will continue to expand indefinitely.

What are we to make of this 'arrow' that might (or might not) point in either direction? The metaphor is presumably a reference to the signs on a map or flow-chart, or on a one-way traffic system, as if there were some kind of track or pipework through which the traffic of events moved, and which existed whether there were any traffic using it or not.

But there *is* no pipework, there *are* no roads. There are only events and processes. Can events and processes proceed in different directions? Some, in the physical world, have a form that is symmetrical in time; they look the same whether they are running forwards or backwards. If you can drive a train from London to Edinburgh you can drive it back to London again, and if you filmed the journey and ran the film backwards and forwards on an editing table you might well be unable to tell without some kind of extra indication (smoke in the slipstream, illuminated arrows on the editing table) which way the film was running. Other processes do not have a form that is symmetrical in time. You can jump off the top of a skyscraper but you can't jump back again, and this for two contingent reasons – one to do with the unidirectional nature of gravity and the other to do with the fragility of the human body – and even if you ran the newsreel footage of the event backwards on a Steenbeck you wouldn't need to look at the indicator on the machine to tell you which way the film was running.

On the Steenbeck or off the Steenbeck, however, with or without an arrow, the train still reaches Edinburgh after it leaves London, and London after it leaves Edinburgh; you still hit the bottom after you leave the top and arrive at the top after you scrape yourself off the bottom. Is this to do with the Second Law of Thermodynamics? No, it's to do with logic! To say 'event', to say 'process' or 'occurs', is to say 'begins', 'ends', 'before', and 'after'.

We are not so easily confused when we come to the spatial dimensions, even though any distance in the world that you can write with a plus sign you can also write with a minus sign. You can perform mathematical operations on these negative numbers with just as much consistency as you can on the positive ones, and you can use negative values in the calculation of the actual physical distances. You can, for instance, locate the position of the Flying Scotsman by treating the distance run northward as positive and adding it to a negative figure representing the distance run southwards. This does not tempt us to believe, however, that objects in the physical world can occupy negative space. If the calculation on the Flying Scotsman comes to a negative overall total then we know that, if it is to have any interpretation at all, it can only be a positive one (the train has crashed through the buffers at King's Cross, perhaps, and ploughed on across Euston Road).

You want to complain that I am seeing all this in terms of psychological time, and that we are trying to understand something much loftier here than the hopelessly limited aspects of the world that can be grasped by the human consciousness. Stephen Hawking engagingly confesses, in *A Brief History of Time*,[8] which probably did more than any other work to promote the notions of reversible time and a future contraction of the universe, that he did once believe that in the contracting phase all the processes, including those connected with human experience, would simply run backwards: 'the contracting phase would be like the time reverse of the expanding phase. People in the contracting phase would live their lives backwards: they would die before they were born and get younger as the universe contracted.'[9] In the book, however, he proposes three separable versions of time – the thermodynamic, the psychological, and the cosmological – each with an 'arrow' that can in theory point in a different direction. The psychological must always point in the same direction as the thermodynamic (on the bizarrely impersonal grounds that psychological processes involve an increase in entropy, since the ordering of knowledge, whether in a machine or its human equivalent, requires the dissipation of calorific energy). 'If and when' the universe stops expanding and begins to contract, however, the cosmological arrow will remain, but there will be no 'strong' thermodynamic arrow, hence no conscious life and no psychological arrow.[10] So now we have three different traffic lanes along the highway, with one or more of them able to carry some kind of contraflow. The arrow which Hawking envisages being reversed is after human beings have left the scene.

The suggestion that the arrow of cosmological time might be turned back to front *after* the end of psychological and thermodynamic time repeats and reinforces the assumption that Hawking himself and other cosmologists always seem to have made – that the contraction follows the expansion, that it is yet to come. The two events are evidently located, like all other events, by reference to our own stunted perceptions of what comes first and what comes later. And if the beginning of uninhabited cosmological time is located ahead of us, then the end of it can only be located even further ahead, and if the end is later than the beginning then all its parts presumably follow in chronological order – an order established on the basis of before and after in time as we are familiar with it.

You dismiss this as of no account. Obviously we can only struggle to make sense of cosmological time by reference to perceived time, just as we can only struggle to conceive of God by reference to inadequate human concepts of fatherhood, morality, etc. But isn't this to say that if we don't view it in much the same way as we view 'psychological' time we don't view it at all?

You airily wave this away once again. If it makes me any happier, you say, I can view the end of the great contraction as coming *before* the beginning of it. Hold on . . . The end before the beginning? – mere words, you tell me, since we're stuck with language. Change the words round, too. Call the beginning the end and the end the beginning – all right . . . But this new end which used to be called a beginning . . . it hasn't happened *yet* . . . ?

Perhaps we are to see time as a dimension of the same sort as the ten (or eleven) spatial dimensions of superstring theory – as a variable which neither can be nor needs to be interpreted in terms of human experience. In which case the notions of before and after, past and future, are merely conventional signs, as arbitrarily chosen as brown for live and blue for neutral. When did the Big Bang occur? When will the Big Crunch happen (if it does)? Nowhen. The questions have no experientially meaningful answers. Each of the two events simply has a numerical index attached to it which is as devoid of physical significance as its counterpart in the series of natural numbers, where the precedence of 72 over 93 has no more temporal implication than the precedence of a duchess over a countess.

So, once again, we're all colour-blind. Or rather, we all suffer from chronic pathological colour vision in a monochromatic world.

Well, if we are not to view the universe through human eyes and human understanding, it would at least reduce some of the sheer comicality of the cosmic process as we now conceive it. You can't help being struck by the egregious oddity of our apparent location at the very centre of the great expansion – and by the indecent haste of all those stars and galaxies fleeing our company in every direction we look, as if we were the hosts at a quite spectacularly unsuccessful party.[11]

And what happens if Hawking and others were right after all – if the great changeover from expansion to contraction *does* occur, and all the processes go into reverse? Spare a thought for all those photons and other particles that have been ploughing doggedly towards us from the edge of things at 186,000 miles a second for the past fourteen billion years. Just as they're about to touch down at last upon a hospitable shore they find themselves turned around, and starting the fourteen-billion-year-long journey back to their point of departure, like refugees shipped home by immigration officers. Or the three heroic Russian aviators in *A Night at the Opera* who ran out of fuel just as they were about to land at New York after their epoch-making flight from Moscow – and had to turn round and fly all the way back to Moscow.

Or have the photons spent any time at all on the trip? The relative ageing of space travellers is one of those hoary subjects which give rise to recurring newspaper correspondences. To go back to the clear and simple formulation by the physicist and mathematician Brian Greene: 'Light does not get old . . . There is no passage of time at light speed.'[12] Well, as the terrestrial observers we may ease our minds by seeing life aboard the speeding photon as ageless. All motion is relative, however. To the crew of the good ship photon it is the photon that is at rest, and we who are moving at the speed of light. During the billions of years of their journey, as *they* experience it, they are able to comfort themselves with the thought that on earth we have not had to endure any great wait for the joy of their arrival, because for us, as they see it, no time has passed at all.

So things here would still be at a pretty primitive stage of development, as the crew of the photon experience it, when they step ashore? No reality television – no main drainage? No earth, for that matter – no solar system – no nothing? Not at all. Everything would be in place, down to the immigration officer and the detention centre waiting to receive them. And would have been there, as they now realise, seeing it for the first time, for the last fourteen billion years.

Another consequence of the great changeover would be a dramatic redistribution of determinability and probability. Not only will the first be last, and the last first, as the Bible says, but the random will become the determined and the determined the random. In the present world the entirely random emission of an alpha-particle from a uranium nucleus produces a fully determined click in a geiger-counter. In the world to come (as we are crudely obliged to think of it), it's the geiger-counter that will be radioactive, and will randomly emit an alpha-particle that will scurry off to reunite with the inert uranium nucleus in an entirely determinate fashion. These random emissions, however, will be entirely predictable to anyone who remembers the forward phase of the universe, because they will occur at exactly the same time as the determinate absorptions before.

All probabilities will be reversed. Where a particular event now has various possible outcomes, one of those same outcomes (whichever one it turned out to be in the forward world) will now lead to that same event with absolute certainty. Conversely, events in the forward world which may be determined by any one of a number of different possibilities will have only probable outcomes. If the outcomes are only probable, though, then the outcome in any particular case may well not be the particular cause of which it was the result in the forward universe, and the reverse universe is no longer a mirror image of the forward one. Perhaps we are to suppose that, in the aggregate, the different causal pathways opened up by these changes will somehow cancel themselves out. The probability of *this*, however, seems low.

Imagine that you've set sail on your improvised raft from the island where you were shipwrecked, and have been navigating for some weeks without a compass by successive estimations of probable bearings. Now you realise you've left your passport behind. You attempt to get back to the island by making a converse series of estimations of the probable back-bearings. Hazard a guess at your chances of seeing your passport again.

So the universe may never find its way back to its point of departure after all.

Then again, is it admissible to postulate even a potential observer of probabilities in a phase of the universe when no observer can by definition exist, since psychological and biological time has ended? If not then there will be no potential points of view in this world. Relativity and quantum mechanics will have no application; there will be no measurement; no physics of any sort. Worse: since before and after are local concepts, dependent upon a standpoint, it seems difficult to say that anything in this world is before or after anything else – that anything is occurring in a temporal dimension at all. The possibility of saying this – or anything else about the universe in this phase – seems to have vanished. Even the prediction that it will terminate *at all*.

Hawking seems to believe that the thermodynamic arrow of time both can explain and is necessary to explain why we remember events in our past and not events in the future. No such explanation is necessary, however – or possible. We remember events in the past because this is what remembering is, and because any attempt to construe it differently makes a nonsense. A nonsense of remembering, certainly – and a nonsense of 'psychological time' and its attendant arrows.

The impossibilities of backwards time are used to dazzling dramatic effect in Martin Amis's novel *Time's Arrow*, which is constructed as if Hawking's suggestion had been realised in its original form, and the contracting phase of the universe were the time-reverse of the present one. But only for one particular human being, Tod Friendly, an American doctor who (in the familiar forward world), has changed his identity from that of a German called Odilo Unverdorben so as to conceal his Nazi past, when he was involved in medical experimentation in Auschwitz. All the rest of the world is continuing to live in forwards mode, anticipating on Monday what will happen on Tuesday, and (more importantly) remembering in the post-war years what has happened during the war. Friendly, however, experiences the events of his life the other way round. He has no idea, as American citizen Friendly, that as he gets younger he is going to evolve into Unverdorben the Nazi.

In fact Amis's rearrangement of reality is even more radical. He has had the brilliant literary conceit of splitting his central character into two. One of the two avatars is leading his life forwards, so that his German period as Unverdorben comes before his American period as Friendly. The other, who is living the same life backwards, is a kind of disembodied internal observer, who knows everything about even the most private events of the character's life, and this includes having some insight into his fears and dreams. What he has no direct knowledge of, however, is the character's specific thinking, or the exact nature of his feelings. He can't share the character's internal world in any real way because his own is totally different. He has no inkling of the events that haunt Forwards Friendly's memory, because for him they have not yet occurred, so that he has no understanding of what it is that his Doppelgänger is struggling to keep concealed.

This works superbly as a literary device. At the tactical level it enables Amis to describe familiar events in a way that makes them seem observed for the first time, with the freshness of complete unfamiliarity, because Backwards Friendly, seeing them in reverse, misunderstands everything about them. For him, Forwards Friendly's seedy sexual liaisons begin in disenchantment and end in anticipation. Prostitutes pay him for the pleasure of his services. All the normal causality is destroyed, so that Backwards Friendly has to invent a completely new causality to explain to himself what happens. (Yet, since he is often surprised by the way things work in the world around him, he has obviously had experience at some point of things happening in the more familiar sequence, with the logic that we understand.) On the strategic level, more importantly, it enables Amis to dramatise evil more effectively by the ironic device of showing Unverdorben the Nazi apparently bestowing not death but life upon his victims, and progressing not from childish innocence to adult experience but from adult experience to childish innocence.

All this works, however, only because Backwards Friendly is the one fish to be swimming against the current. If *everyone* (including his own self as a practical participant in events) were living backwards, none of their actions would make sense *to themselves*. Or at any rate not the sense they made when the universe was in forwards mode. The

fear that haunts Forwards Friendly would turn out in the backwards
world to be caused not by specific memories of particular events (of
the lies he has told to conceal his past) and the anticipation of unspecified possibilities (that he will be unmasked), but by the anticipation of
telling the lies, and by the memory of . . . of their *not* having been
uncovered. In other words, the causality that drives the backwards
world is not the causality that drives the forwards one. The thoughts
and feelings that backwards people have are not the thoughts and feelings that forwards people have. We are in a different world, not in the
same one. The clock has moved not backwards but (as ever) forwards.

Whatever view you take of time, you can't help being struck by one
glaring disparity between it and space: it seems to be unidimensional,
and only linear measurements of it are possible.[13] Space, however, has
more than one dimension. In fact it has on oddly contingent number –
three. (Or ten, or eleven, in string theory; but in any case more than
one, and bizarrely happenstancical.)

These three dimensions in their familiarity seem to be natural features of the physical universe. Plainly, though, they are not themselves
numbered among the objects that make up the universe – they are
made manifest by them, and by our situation among them. And when
you begin to consider them for a moment you might wonder whether
they are not, like so many other apparently natural features, really
human constructs, created to make the natural features themselves (in
this case the solidity and disposition of physical objects) accessible to
our purposes.

To say that the universe is three-dimensional (as the term is usually
understood) is to say that any object can be located by making a linear
measurement along no more than three straight lines drawn at right
angles to each other. Now in the first place, as we all understand, there
are no straight lines in the physical world, only conventional representations of a geometrical abstraction – the shortest distance between
two points.[14] These points in their turn are geometrical abstractions.
So is the notion of a shortest distance between them. In fact you might
think that there is an inherent circularity in the notion of a straight
line, because it's difficult to see what 'the shortest distance' can mean
except the length of the straight line that embodies it.

When we talk about three dimensions we have a vague picture of
some kind of standard grid imposed upon the universe, something like

the lines of longitude and latitude on a two-dimensional map, with lines of altitude added. The lines on a map, though, are oriented north and south and east and west; they are drawn from a particular starting point on a particular great circle. The orientation of the three straight lines that locate objects in space is entirely arbitrary. So is their point of origin. If you don't like lines drawn from Greenwich to the North Pole, or the centre of the earth to the centre of the Sun, you can base your dimensions on a line drawn in the direction of Jerusalem from a starting point at the eel-and-pie shop in the Balls Pond Road.

None of these imaginary lines locates physical objects in any absolute sense. They all simply relate physical objects to each other. This is the only way in which the position of any one object can be specified – in relation to other objects. In unoccupied space the concept of dimension (as I argued earlier in the case of measurement) is vacuous, except in so far as it is bounded by objects, or might have objects located in it. The concept of lines and orientation has after all no meaning without the concept of a starting point. A starting point is any point we like to nominate, but, until we nominate it, there is no starting point.

When we locate things in practice, of course, we often don't use the whole three-dimensional apparatus. As in all our representations of the world, we disregard whatever is inessential for the purposes in hand. We make things as simple as we can for ourselves. If it's a question of framing by-laws to control the fouling of the footpath, we treat all dogs simply as dogs, and leave out of consideration all the many other ways in which dogs can be categorised, striking as they are – size, colour, marking, length of tail, shagginess of coat, ferocity, lovability, trustfulness and individual name. Likewise, when I ask where the door keys are, you don't point first at the North Pole and say '3.7 metres in this direction', then point to the place on the horizon where the sun rises at the equinox and say '1.7 metres this way', then point to the centre of the earth and say '0.9 metres this way'. You point straight at them. You indicate the single imaginary straight line that could be constructed from your finger to the keys. And I in my turn make the beeline for them which is the basis of so much of our actual measurement.

You ask me where Haltwhistle is. I point to it on the horizon, or on the map – one dimension again. If we're not near it, and haven't got a map in front of us, though, I tell you it's twenty-two kilometres due west of Hexham. At grid reference NY 705 640. So now I'm using two dimensions. Not one, because I can't point, and not three, because I know that Haltwhistle is on the surface of the earth, and although I understand perfectly well that there are no plane surfaces in the physical world, the flat map treats the surface of Northumberland as a plane, and I'm assuming that for your purpose I can do the same. If you tell me that you're thinking of building a sewer between Haltwhistle and Hexham, though, whose function will depend upon the natural fall of the land, then I might need to add that Haltwhistle is also seventy-nine metres further from the centre of the earth than Hexham.

In mid-town Manhattan, of course, it might well be practical to specify a route between two points by specifying the distances in each of the three dimensions consecutively, because this is the course you will in practice be constrained to take – five blocks uptown, say, three across town, and thirty-seven floors up. The three dimensions are physically inscribed upon the city. Not that our three dimensions are necessarily at right angles to each other, as they are in Manhattan. They could run slantwise, as the avenues do in Washington DC, and all the skyscrapers could be thirteen degrees out of plumb, as they might well be one day if certain modern architectural fashions become general.

Do the lines that represent our dimensions even need to be *straight*? If you measure the distance from Manchester to Newcastle on the map with a ruler it is 178 km. According to the distance chart in the road atlas, however, it is 224 km. This is because the publishers of the road atlas imagine that few motorists driving from Manchester to Newcastle will follow an imaginary straight line over the top of the Pennines, and that most of them will stay on the roads, which are not straight at all. For motorists the irregularly curling road is a better measure of the distance from Manchester to Newcastle than the imaginary straight line. You want to protest that the *real* distance is the one measured off along the straight line. Are you trying to tell us that the broad ribbon of well-drained tarmac, carrying its endless stream of traffic, hooded in exhaust fumes, is *less real* than a notional line which ceases to have any visible form, even on paper, as soon as you lift the ruler away from the map? Or do you mean that, while the distance in the chart makes it easy to bring it into comparison with other distances between cities for the practical purpose of driving, the imaginary straight line across the map, being constructed on an easily reproducible, universal principle, makes more general comparisons possible – with distances by air, with distances between planets and molecules? That the imaginary line is simply a tool like any other, convenient for some purposes but not for others?

You still believe that the straight line, and the 178 km measured along it, have some kind of logical primacy, some kind of archetypal, eternal ... something or other. Some profound quality, certainly, that the arbitrary assemblage of meandering trunk road and motorway, and the 224 km they measure, most signally lack. Then you recall (I hope) that your imaginary straight line is only a straight line at all because you are ignoring another imaginary straight line that runs through the surface of the earth from its (notional) centre. If you *really* want to construct an imaginary straight line between Manchester and Newcastle (and you might, for some arcane purpose involving beams of particles inside tunnels) you are going to have to allow for the curvature of the earth, and to treat the first imaginary line, the one you visualised with the aid of a ruler, as the arc of a circle. This, the straightest line allowed by the geometry, is the geodesic, and all measurements of distance on the surface of this earth are necessarily geodesics. If you don't like the idea of a curve as the true straight line between Manchester and Newcastle, of course, you can still calculate its even less tangible subterranean chord. Now the distance comes out different again.

After you have calculated a few distances as subterranean chords you may begin to think that using three straight lines to represent dimensions is going to be a remarkably cumbersome and unhelpful way of making measurements here on earth. For the purposes of this wicked world you may feel ready to accept that physical objects have the inherent characteristic of being specifiable not by three straight lines but by one straight line and two arcs of a circle.

You might say that the arc of a circle is itself a two-dimensional conception. Well, it is if you consider the straight line as the gold standard of dimensionality, to which all other constructions have to be reduced. But then you might equally well insist upon the converse – that the arc is the fundamental measure of dimension, so that a straight line is defined as the limiting case of an arc – part of a circle whose radius is infinitely large – which means that now a straight line has to be specified in two-dimensional terms.

In the universe at large, however, you stick by the three well-tried straight lines. Except that now you are beginning to worry about the gravitational curvature of space (i.e., as I am arguing, the curvature of all objects and paths of objects), which surely means that even the *concept* of a Euclidean straight line has no application in the world of solids unless we build in the notion of an observer, looking at it from a particular point of view. Can the universe still be seen, since relativity, as spatially three-dimensional? *Was* it, for that matter, before Euclid invented the straight line?

It goes two ways, though, like everything else in the traffic with our surroundings. The world around me can be given form only through my spatial and temporal sense of it – but my spatial and temporal sense is formed in its turn by my dealings with the world around me.

If we have a sense of time as an inescapable aspect of our lives, it is because our lives have always been marked by change, around us and within ourselves, and governed by events with a clear periodicity – the alternation of day and night and the procession of the seasons; and in so far as our sense of time is one of relatively stable progression it is because this periodicity is a stable one. Our perception of time changes as other periodicities take their place among these ancient ones, and become important to us. Our temporal discrimination has become more acute since the invention of clocks. There was no clear conception of half- and quarter-hours, I read somewhere, until church clocks began to mark and sound them – and even then there was still no feeling of the minutes that made up the halves and quarters. Blindfolded hostages lose track of time – the fundamental sense of before and after itself can become confused. Yet even in the hostage's cell – even in the conditions of total sensory deprivation set up for psychologists' experiments – certain approximate internal periodicities remain: of hunger and thirst, of ingestion and digestion and excretion, of breathing in and breathing out. If we didn't eat or drink – if we didn't even *breathe* – then time would indeed have a stop. As long as we have breath in our bodies we are bound to the cycle that sustains our sense of time.

Once again, as in our perception of objects as objects, as in the construction of language, we have picked upon some natural feature of the world which turns out to be discriminable (which is to say that we have learnt to discriminate it). We can't go round the back of this and say in other terms what it is that we are discriminating. Daniel Dennett points out: 'The only way to answer questions about what quality it is which is detected by an m-detector is to point to the detector and say it's the quality detected by this thing here.'[15] Sometimes, as Dennett says, new properties can be brought into existence because a context and a purpose for them has been established. As an example of this he offers the password system devised by Julius and Ethel Rosenberg, the atom spies: 'A cardboard Jell-O box was torn in two, and the pieces were taken to two individuals who had to be very careful about identifying each other. Each ragged piece became a practically foolproof and unique "detector" of its mate . . .'[16] What quality did it discriminate in its counterpart? The quality of being the negative image of itself.[17]

Dennett also quotes Jonathan Bennett on the conceptual challenge posed by the chemical phenol-thio-urea, which three-quarters of the human population finds bitter and which the rest find tasteless. Is it bitter or is it tasteless? The capacity to discriminate it is genetically determined. 'Suppose . . . ,' says Bennett, 'that a dynasty of world dictators begins intensive breeding of non-tasters and gradually allows the tasters to die out . . .':

> After a few dozen generations, phenol-thio-urea is tasteless to everyone living, so that there are as good grounds for calling phenol-thio-urea tasteless as for calling water tasteless. This describes a course of events in which something (a) is bitter at one time, (b) is tasteless at a later time, and (c) does not itself change in the interim. Similar arguments could be developed for the taste of any given kind of stuff, and also for colors, sounds, and smells. A simple genetic control would not always be available; but mass microsurgery might bring it about that no human could detect any difference in color between grass and blood, and to do this would be to bring it about that grass was the same color as blood.[18]

I suppose this is the physical equivalent of all the questions in aesthetics and ethics that really do exercise us, often for urgent practical reasons. Among all the patients with variously altered perceptual capacities studied by Oliver Sacks and other neurologists may well be some with the inability to discriminate spatial or temporal relationships. If so, the categories of space and time, which we had once been inclined to accept as loftily antecedent to human experience, might well begin to seem as insecurely rooted as the taste of phenol-thio-urea, as elusive as the aesthetic qualities of a forgery or the moral ones of enlightened self-interest.

Colour vision and the colours of the natural world have developed in tandem, as Dennett points out, in rather the same way as the two halves of the Rosenbergs' Jell-O packet. Like the halves of the Jell-O packet, they have come into being for a good practical reason, though in this case that reason has to do with natural selection. Birds can see the redness of apples by selection because it enables them to identify a valuable food; apples are red by selection because it gets them noticed by the birds, which then eat them and distribute their seeds. The spatial and temporal nature of the inanimate world is not developed by selection, because the inanimate world does not reproduce itself genetically. But the human side of this interlocking relationship has surely been shaped by selection. We have to discriminate spatially in order to eat and not to be eaten, and in order to locate our mates. For the same reason we have to understand the rudiments of cause and effect, and a necessary precondition of understanding cause and effect is the discrimination of before and after. (Even this primary discrimination can break down, as has been demonstrated in certain psychological tests – and without any question of sensory deprivation or neurological abnormality.)

These abilities (and the explanation of them) are of course not restricted to human beings. The same holds for the lion that will or will not eat the antelope, and of the antelope that will or will not be eaten by the lion. It holds for the bird in its relation with the apple, of the fly in its relation with the lizard and the lizard with the fly. We don't need to suppose any conscious thoughts in the mind of fly or lizard. Their behaviour testifies to a working grasp of distances and solid objects, of the causality of if-eat-live, if-eaten-die.

Is the same testimony offered by the behaviour of mitochondria, as they strip electrons from hydrogen molecules, on the very lowest link of the food chain? Yes! Not that their behaviour is purposive – they just do what they do (and prosper thereby). But the first shadowy beginnings of the concepts of causality, of space and time, are present even here.

You might object that this explanation is trivial, that *obviously* we discriminate things spatially and temporally. How could we not? They *exist* in space and time! – they also exist in gravitational and magnetic fields, both of which may be blindingly obvious to Martians.

But *we* don't in everyday life discriminate the gravitational field until its source gets to be the size of the earth, or at any rate of the moon, and the magnetic field we didn't discriminate at all until the twelfth century, when someone invented an instrument for detecting it.

Perhaps, before we confuse ourselves in our efforts to think about the beginning and end of the universe, we should see if we have any very firm grasp upon the beginning and end of the present week.

Difficulties arise even here. The beginning of the week is in the past, and the past has gone beyond recapture, approachable only through the shifting lights of memory and the ambiguous traces left in the sand in front of us. The end of the week is in the future, and the future is beyond our grasp, approachable only through the uncertain devices of projection and conjecture, of plans and intentions. All we have actually available to us, in our hands and before our eyes, is the present moment. And yet hardly have we begun to think about its being the present moment than the thought is no longer being thought now. The more finely we try to consider it the more elusive it becomes. It's like a fly that's always one jump ahead of your slap, blown to safety by the descending hand itself. So let's lie in wait for it. Let's designate some precise temporal address in the future – a mathematical point in time when the runner will breast the tape, or the hand of the clock will sweep across the line . . . It's coming, it's coming . . . ! And once again it's gone.

It's impossible even to find a way of thinking about it. Every analogy breaks down as soon as we put it forward. We try considering the present as a kind of mathematical line that divides the past from the future; but a line has no width; how could we experience anything without some stretch of time to experience it *in*? We think of it as the edge of the past as it advances over the future, like the edge of a wave sweeping across the sand; but the sand exists and the wave leaves standing water upon it, while the sand of the future has yet to come into being and the water behind the wave has already evaporated. We try the flying dot of the cathode ray tube, that leaves complete moving pictures behind on the television screen. Once again, though, the dot needs *time* to construct a picture for you, as you cannot help but realise if you glance briefly at a television screen, and catch not a complete moving picture but a part of one frame – narrow bands of game reserve, thin slices of game show. And already, you note, the stripe of picture with half a lion in it is complete, even though the other half of the lion is missing.[19] The establishment of the picture, and of its movement, depends upon persistence of vision – the momentary lingering of the image on the retina after the flying spot has gone on its way. The past, as we look at a television screen, comprises almost the whole of what seems currently present to us. But then to see *anything* the eye has to scan the scene in rather the same way. Our ability to experience the present in any way at all requires a general persistence of consciousness.

The present is already the past. Even if we could lay some hold upon this ungraspable instant in which we are condemned, or privileged, to live, we should still know that we were looking at a world that had already ceased to exist. Any piece of knowledge that we have, however small, takes time to assemble. It's knowledge of a world already in the past, in any case, even as we begin to assemble it; every signal that reaches us, even the words being spoken by lips an inch from our ears, even the light shining from eyes an inch from our eyes, comes out of the past. Each of us is alone in time. And when those lips and eyes are not an inch from ours, but twelve thousand miles away, and every word has to be echoed back from space, or written down and kept in the hold of a ship for twelve weeks, when every glance has to be frozen and processed before it can even start on its journey towards us – in what real sense do those lips and eyes now exist at all?

At this very moment, I think to myself when I wake in the night, someone somewhere is giving birth, someone is being tortured, someone is dying . . . But what content do these statements have, except that I can imagine the situation, and can suppose the statements to be true? *At this very moment* . . . It sounds so immediate and urgent – it challenges me to some practical response. But it relates the events to my present experience no more than does 'At 11.37 on 12 May 1704 someone was giving birth,' etc.

My idea of you as existing *now*, whether you are in Australia or here, is largely a construction from recollection, report, and pure imagination. Your existence is for me only intermittently a matter of immediate personal experience. (So in this sense the difference between your being alive and dead is not so absolute; the recollection, the historical narrative that I have constructed, remains. For me. Though not for you. Not that this seems all that much of a consolation even for me.) And even when I am seeing and hearing you *now* (yes, and touching you – *that* takes time, as well), my sense of you as being presently alive is a projection forward into your future. But then my own present, too, is already my own future. This is what the present is *for*. (By which I mean, this is why our conception of the world in this form has been selected.) Not so as to draw us into some kind of passive embrace with the world as it is, or recently was, but so as to enable us to anticipate and shape events which have not yet occurred. We spend our lives not being and doing, but being on the verge of being, being just about to do. And what we are on the verge of is . . . being in the past. What we are just about to change is . . . what's already history.

And the best moment of all, as a girl once shyly but perceptively told me, as we sat in her room drinking tea and thinking over the events of the afternoon one long-lost summer's day, is the one just before the delight that is about to burst upon us becomes the delight that already has.[20]

In scientific experiments times are often measured with extreme accuracy, in thousand-millionths and million-millionths of a second. This precision, however, relates to the comparison of external events and processes one with another. Any attempt to anchor these external conjunctions in some observation taking place *now* is only notional – as notional as every other attempt to anchor the precise structure of external reality in human observation.

The implications of quantum mechanics, in any case, fatally erode the status of the present in another way. Its general laws, as one physicist puts it,[21] 'imply that, as a matter of first principle, one cannot know all the determining elements of the present with unlimited accuracy'. If a potential observer is required by quantum mechanics and relativity, then this introduces a further element of indeterminacy in itself. Any real observation by any real observer is an act that is extended in time. All completed acts of observation lie in the past.

A real observer is for that matter extended in space. I am *here*. By which I might mean, 'in this country', 'in this town', or 'where I am standing'. I suppose I might, with an appropriate gesture, indicate that the act of observation is centred in my head. I can't mean anything very much more precise than this, however, because I, and even my eyes and brain alone, take up an irreducible minimum of space. 'Here' and 'now' are both concepts that are visible only when they're marked not with a mapping-pen but a distemper brush. The more precisely you attempt to locate them, the less application they have. It is a familiar limitation. The more precisely you define 'red' – the narrower the waveband to which it refers – the less use it is for referring to any discernible quality, and the more we obscure the relationship between red and apparently red.

I might, of course, use 'here' to indicate a point outside myself, with the help of another suitable gesture or reference – resting the point of a pencil on a map, for instance. (Though in this case it becomes a synonym for 'there', and wherever *there* is, the one place it is not is *here*, where I am.) You can imagine that my pointer gets finer and finer, until at last we are in the sub-atomic realm. But now any imaginable indicator is at any particular moment entirely indeterminate. At its finest focus the notion of a particular location has finally disappeared into a blur of probability. In any case, specifying the location of our pointer depends upon an observer, the great clumsy brute that we were trying to escape from.

There is an inherent fuzziness about all our measurements and all our physical concepts. Not that this deters us, of course, any more than the failure of the world to provide a single example of a Euclidean point or line deters us from Euclidean geometry. Just as we understand that a geometrical point is not a physical object but a mathematical fiction, so we understand that the notional observer upon whom all our measurement of space and time depends is a fictitious character whose magical smallness and ubiquity are called into being, like the magical power of invisibility possessed by the hero of a legend, purely by the words of the story.

But this in turn means that the precisely measurable universe he supports is an abstraction, a construct as fictitious as the deeds that the wizard's invisibility makes possible. It bears no closer a relation to the actual universe than a triangle as defined by Euclid does to even the most neatly cut cucumber sandwich. To locate entities that really are independent of human consciousness and understanding we have to go behind the categories of time and space that once seemed so autonomous and secure, and enter a realm more abstract and fundamental still.

Fingerhold

The world as numbers

The kingdom of numbers is a strange place, and gets stranger the further human exploration of it goes. It is entirely abstract and self-contained; and yet it is the key to the structure of the physical world. It is not, however, validated or verified by anything in that world. Mathematicians explore it not by sight, sound, touch, or taste; not by experiment; not by measurement; but by the pure cold light of reason. They advance from one discovery to the next by the recognition of what will be consonant with what has already been accepted. The criteria for this recognition are unstatable and unexaminable. It is as if speleologists were to explore some great system of underground caverns, and to do it not by the light of torches but by a sixth sense of which they could give no account – a sixth sense whose promptings seemed to be justified by the fact that they were identical in most of the speleologists most of the time (though not in all of them all of the time).

The explorers of these caverns map stranger and stranger configurations of stalactite and stalagmite. Some of these deposits, like the cathedrals and Swiss villages identified by the guides in real caves, seem to model familar objects in the world outside. Others have fantastical and impossible shapes with no possible counterpart in the physical world. And yet when the party gets back to the open air, some of the things they see above ground now seem to embody aspects of those subterranean fantasies, rather as we catch an echo of our dreams in the waking world.

One of the most surprising things about this city of invisible palaces, though, is the way you get into it. Once again it's like the entrance to a hidden cave system – a familiar spring emerging from a hillside, perhaps, used by generations of local people for watering their animals and washing their clothes; a portal as modest and unremarkable as a gap in a hedge. In mathematics it's some configuration like the ten fingers of the human hand, folded over one by one, or a few groups of scratches on a wall, struck through one after the other – simple dodges to help keep track of the size of a flock of sheep or deliveries of grain.

Now we've realised that there's an even simpler, even less conspicuous way in. You don't need to get mixed up with the shepherds and washerwomen – you don't need all those ten fingers. A single finger will do just as well, provided you can crook it or straighten it. Or a leaf, dark green on one side, pale on the other. A wisp of thistledown that is here one moment and gone in the next. Anything that can be now like this, now like that.

And from these humble beginnings stretches a line of entities that will never end, each of which can be split into smaller and smaller entities without ever reaching either some irreducible quantum or vanishing completely – which can at the same time jump effortlessly over the non-existent vanishing-point and become less than nothing, and lesser still than that less, and lesser yet without end – which can at the same time be *both* greater and less than nothing – which, though their supply is inexhaustible, will march on beyond the point of inexhaustibility – which can express states of affairs that cannot be expressed in language, nor yet in stone or steel or cells or molecules or quarks.

The oddest thing of all about this world, though, is that (like the universe itself) it is not entirely regular. The whole system has been developed by the rigorous application of uniform principles to a simple symmetrical model, as in some ideal society based upon absolute equality and entirely rational principles. And yet, as in all attempts at such a society, among the neat marching squadrons of contented citizens various dissenters and malcontents can be found. Most numbers are happy to be seen as multiples of their colleagues, some remain obstinately irreducible. Some are 'sociable numbers' – rare chains of conspirators, each of which is the sum of the divisors of the preceding one, and where the last number of the chain is the same as the first.[1] And these oddities, the primes and sociable numbers, don't occur regularly or predictably. The only way their nature can be discovered is to put them to the test. Their occurrence is as contingent as the occurrence of their human counterparts. No, it's worse than that. For the occurrence of society's exceptions reasons can always be found, at any rate after the event. For the distribution of primes and sociable numbers no explanation has yet been discovered, even with the benefit of hindsight. They are dotted about among their fellows apparently at random. How can any element of even the contingent, let alone the random, survive in this perfectly rational world?

So what *is* number?

A simple enough question, but the answer seems to slip through our fingers as we reach for it. You might produce a similar sense of mystification, though, by asking 'What is redness?' or 'What is human-being-ness?' Robust answers, of course, at once suggest themselves. Redness is the characteristic shared by red objects. Human-being-ness is the property of being a human being. If you then ask what the characteristic *is* that red objects share, what *constitutes* the property of being a human being, I can only point at a ripe apple and then at the setting sun, or introduce you to my friends Darren and Laetitia. You turn from the apple to the sun, the sun to the apple. 'You mean round?' you say finally. 'You mean nice? Health-giving? Sources of energy? Shortly to get eaten, either by the horizon or me?' You covertly size up my two good friends. 'You mean slightly overweight? Inarticulate? Brown-haired . . . ?' I point out a letter box, a number 13 bus. I introduce you to more of my friends, some anorexic, some loquacious, some silver-haired, some completely bald . . . You get it now?

In the end either you do or you don't. How else can I help you? You want me to tell you that redness is the quality of objects that reflect light with a wavelength of around six hundred millimicrons, or talk about featherless bipeds? You look hopeful for a moment. Then bafflement returns. What, you ask, *is* this quality possessed by objects that reflect light with a wavelength of around six hundred millimicrons? What *do* featherless bipeds have in common, apart from featherless-ness and two legs? In any case, you go on, doesn't that one-legged friend of mine whom you met, and who evidently has some rare medical condition involving a plume of feathers growing out of his scalp, but who otherwise seems so charming – doesn't he have the property of being a human being?

I set up a spectroscope and point to the light around the six-hundred-millimicron mark – '*That* quality!' But we're plainly going round in circles. I tell you to ignore Theodore's wooden leg, and the curious developments on his head, and to concentrate on his charm – yes, you noticed *that*! – and also his ability to think, his habits, his parentage, his . . . I don't know! Everything about him!

But you do not, in practice, drive me to such lengths. You *do* get it, and you get it quickly. Almost always, unless you have a specific perceptual problem such as colour-blindness. Because this is what living creatures are good at – what they *have* to be good at in order to survive. They see analogies.

So what is the analogy between a ripe apple and a letter box? Their both looking in some way like a number 13 bus. Looking in *what* way like a number 13 bus? In the way that a number 13 bus looks like a blush, and a blush like a tomato, and a tomato like a tail-light . . .

It's the story of phenol-thio-urea and the torn Jell-O packet once again.

What is number, then? Number is the property that particular numbers have in common. Number is the way that 3 is like 14, and 14 is like 7 billion, and 7 billion is like the square root of minus 0.0000000000000001.

Y-e-e-e-s . . . So what are *numbers*?

To think about this straightforwardly we need to get out of the dark and mysterious vastnesses of mathematics and back into the light of day. Back to the village spring where the shepherds are watering their sheep and the women are washing their clothes, in complete ignorance of the extraordinary subterranean system that has delivered this simple supply of plain water to meet their needs.

How did human beings first make the acquaintance of numbers? Quite literally, surely, in activities of precisely this kind – the tending of sheep and the washing of clothes. The beginning of arithmetic is simple counting. It still is, in the childhood of each of us now. What happens when a child learns to count? It is coached to recite a series of words, 'One, two, three,' etc., much as it is coached to recite 'Hey diddle diddle, the cat and the fiddle.' Like 'Hey diddle diddle' the words are meaningless sounds. But like 'Hey diddle diddle' they are to be recited in a certain fixed order. Now the child is coached to point to various objects as it recites the rubric. The game for a start has no meaning except to please adults. The word order keeps going wrong – 'One, two, three, five, eleven . . .' The child's finger jumps meaninglessly about among the scattered toys, and returns to toys it has already touched. Then, gradually, the word order becomes more reliable. The child is persuaded to touch not random objects in the group as it recites, but each one in turn, and to touch it once only. Until, by dint of repetition, the child notices that the recitation performed over the heap of wooden bricks has something in common with the recitation performed over the cotful of dolls – they each end at a particular word in the sequence. And eventually it makes the great analogical leap: it sees that the heap of bricks and the cotful of dolls, which look and feel and taste so remarkably unalike, have something in common with the words of the rubric – and therefore something in common with each other.

Now we go back to all our beginnings, when there was no established sequence of number-words for anyone to be coached in. Abel wants to know whether all his sheep have been folded for the night, whether all of them have survived the winter, whether all of them have been rounded up for slaughter.[2] Difficult to tell by looking at the flock as a whole whether or not this shifting mass is the same size as it was yesterday. So, as the flock files out of the fold in the morning, he takes a stick in his left hand and a stone in his right, and he makes a scratch on the stick as each animal goes by. He ends up with a lot of marks. *How* many he has no way of telling, because he is still only in the early stages of inventing counting. All he knows is that, since he made a mark each time a sheep went through the gap, the flock of sheep and the marking on the stick have something in common. The something that they have in common can be put to good use. When the flock comes back into the fold that evening he takes the stone again and he crosses off one of the marks on the stick for each sheep that passes. If he ends up with all the marks crossed off he knows he has got his entire flock back.

Has he counted the sheep? Yes, in the sense of capturing their all-ness, or their deviation from it. No, if you mean being able to tell you the result in a word that could be used to compare the size of the flock with the quantity of corn being harvested by his brother Cain, who is a tiller of the ground. He could use the same process, though, to establish a comparison. He could cross through each of the scratches on his stick a second time as each stook of corn is thrown on to the cart, and see whether there are any single-crossed scratches left after the last stook has been thrown, or whether there are any stooks left after he has re-crossed the last scratch.

So has he compared the number of the sheep and the number of the stooks? 'Number'! These long words! Number is still in the process of being invented! What he has compared is the size, the bigness, of the flock and the harvest, on an arbitrary scale of one sheep to one stook, as you might compare the length of two sticks, without any resort to rulers or centimetres, by laying one against the other.

Another morning dawns, and this time, as the sheep go out to pasture and Abel picks up his marker stone, there seems to be no stick around to scratch with it. He has to improvise with whatever alternative material comes to hand. Hand! Yes! He has the stone in his right hand, but his left hand is free – he could make the scratches on that. It turns out to be remarkably painful. But there's another and simpler way of marking his hand, he realises, because it has various projections sticking off it, and these projections can be bent over. So he drops the stone, and instead, each time a sheep passes, he bends down a finger on his left hand.

He hasn't got very far, of course, before he's used up all the available projections. He has a few more projections on his right hand that can be pressed into service, though it's awkward at first bending them over with his left hand. And now he's completely run out of projections because he's run out of hands – but still the sheep keep coming! There are more projections on his feet, but bending over is painful – carrying all those lost sheep back to the fold on his shoulders over the years has done his lower back no good at all. So he returns to his left hand and starts over again. By the time the fold is empty he has gone all through the fingers on his left hand and right hand again – and again – then started yet again and got as far as finger, finger, finger. So hands, hands, hands, finger, finger, finger of sheep have gone out of the fold.

Every morning and every evening he goes through the same performance. Sometimes he finds that finger or finger-finger sheep have been lost to the wolves. When the lambing season starts he turns down more fingers – more hands – more *pairs* of hands – every day.

He offers up hands-hands-finger-finger lambs as an offering unto the Lord. He asks his brother Cain, the tiller of the ground, how much *he* is proposing to offer unto the Lord. Oh, about so much, says Cain, who has not discovered the new science, making vague gestures to indicate a little pile of turnips.

The Lord has respect unto Abel for his hands-hands-finger-finger lambs. Unto Cain and his vague unquantified heap of turnips He has not much respect. Well, to be precise, for Cain and the turnips He has (being a just Lord) a vague unquantified little pile of respect, and for Abel and the lambs no less than hands-hands-finger-finger units of respect.

Abel begins to prosper gratifyingly from this trade of lambs for divine recognition. He acquires more hands of sheep, and even begins to employ a few fingers of extra shepherds. First a few fingers on the left hand – then the workforce increases to finger-finger on the *right*. He is able to keep track of his employees fairly easily because they all have names, and it occurs to him that it would make life easier if he did the same with his fingers. So that instead of having finger-finger-finger-finger-finger-finger men on the day shift, and finger-finger men off sick, he could simply say Little-Finger-on-the-Right-Hand employees at work and Index-Finger-on-the-Left-Hand employees off; or, to use the nicknames he gives the fingers for the sake of brevity, Six and Two.

Soon Abel has so many sheep and so many shepherds that he finds it difficult to remember how often he has to go through both his hands to match them. Keeping track of his hands is becoming almost as difficult as keeping track of the stock and staff.

So he adopts the same system with his hands as he has with them. He is able to tell Cain that he now has more sheep than match a pair of hands counted several times over – counted a pair of hands times over, in fact – and then the whole lot counted Ring-Finger-on-the-Right-Hand times over again. Or, as he puts it in his special new technical language, ten counted ten times over, and the whole lot counted seven times over again. Which brings in three-times-ten-and-then-five-more units of respect per annum from the Lord – a solid return of five per cent on capital. How this compares with Cain's muddled little piles of turnips going out and vague dribbles of divine respect coming in, Cain himself has no way of assessing. But there is something about all these clever new words that Abel uses – three, ten, five, and so on – that makes him want to get his unquantified multiplicity of fingers round Abel's throat.[3]

Or maybe most of Abel's fingers had been bitten off by a wolf, so he had to work in fours, with the four that remained. Or maybe he used seven beans, or the first sixteen words of a prayer. Maybe he went straight to the words one, two, three. Whatever it was, he must have used a counterpart of *some* sort, whether physical or abstract. My point is that counting, like so much human activity, is the recognition of an analogy between one thing and another.

This, after all, is the *point* of counting. To bring the sheep into relation with your fingers (or the marks on your stick, or the figures in your ledger, or the magnetism in your computer). And the point of *that* is to enable the sheep to be compared in this respect with other things, to set up a standard for the same lot of sheep at another time, or with the sacks of flour for which the sheep are exchanged. It is not an abstract exercise. It is a practical one.

What is this supposed analogy between the third sheep and the middle finger of your left hand? – None, except in the context of counting. Sheep, sheep, and sheep have walked through the gate, and at the same time you've turned down thumb, index, and middle. Number is not an inert aspect of the natural world. It is derived from a human activity. It is part of the process, brought into being by the forward movement of events, as sound is by the movement of the tape or the disc under the head. It is an aspect of our traffic with our surroundings, no less than language.

If we are going to count sheep we need to use our powers of analogy in an even more basic way, long before we reach the sophisticated stage of seeing the analogy between the sheep and our fingers. To be able to bend down your finger each time a sheep passes, you have first to be able to recognise a sheep as a sheep – you have to see the analogy between one sheep and another. You have to identify a class of things to be counted. It would defeat the purpose of the undertaking if you bent down another finger when a wolf instead of a sheep walked into the fold, or if you included your brother and some of his miserable turnips.

You object that you can perfectly well count sheep sheep Cain turnip, and that the answer is left-hand-ring-finger, or four, just as much sheep sheep sheep sheep is. What would be the point, though, of counting brothers and turnips as well as sheep if you want to establish the size of your flock? Well, you might want to establish something else – the sheer multiplicity of separate creatures and vegetables provided by a beneficent God for his subjects east of Eden. But you have to be counting *something*! You have to have identified a class of some sort for the operation to have any imaginable point or meaning. I tell you that I've counted nine, but nine what I don't know – I just know that there are nine. What information could I possibly be giving you? Well, I'm giving you the information that I've counted nine somethings. I suppose that might have some minimal meaning, though it's difficult to imagine in what context.

You tell me that people are always counting up to ten, or whatever, in an entirely abstract way. In games of hide-and-seek, for example, or to control their anger before they respond to something that provokes it. But what they're counting is ten *seconds*, or at any rate ten indefinite units of time of around that length. And in the case of counting to control their anger they're probably also simply reciting a rubric, in the way that they might equally well murmur some entirely non-arithmetic mantra to themselves at moments of stress.

So what's the analogy if you count your *fingers*? Am I telling you that you're seeing the analogy, not only between one finger and another finger, but between all your fingers and all your fingers? Well, it could be the analogy with sheep, or your toes. Nowadays, since the human race has invented the series of terms that make up the natural numbers, it's more likely to be with that. Fingers and numbers have gone their ways. Although threeness may have originally been a characteristic of three fingers (just as headship, before it signified the position occupied by the leader of a tribe, was simply a characteristic of the head on your shoulders, and nosiness was the quality of being the nose on your face) it has long since become detached from the fingers, and established as a freestanding concept (just as headship and nosiness have).

When in fact might you count your fingers? Just imaginably, I suppose, when you emerge from the anaesthetic after a road accident and suspect that you have a shortfall in the finger department. Much more probably when you're learning to count – when you're being taught as a child to use the recitation of one-two-three that you have learned for practical purposes. Or, later in life, perhaps, if you are learning to count in binary or hexadecimal.

The mention of *binary* prompts you to another objection. Isn't all this laborious palaver with fingers and sheep beside the point, if we can derive the whole world of number from just two concepts, zero and one? Not at all. The breathtaking simplicity of binary, like so much simplicity, whether in art, manners, or the gear-change in a car, is a late and sophisticated discovery. The concept of zero (the mark of the empty column, the empty class) arrived in Greek arithmetic only around the end of the first century AD, and in Western European arithmetic not until the seventeenth century. The history of *one* is not so well recorded, but the concept is almost as abstract and elusive as zero. How would you ever see the analogy between your head, for example, and a sheep, if you hadn't already seen the analogy between (say) five sheep and five fingers? Even the twoness of your two hands would be difficult to understand if you hadn't already grasped the more-than-twoness of so much else.[4]

Number, in short, is not something logically and mysteriously anterior to space and time, or to cause, or to the human presence in the world. It is, like distance and duration, like redness and sheephood, an aspect of the contingent world which is given form just as distance and duration, redness and sheephood are, by becoming part of our purposes and activities – by being drawn into the traffic and becoming currency in our dealings with the world.

Couldn't you, though, represent any number *topographically*, without naming it, without comparing it with anything else, without any counting procedures? Couldn't you simply lay it out as . . . well . . . dots? This seems straightforward enough if you think of a small number like three or five. You can at once, in one glance, grasp its threeness or fiveness. You can observe its numerical properties geometrically – see the four points of a square emerging from the five, and observe that each side is two points long. You can see the extra point that makes the collection odd instead of even.

But when you think about it, even this isn't quite as straightforward as it seems at first sight. We need to have some information about these dots. We need to understand that they are to be read as a single group, and not (for instance) as one group of three and one of two. We need to understand that this group is to be read as a number, and not as (for instance) two colons and a full stop.

Now think of a slightly larger number – thirteen, say. Its thirteenness is not immediately apparent to the eye – not at any rate to mine. We need to see it as (say) two groups of five and one of three, and to understand that the three groups are to be summed. And if we are to notice its numerical properties – the fact that it is odd, and a prime, and the square of three plus the square of two, then we need to organise those dots into ranks.

Now think of a vast profusion of dots on the paper. It may represent a number (if this is what its purpose is stated to be) just as unambiguously as any named term does, but without counting them we've no idea what that number is. So it's not a picture or a diagram of the number, because it doesn't do what a picture or diagram does – enable us to grasp its subject and operate with it. What use could this profusion possibly have? How can it be related to any other collection in the world? Well, perhaps by drawing up the dots in ranks and squadrons once again. Now we observe that there are regular groups, from which some of its numerical characteristics (its parity, for example) can be read off geometrically, and which can be aligned with other groups of objects, and found to match, or to outnumber them or be outnumbered. But this process of arranging in ranks and squadrons is based on analogy just as surely as the comparison between sheep and fingers. The only difference is that the analogy is internal. The analogy is between this rank and that rank, between this squadron and that squadron.

You feel very strongly that I've got the whole thing back to front. I've sidestepped the obvious question of what the analogy *is* between five sheep and the fingers of the human hand. You want to shout in my ear that the analogy between them, waiting for us to see it and state it, is surely just this – that both the sheep and the fingers have five members.

We are taking part in an ancient philosophical debate here. It is the problem of universals. Does red exist as in some sense a constituent of the universe over and above all the red objects? Is it something which is partaken of by all red objects – which explains why we see them as red and call them red?

I choose red as a candidate for election as a universal because it is one often mentioned by the philosophers who consider the question, and who usually stick to rather familiar qualities which do seem – well, yes – universal. Red seems to be an especial favourite. (Why do blue, green, and yellow clamour less for explanation? Perhaps there's something about the sheer flamboyance of red that does suggest a dashing independence of the particular.)

But since you may not have a bowl of strawberries or a pair of scarlet lips before you as you read this, and since you almost certainly still have the well-worn hand which has served us so well for so many purposes already, take another look at that. At the colour of it, this time. Is *this* colour a universal?

The first difficulty in answering this is to know what colour it *is*. Brownish, perhaps, or blackish. Pinkish, probably, if it's like mine. Pinkish somehow doesn't seem to shout its ontological independence aloud quite as stridently as red. In any case 'pinkish' is an over-simplification. Its general effect is pinkish, but it's not uniformly pinkish. Pinkish-in-general as a quality seems even less forthcoming about itself than pinkish. Particularly since this pinkish-in-general effect is built up from many different shades of pinkish side by side, in a pattern of mottling of a complexity that defies description.⁵ You suggest that you could in theory analyse all this confusion into pure flamboyant red and pure innocent white (with of course a bit of cyan and yellow thrown in). But pure red and pure white are not the quality – not even the qualities – that you're actually *seeing*. The quality that I think you're suggesting now is an abstract one – analysability into pure red and pure white. Is *this* the universal we're trying to identify? If so a whole new world of universals is opening up before us, as boundless as the Pacific in front of stout Cortez. Your hand shares with an artichoke the universal of analysability into cells; with any other physical object the universals of analysability into elements and into molecules. You're happy to accept this. We've scarcely started on the list yet, though! With the Royal Standard floating over Buckingham Palace it shares the universal of flappability. With a pancake flippability. With a cork the ability to stop the contents of a jar of hundreds and thousands from cascading out over the floor . . .

In any case, what we are trying to explain, whether by analysis or not, is not pinkish mottling in general but precisely *this* distribution of pinkish mottling. Each of us looks at his own hand, and then at the other's. They are not exactly the same range of pinks, or the same pattern of mottling. You evidently have one universal, I have another. Does each of the millions of people in the world with pinkish skins have a different universal? Can you have a unique universal? Isn't that a contradiction in terms? So, all right, this particular colour distribution isn't a universal. But, then again, there may be someone somewhere who has hands with exactly the same arrangement as you, or as me. So what we are looking at here may be a universal and it may not be?

And, even if we can establish that all the various arrangements of pinkish mottling on hands are as unique as fingerprints, and even if we therefore feel safe in deciding that they are not themselves universals, do whatever they are share *another* universal – the universal of being an arrangement of pinkish mottling which is different from everybody else's?

So much for the *colour* of your hand. Now we come to the texture of the skin, the shape of the knuckles, the termination of the fingers in nails, the gradations in finger length, the triple inward articulation . . .

The triple inward articulation, I can't help noticing, is a universal shared by my hand and my reading-lamp. Triple articulation is a universal shared by triple inward articulation and triple backward articulation, if it exists . . . And off we go again. Tripleness is a universal shared with the three persons of the Trinity and the three French hens of the carol and the three fingers I have bent over by virtue of their triple inward articulation . . .

Universality itself, for that matter, is a universal shared by the Catholic Church, a universal joint, the Great Universal Stores, and the great universal Sea of Generality which once again is closing above our heads.

Our use of number, in other words, calls upon the same fundamental human ability as language does: the ability to see an analogy – to find ways in which, for the particular purpose we have in mind, one thing is like another thing. We do not have to suppose the pre-existence of kingdom upon kingdom of common properties in order to explain what this ability seizes upon, any more than we have to suppose the pre-existence of a realm of possibilities for houses and bicycle pumps in order to explain our ability to manufacture houses and bicycle pumps. We need a house to shelter from the winter storms, and see that we could do it by scrabbling together this bit of tree and that heap of straw – and that's all there is to it. We need to find a way of telling a ripe cherry from an unripe one to avoid indigestion, and we see that there is a way in which we could do it by matching it with either a leaf or a gashed finger.

Am I saying that an analogy can be found between anything and anything? I'm saying, in fact, that the pressure of our purposes makes us as ingenious in finding material for analogy as rats are in finding things to eat. However unpromising the parallel, something can probably be found if needs must. I pluck an improbable pair of candidates out of the air at random: a galvanised nail and the Church of England. A clergyman with a sermon to write might well seize upon their shared upstandingness and resistance to change, their possession of an acknowledged head and a clear point. How about – I don't know – a pang of hunger and the Northern Lights? Already I begin to see the nagging cold elusive unavoidability that they share.

There is an analogy (of course!) between seeing an analogy and . . . and almost anything I care to name, on my own submission – baking a cake, running the marathon – certainly, certainly, all of them, no doubt, for somebody's purposes; but for *my* purposes here and now – the assigning of value. We light upon something that will serve as a currency to draw different objects, different activities, different purposes, into relation with each other.

Can I really expect you to believe that the astonishing fairy palace of mathematics, with its unreal and irrational and transfinite numbers and its strange resonances and recurrences, is built out of nothing but a few sheep and a few fingers, and some otherwise unstatable relationship between them? Even if we go on to imagine fingers reflected in mirrors, fingers chopped in pieces, spaces where fingers were but are no longer, imaginary fingers? – Well, can you really expect me to believe that a Riemenschneider altarpiece is nothing but sticks, and the church it stands in nothing but stones? – You don't expect me to believe any such thing, you reply. The significance of the sticks and the stones is the way in which they have been shaped. – Of course. And the significance of the sheep and the fingers is the traffic in which they have become involved. Anyway, if you don't like fingers and sheep, try the Turing machine (the forebear of the digital computer), the imaginary device on which, as its inventor Alan Turing showed, *any* mathematical process could be reproduced (and – this was the point of the machine – its decidability or undecidability determined) simply by moving a paper strip forwards or backwards one place at a time. (And maybe this is what Abel used: not fingers bent and straightened, but a segmented stick slid forwards and backwards.)

But does all this do anything at all to explain some of the features of numbers that we find most puzzling – the irregularities, the contingency and randomness, that seem to lurk among them? It doesn't explain them, no. Some of the anomalies, though, arise from the demands that we make upon numbers. Is being a prime, for instance, a natural feature of the numerical universe? Or is it another of the categories that we pick out from their surroundings and establish for some particular purpose of our own? Isn't it created by the definition we have constructed: 'a number exactly divisible only by itself and one'? And isn't that definition really quite arbitrary? Are we surprised, when we use an algorithm which generates a number with infinitely many decimal places (to calculate the value of π for instance), that the resulting digits may (as in the case of the value of π) appear to be no more regularly distributed than primes among the series of natural numbers? Or chaffinches in Kent and thunderstorms in Estonia?

Arithmetic has a self-contained and self-sufficient air, like an absolute monarchy, which seems to excuse it from any requirement to explain itself, or any entanglement with the vulgar contingencies of economic activity and public opinion that beset more workaday governments. Geometry has the same style. It stands beside arithmetic like the old Austrian Empire beside the Russian Empire, foursquare against all the disorders of nationalism and democracy. The existence of more than one such autocracy reinforces both, and seems to offer a justification that puts them beyond the need for rational examination. The two monarchies are different expressions of the same divinely ordained principle; the two disciplines sit naturally together as part of the mystical underlying harmony of the universe.

For all their common front against religious and political dissent, however, the Tsar of Russia and the Emperor of Austria have different territorial and hierarchical interests to pursue. Arithmetic and geometry, likewise, are quite separate conventions, which march together in places and not in others. Numbers and the geometry of even the simple circle, for instance, don't relate to each other *at all*. There is no completely writable number for relating the straight line that bisects a circle to any of the measurements related to its circumference. There is an entirely definite and unambiguous relationship between the two, but it is a geometrical one. It has a name, π, but no exact numerical counterpart. 'π', in fact, is a proper noun, the name of this one single specific relationship, and it cannot be defined or explained numerically any more than 'Queen Victoria' can. It is given significance by its geometrical context, just as 'Queen Victoria' is by the context of British political history.

(You could of course *identify* Queen Victoria without recourse to a context by – discreetly – pointing at her, whereas you can't point to the object named π. But then you can't point to the person named Jack the Ripper. All the same, 'Jack the Ripper' names a certain supposedly single specific person, even though we don't know who that person is. All we know about Jack the Ripper is that he is the – supposedly single – author of certain particular crimes. All we know about π is that it is the constant that relates straight lines to circles.)

The numerical indeterminacy of π is not merely abstract. It means that the area and circumference of the actual circle in front of me are numerically indeterminate and undetermined. In theory, I suppose, though I think not in practice, the area and circumference could be given, and the diameter calculated from them – in which case the diameter would be numerically indeterminate and undetermined. So there is a suggestive analogy with physical indeterminacy here: the measurement of the rectilinear and curvilinear variables of a circle cannot be determined with equal accuracy; and, where one is given, the other necessarily remains indeterminate.

Of course, a translation of π into numerical language can always be undertaken, for all practical purposes, and made as arbitrarily close as those purposes require. But it is a little like translating between two different spoken languages. There, too, practical equivalents can always be found, but there is no way in which colloquial usages can be mapped on to each other one to one (as translators and interpreters soon discover, even if they are completely bilingual – perhaps especially if they are completely bilingual), because each language has developed in different circumstances, and is used by its speakers to come at the world in different ways.

But then even paraphrase within a single language doesn't always work out precisely. Not only is there no numerical equivalent for π, but there is no geometrical interpretation for it, either – no axiomatic procedure that will relate the rectilinear and the curvilinear. Nor, within the confines of arithemetic, do vulgar fractions (the approximation of π as $22/7$, for instance) necessarily have any exact equivalent in decimal form. The relations between numerical terms, in other words, are not always numerically formulable. Often the simplest arithmetical operations, when performed upon the simplest numerical terms – determining the value of $\sqrt{2}$, for example – open trails of digits which pursue numerical solutions through all eternity without ever quite locating them. The absolute monarchies are not as absolute as they appear.

Pure arithmetic is not about particular things but about particular relations between things. Algebra preserves an even loftier detachment from the world. It is not about relations between things – it is about relations between relations. Numbers are variables that can be quantified in the physical world. Algebraic terms are variables that can be quantified in the world of numbers. To use an algebraic equation to solve problems in the physical world (as we do) we have to go through two stages of quantification – the letters of the algebra in terms of numbers, and the numbers in terms of objects and units of measurement in the physical world. But then we can apply non-mathematical abstractions in the same way. We can start with '(A is married to B) + (B is child of C) + (C is female) entails that (C is mother-in-law of A)', and we can use this formulation to supply the identity of Jack the Ripper's mother-in-law – provided we replace A with 'Jack the Ripper', and quantify 'Jack the Ripper' as the Archbishop of Canterbury (or whoever).

But the noble purity of arithmetic, algebra, and geometry becomes compromised at once when they have to come down from their palaces like this to dispense practical administration in the marketplace. The relations of sides and angles of a triangle have a perfect clarity in the pages of the trigonometry primer, but when the surveyor has to use them to calculate the height of a hill, no base-line can ever be definite enough or any reading on a theodolite sharp enough to translate those absolute relationships into more than approximate quantities.[6] Two and five can be added or multiplied together and divided into each other most elegantly when they are pure numbers, but when these unworldly variables have snotty-nosed words tugging at their skirts – substantive values in the physical world such as metres, angstroms, shoes, or batches of shoes – life gets messier.

At once all kinds of contingent limitations apply, just as they do to all the other uses of language, and the fact that something is mathematically formulable doesn't necessarily give it employment down here in the real world. 'Two left feet' – yes. 'Two left-footednesses' – no, or not without further explanation. 'Two shoes × four = eight shoes' – certainly. 'Five shoes × thirty cm = 150 cm' – yes, provided we understand that the thirty cm is intended to refer to some particular dimension of each shoe – the length, say, or the width – and that the sum is to be understood as indicating an actual measurement in the world only if the shoes are placed contiguously in the direction of this dimension. 'Five cm × thirty cm = 150 sq cm' – impeccable. But 'Five shoes × thirty shoes = 150 square shoes' – I don't think so. I suppose that once again you might make the equation interpretable by giving it a particular use – perhaps to demonstrate how much of the floor a rectangular arrangement five shoes wide by thirty shoes long would occupy . . . Now let's try multiplying five shoes by thirty ice creams . . . There's probably some context in which even this formulation could be found some application. It's difficult to think quite what it could be, though.

And we're still in the realm of hypotheticals! 'Two shoes × four = eight shoes' asserts nothing about the world except that *if* you have two shoes and multiply them by four you will end up with eight shoes, and we can say this quite safely even in a land where everyone goes barefoot. The expression 'two shoes', and the assertion that multiplying them by four will turn them into eight shoes, have no more actual standing in the world of things than '$2x$' and '$4(2x) = 8x$' have in the world of numbers, or '2' and '$2 \times 4 = 8$' in the world of language, until they are assigned a specific value there – *these* two shoes (or, any two of this batch of shoes), *this* fetching of *these* two shoes and *those* two shoes, and *those*, and *those*.

There is a parallel between the standing of the expressions relating to the shoes and the shoes themselves. These two shoes in front of us, real and wearable and multipliable as they are, have no standing in the world of commerce until they are assigned a monetary value. They have no standing in the world of nutrition, for that matter, until a food value has been determined for them (which may of course be zero, so that eating five batches of two shoes each is no more nutritious than eating one, or none). No standing in the world of aesthetics until the question of their visual appeal arises, and judgements are made about their beauty. No standing in the world of morals until questions are asked about their influence for good or bad.

Can shoes have an influence for good or bad? Of course – *anything* can. Certainly shoes. Ask any head teacher who is struggling to ban trainers or high heels in their school! Trainers, they will tell you, imply an allegiance to a system of commercial and popular values which is at odds with the sense of community that the school is attempting to inculcate; high heels express an unworthy readiness to sacrifice mobility and health in order to gain male approval. Or ask the rebellious pupils who are resisting the school's edicts. Trainers, you will discover, express the wearer's rejection of the values of the cultural elite and the exploitation of animals to provide shoe-leather. High heels enable a woman to walk taller and prouder.

You may feel that we have drifted a long way from the mathematical by this time, that aesthetic and moral values cannot be arithmetically quantified and calculated in the way that more naturalistic values such as nutritional and financial value can be. Five beautiful shoes are not five times as beautiful as one beautiful shoe; a pair of trainers is not 0.79 times as corrupting as a pair of high heels. But then the mathematics has already faltered even with the supposedly more naturalistic values. The food value of twenty bananas for a single banana-eater is unlikely to be twenty times as great as that of one, because the body cannot go on extracting nutrition in a simple arithmetical progression. Nor is the monetary value of five hundred shoes likely to be five hundred times the price of one. Discounts are expected for bulk purchases. Markets become increasingly saturated, and prices fall. In any case the monetary value of things, just like the aesthetic and moral value of them, depends not directly upon the trade in them that has already occurred, but in the expected future trade, and this is affected not only by the prices recorded in the past but by other factors as well – the larger context of political prospects, of changing taste and opinion.

If the world seems to behave tidily in mathematical terms it's because the untidy work of fitting the world into numbers, of assigning numerical values, has already been done. The problem in clearing up a room is to know what to put in which drawer. Once you've done that the room looks tidy. But if you can't retrieve everything again when you want it you may have destroyed the utility of both room and contents.

The results of even the best-conducted experiments in the most controlled of laboratory conditions are nearly always very approximate. They have to be averaged out and rounded off before they fit into the neat schemata of the equations that prove the elegant hypothesis written on the blackboard.[7] Battles, for that matter, can be fought and won in fine style, once all that confusion on the ground, all that mud and muddle, all that misunderstanding and mechanical failure, all that sudden panic and loss of morale, have been reduced to boldly sweeping arrows on a map.

The so-called 'fuzzy logic' that was developed some years ago by, among others, Bart Kosko and Lotfi Zadeh, is a brave attempt to take account of the lack of fit between the precision of number and the indefiniteness of the world it is used to quantify. The intention is to render the discrepancy tractable by accommodating it to arithmetic –

by quantifying it in its turn, and establishing a fit between numbers and the discrepancy itself.

'Words,' says Kosko, in *Fuzzy Thinking*,[8] 'stand for sets of things. These sets are fuzzy. The things belong to them to some degree. That means the things also belong to their opposites to some degree ... The man who is five feet seven inches is tall and not tall, maybe more tall than not. The man who is thirty years old is old and not old to some degree, maybe less old than not ...'

Tall and not tall? Old and not old? These seem strange ways of expressing it, but what they are to mean in practice is made clear by Zadeh[9] in the case of tallness, a straightforward concept that we all understand and find useful – and a characteristically indeterminate one. It is difficult to accommodate to ordinary logic, and difficult to quantify in ordinary arithmetic. Logic deals with your being either tall or not-tall; arithmetic with your being five feet ten inches or six feet two inches, or any other specific height, or within any specified range of heights; but when I remark upon your son's tallness I don't mean that he is any one particular height, or even that he is within a specified range of heights.

Here's how Zadeh copes with it. The membership curve of tall men, he says, 'for each height ... gives the degree or fit of membership in the set of tall men. Tallness is a smooth function of height. Every man is tall to some degree. Every man is not tall to some degree.' From the accompanying graph you can work out that if you are five feet tall then (on a scale of 0–1) you belong about 0.2 to the set. Is a six-foot man tall or not? 'On the fuzzy view,' says Kosko, 'he is tall, say, to degree 80% and so he is not tall to degree 20%.'[10]

Kosko applies this technique not only to continuously variable quantities such as height and age, but also to sets of unquantifiably disparate objects such as houses:

> *House* stands for a set of houses, a list of houses, a group or collection of things, and each thing we can point to and call 'house'. But which structures are houses and which are not? You can point out some things as houses more easily than you can point out others. What about castles and trailers and mobile homes and duplexes and time-share condos and teepees and yurts and lean-tos and caves and tents and cardboard boxes in alleys? It's a matter of degree. Some structures are more 'a house' than others are. They are to some degree a house and not a house.

This, it seems to me, is quantification run mad. Certainly I don't mean, by my admiring remarks about your son's tallness, that he is any one specific height. Still less, though, do I mean that his height has any specific relationship to the mean height of some specific group (selected, presumably, by age, race, nationality, gender, class, etc.). His embarrassed shuffling about and hanging of his head may be intended to mean that he's not as tall as all that, but it certainly isn't intended to mean that he's seventeen per cent not-tall.

The application of the idea to the identification of objects moves even further into unreality. A cardboard box isn't ten per cent a house! (Let alone ninety per cent 'not a house' – whatever that might mean.) It isn't even vaguely 'to some degree' a house. It's something that might actually *be* a house – a hundred per cent a house – if circumstances compelled you to live in it. But if you use it to pack up your belongings in, or to catch hailstones to ice your drink, or to crush snails with, then it's not a house at all.

Kosko believes that we reason in rules, and that when we use our common sense in everyday life we are reasoning in fuzzy sets. 'If the man is tall he stands at the back. If it rains, you get wet. If the air is warm, turn up the air conditioner.' In everyday life, though, in the exercise of our common sense, we don't need any special new numerical interpretation of 'tall' to stand the tall people at the back of the photograph, or any statistical assessment of whether something is rain or not-rain to tell you to come in out of it, or of warmth to know when to turn up the air conditioner.

Why don't we? Because in everyday life, when we are exercising common sense, we are not (usually) reasoning from rules. If we needed a rule to tell us to come in out of the rain, and for every other situation in life where we proceed by common sense, we should need as many rules in our head as there are moments in our life. Extra rules, for a start, about coming in out of hail . . . sleet . . . snow . . . Subsidiary rules under 'snow' allowing us to stay out if engaged in snowballing . . . discovering the South Pole . . . assisting a reigning monarch to take provisions to a distressed peasant . . . Special one-off rules about what to do in icy rain if looking for a lost five-pound note . . . child's beloved teddy-bear . . . Father Christmas . . . And where would we have acquired these rules? Are children taught in school or at their mother's knee when to turn up the air conditioner? Did we have to study the instruction book that came with the machine? Have we formulated it for ourself by testing out and rejecting other hypotheses in the spirit of Popper: 'If too warm, turn the air conditioner down . . . off . . . back-to-front . . . Leave the air conditioner as it is and turn the television up . . .', etc? The first time we felt too warm and discovered there was an air conditioner to hand, didn't we work out what to do by . . . well, by using our common sense?

We shall return to the subject of rules later. And of course there *is* a need for a rule about when to turn up the air conditioner: the air conditioner needs it. It needs it precisely because it hasn't got any common sense, or any feeling of its own for comfort. This is why fuzzy logic, and fuzzy reasoning with rules, have an application – to adapt the procedures of common sense to the bivalent logic of servo mechanisms and computers, and to their lack of human affect. As Kosko says: 'You can't hand a fuzzy set to a motor. You have to tell the motor a speed. You have to give it a number.'[11]

But you *don't* have to give *yourself* a number. That's the difference (or one of the differences) between you and the machine. The numbers are simply a device to make what you do workable on the machine – to digitalise the analogue processes of human thought and behaviour. All this is entirely helpful and constructive. What's mad, though, is to try to map the model *back* on to human thought. It's as if the bronze figure of the Prince Consort in the Albert Memorial made us think that the living flesh of the Prince himself had been composed of a mixture of copper and tin, and his noble features had been shaped by melting the flesh down and pouring it into a mould.

Kosko asks how you, as a human being, make a practical decision, and notes percipiently what an apparently indeterminate procedure it often is:

> You do not solve math equations or draw rules patches on a page. You do it by feel . . . Suppose you get a new job offer. And you take it. What do you tell your friends when they ask why you took it? You will say a lot of things but each thing will be only part of the answer: Better pay. Liked the people. More room for growth and advancement. More interesting work. Better benefits. Bigger office and better view. Shorter drive to work. Can't stand my old boss. You can *defend* your choice with these reasons. But you cannot fully explain it.[12]

Admirable. But when it comes to relating all this to the decision that emerges, the account begins to go off the rails at once. 'Each reason you give is a matter of degree. Each has a fuzzy or grey weight. How much better is your new pay than your old pay? How much more do you like the people? How much shorter is the drive? No one reason throws the decision. They all add up to the decision.'

Well, for a start, as we know only too well, the list of rational considerations you offer to others (and possibly to yourself) may have only the most tangential bearing on the final decision – or none at all. The pay's worse, the drive's longer – everything's against taking the job; but you take it nonetheless because of that view out of the window; or because of some quite irrational factor that never even figures in the list you offer yourself or your friends because you weren't consciously aware of it (the ghost of a familiar scent in the corridor, the ghost of a promise in the look of the woman in the office opposite). And even if all the rational factors in your list do really play a part in your decision, how do you bring them into relation with each other? How do they 'add up to' the decision? And here Kosko's account loses all contact with observable reality. He suggests that we *mathematicalise* the various factors. The 'fuzzy or grey weight' of each, which seems reasonable as a metaphor, becomes alarmingly literal – literal enough for the weight to be in some sense measured and computed with. 'You pick a fuzzy weighted average,' says Kosko. 'You add up a lot of things and weight each thing to some degree. Then you go with the average or "centroid" or center of mass.'

At this point we have moved into pure metaphysics. Averages, whether weighted or not, and however fuzzy, can be meaningfully established only between different quantities, and there is no way of quantifying these feelings, either in our own experience or in what can be observed of neural functioning. This is an explanation that is entirely without empirical evidence, derived purely from the principle that it is trying to establish, by mapping back the mechanics of the model on to the original.

The making of decisions is another subject that we shall return to later. The point at issue here is that fuzzy logic, useful as it may be in designing machines, doesn't help in understanding the way we function ourselves, because so much of our categorisation is inherently unquantifiable, and its utility resides precisely in its imprecision. Even with entirely quantifiable concepts the *act* of quantification cannot itself be made numerical. The *application* of arithmetic cannot itself be arithmetised. These things involve the powers of recognition and judgement, not calculation.[13]

You feel understandably impatient with all the tales of hypothetical shepherds and fingers. However human beings stumbled upon mathematics, it must surely have been there before fingers and sheep were. If we need it, as we do, to model the physical structure of the universe, it is because that structure is itself inherently mathematical. The mathematics is in the world, not in us, and always was there, from the very beginning of things. We have discovered it, not invented it.

Is the world inherently mathematical, though? It is inherently *susceptible* to mathematics, because of its division into elements that are identical – or at any rate sufficiently similar – to be classed and as a result capable of being matched up with other classes (i.e. counted). Is nature doing mathematics when particles combine in regularly sized groups to form atoms and molecules, or when cells divide? Or is it going simply about its business in a way which, as it happens, human beings have been able to model by combining fingers and numerals according to rules that echo the structure of the original processes?

You tell me that the mathematicality of the world is far more subtle than this. Length, time, and mass can all be measured in so-called 'natural' units – the Planck units – which are defined as relationships between different constants,[14] and which, as Planck said, 'retain their meaning for all times and for all cultures, including extraterrestrial and extrahuman ones'. These units, though, and the constants on which they are based, are derived from various contingent aspects of the universe.[15] They are what they contingently are. They have a precise value in terms of the arbitrary units with which we are more familiar – seconds, metres, and kilograms. The 'Planck length', for instance, could be modelled by a contingent analogue – a piece of platinum, like the platinum rod kept in Paris as the standard of the metre (or it could if only it weren't so much smaller than a platinum molecule). And, if physicists find – as they do – increasingly fundamental common equations to model apparently different aspects of the world, these underlying similarities are what they are, what they contingently happen to be. If they can be modelled mathematically it's because countable discrete units can be found in which they can be marked off.

And in fact there are all kinds of continuous variables in nature – temperature, for instance, at macroscopic levels – for which arithmetical models based upon digital units can only ever be approximations (which is why we are driven back upon expedients like fuzzy logic). At the quantum level, of course, these things are not continuous variables at all – they have discrete values. But these discrete values are probabilistic, and mathematical representations of probability at the quantum level are also unsatisfactory approximations, because quantum states cannot be discretely counted. Quantum probability would be better modelled by some kind of continuously variable analogues, if anyone could think of suitable ones. Perhaps even now one of Abel's descendants is looking at some aspect of his bodily functions and seeing a useful parallel with the flock of particles he is trying to keep track of . . .

Any mathematician would protest at my account. After all, plausible-looking attempts have been made to derive mathematics from a handful of axioms which depend purely upon logical operators – to show that its roots are in logic. (Though this has not proved to be entirely possible, since Gödel proved that no conceivable axiomatic system can ever be complete.) This is plainly not its *historical* provenance, any more than Chomsky's supposed deep grammar is the historical source of language. In both cases the practice comes first and its theoretical structure is deduced later. Abel bends over the thumb and index finger of his left hand as two sheep go through the gate. Another sheep jumps over the wall, and he bends over the thumb on his right hand. Now he puts left hand and right hand side by side, and sees that if he wants to know how many sheep have left the fold altogether he could simplify the use of his hands by bending down the middle finger of his left hand to stand in for the thumb on his right. Has he taken the first step in discovering a rule of addition which has been lying around in logical space, waiting for him? Or has he invented a useful practical technique which can later be extended in all kinds of unforeseen ways?

Well, when he first puts a few sticks across the exit to the fold to stop the sheep getting out, has he discovered a pre-existing rule of gates? When he winds a strand of vine round the sticks, and tucks the end under a previous turn to stop it unwinding, has he discovered a rule of knots?

There turn out to be standard ways in which gates can be made to function well, involving hinges and latches, and standard configurations of knots. These all depend upon a combination of the qualities of the materials available and the human purposes to which those materials are being put, and they are standard because the materials and the purposes are standard. We can go further, and generalise the qualities of the materials to universal principles of friction and gravity and turning moment, and the purposes of their users to survival and reproduction. But the principles remain rooted in the contingent particularities of the traffic in which they occur.

Doesn't Gödel's Theorem make it impossible to use derivability from an axiomatic system as the *source* of mathematics? An incomplete and incompletable formalisation of a functioning system is a vastly more plausible proposition than a provably defective *derivation* of it. *All* our formalisations of reality, after all, are incomplete.

Then again, the common ground with logic has come to seem a less self-evidently secure foundation than it once did. The whole of number, as we know, can be reduced to permutations of just two alternatives (o/1, on/off, yes/no), and the attraction of propositional logic was that it seemed to be based on just such a fundamental dichotomy, true/false. Sweet True and False! The Darby and Joan of the syllogism! The Romeo and Juliet of the propositional calculus! They have such an air of archetypal coupledom as they stand holding hands there side by side that they seem to embody some irresistible natural principle of life. Our simple picture, however, has been as rudely challenged as our simple picture of the human couple. Lukasiewicz and others have long since demonstrated that systems of logic can be constructed perfectly well with three values – or four, or five, or five thousand (though I don't think that anyone has yet suggested the viability of a single-parent logical family). The idea that all these arbitrarily different selections of values have some objective and self-explanatory standing in the universe seems markedly less appealing.

The standing of logic, in any traditional sense, as an inherent component or condition of the natural world, was yet another casualty of quantum mechanics. If a particle can be in two mutually contradictory situations at the same time, then this is the end of the Law of Excluded Middle; it means that there are occasions on which it is simply not true, as for centuries seemed self-evident beyond any shadow of doubt, that [Either P or not-P]. Logic, it turns out, is a theoretical construction, something we have built ourselves, and which models some aspects of the natural world well, others not.[16] In fact the Law of Excluded Middle was surely subverted much earlier by the discovery of imaginary numbers,[17] because the square root of a negative number is both a negative and a positive one – the earliest known identification of what quantum physicists would call a superposition. Only this implication doesn't seem to have been noticed.

Even as the traditional couple, True and False are not as innocent as they look. They lead double lives. They mean one thing when they are applied to the propositions of logic and mathematics, and something quite different when they are applied to contingent propositions. Using the same words for the two different purposes confuses both. It makes mathematics and logic seem to be saying something about how things are arranged in the world, and it makes factual statements seem as if they ought to have the cut-and-dried absoluteness of mathematics and logic.

In fact the truth-value of mathematical and logical propositions depends purely upon consistency within a formal system (a consistency which is supposed to be self-explanatory, once everything has been clearly stated – though, like everything else in life, the reality is slightly more complicated). There can be no formal procedure for establishing the truth-value of a contingent proposition, however, and this has been a source of anguish to philosophers. Many philosopher-hours have been devoted to finding a way of formalising the relationship between such propositions and the states of affairs that they purport to represent. 'The statement *Snow is white* is true,' they propound wisely, for example, 'if and only if snow is white.' This[18] seems a cosy enough arrangement, particularly when the union is sanctified by the grave 'if and only if' of formal logic. But its snugness is illusory. It has the appearance of plausibility only because, in this formulation, the state of affairs against which the statement is being tested has already been smuggled into language in the formulation itself. We are holding up the statement to be examined not against the world which it describes but against another statement; we are finding to our satisfaction that 'Snow is white' corresponds with 'Snow is white', and that 'Snow is white' is true if 'Snow is white' is true. All the tangle of human experience that gives rise to this observation has been tidied away in the cupboard; all the simplification and generalisation involved in generating snow's familiar iconic whiteness; all the soot-black snow falling on industrial towns, the mud-brown snow lying at the side of busy main roads, the inexplicable freak falls of blood-red snow that may or may not have occurred as presage or lament for historical disasters; and by extension all the laborious researches of historians and scientists, all the confrontations of lawyers and witnesses.

You may feel that we have created unnecessary difficulties for ourselves here because the statement we have chosen, simple as it appears, derives from such a large and indeterminate generalisation of our experience. Let's take a simple specific statement about a simple individual case, where we can establish its truth-value, if at all, not by any examination of other statements or review of human experience in general, but by observation of the single specific state of affairs which it purports to express. 'The cat sat on the mat,' you tell me – and not because anyone else told you, but because you've been watching the cat yourself. True or false? A glance is all I need, one single unproblematic act of observation, and at once I know which it is . . .

Well, no. Sorry to be tediously nitpicking – but I can't look now to see if it *sat* in the past. You graciously agree to spoil the internal rhyme. '*Is sitting*', you tell me. 'Now. In broad daylight. No fog in the room. You've got your spectacles on. Look!' One or two tedious indeterminacies remain, even so. The wretched creature seems to me to be sitting half on the mat and half off. You patiently take a ruler to it. 'Fifty-three per cent on the mat and forty-seven per cent off,' you announce. So is that on or not on? We have to discuss this and decide! And in any case it moved slightly further on as you laid the ruler beside it; does *is sitting* apply to when you made the statement or when you checked it? Or both? Or some combination of the two? I might add that what it's sitting on is actually not a mat but a table napkin that happens to be lying on the floor . . . I see that your patience is beginning to evaporate. All right, all right, if you want to *call* it a mat . . .

All these attempts to establish a simple truth procedure for contingent statements come up against the profound difficulties that we encounter (as with scientific laws) in saying what ontological status a fact or a state of affairs has outside the proposition that gives it expression. There's *something* out there, certainly, that the proposition is attempting to lay hold upon, but the question is what it is and how we operate with it, if we are forbidden the circularity of expressing it in language. The one thing we know for sure is that our procedures for determining the truth-value of the result, for deciding whether the attempt has succeeded or failed, are indefinite, arbitrary, and always open to challenge.

So we really need two different words for 'true': one to mean 'consistent with the rules and theorems of the system' and another to mean something more like 'usefully representative of how things are for the purposes in hand'. The two applications are so different that it's astonishing they have ever been conflated; and yet, as with the closely connected anxiety that induction doesn't have the same rustproof qualities as deduction, generations have succumbed to philosophical despair because contingent truth has seemed to be merely the incorrigibly delinquent younger brother of its unimpeachably virtuous older sibling.

They do have *something* in common, certainly; but it is extremely difficult to say exactly what it is. We feel that the truth of a proposition of either sort makes some practical claim upon us. A true proposition, however arrived at, *demands our assent*. And of course as soon as we have said it we feel that this misses the point completely; the point is *why* it does. After all, plenty of things with no truth-value at all make the same demand: just proposals, for example, fair arguments, reasonable offers. So, in their way, do unjust proposals, unfair arguments, and unreasonable offers – or, at any rate, this is the intention behind them. Not to mention bribes, handcuffs, and thumbscrews. Why should anyone bother to utter an untrue statement, for that matter, whether deliberately or inadvertently, except with the intention of seeking our assent to it?

In any case, 'assent' is as two-faced as 'true'. The kinds of assent we give to the two different sorts of truth are different. To analytic truths we yield completely but offhandedly. To contingent truths we offer a less absolute but more substantive allegiance. Even to the latter, of course, we frequently have to assume or feign absolute commitment for practical purposes. But we understand (usually) how arbitrary and unnatural this is, how forced upon us by the necessity for practical action, simply because something has to be decided one way or the other, with no room for compromises or half-measures.

Neal Ascherson remembers[19] a journalist covering the troubles in Belfast telling him excitedly, 'I've got a wonderful story, and it's truish.' Ridiculous, of course; yet there *are* truish things, whether or not the term could ever be found a place in logic (even in multi-valued logic). It's the same with guilty/not guilty, which splits the great muddle of human activity and intention, once it has been dragged into the courtroom, into two neat halves because there are only two doors out of the courtroom – one to the street and one to the cells. There is no shortage of guiltyish and not-guiltyish defendants, all the same, or other defendants with even less pleadable pleas to offer. When I was on jury service once we convicted a docker of theft. He had been caught 'sucking the monkey' with a cask of sherry – boring an airhole at the top and another hole at the bottom, and drinking the contents as they jetted out. The case was open-and-shut, and he only got a £25 fine, but he was furious as he left the dock. I could guess why. The entire shift had been doing it, all sixteen of them – and he was the only one the docks detective had managed to catch. What he would probably have been prepared to accept – perhaps even to plead – was 'Guilty – but everyone else was doing it too'. Juridically – logically – being guilty on your own is no different from being guilty with fifteen others. But in life (if life is of any great concern to justice) there is a difference. There is. There truishly is.

Fuzzy logic, incidentally, wouldn't be of much help in either of these cases. The journalist didn't mean his story was twenty-seven per cent true. The stevedore didn't think he was sixty-four per cent guilty.

Really we could do with *more* than two words to replace 'true', because there are various forms of contingent truth, and they are quite different from one another. The statement 'The Queen is the wife of the Duke of Edinburgh' is not true in the same way as 'the Queen is the darling of Conservative matrons', nor is 'The Queen is the wife of the Archbishop of Canterbury' false in the same way as 'The Queen is the role-model for disaffected black youth'. The statements about the Queen's marital arrangements are contingent, but they are true and false in something like the same kind of absolute sense as analytic statements – because they derive (or fail to derive) from another proposition, written in this case on a certificate of marriage drawn up by a properly appointed Registrar of Births, Marriages, and Deaths. This proposition, unlike the supposed derivation of 'Snow is white', cannot be falsified, by observation or in any other way, because it is not a statement of fact but an *enactment* of it – the fact itself.

Now consider the assertions about the Queen's standing with Conservative matrons and black youth. These, in theory, are to be verified or falsified in the same way as your claim about the cat and the mat – by observation of a world which is not yet represented in language, on judgement and generalisation. Is that in practice, when you make such statements, how you derive them? I suspect not. I imagine you do what everybody else does, and rely on some vague congruity, or lack of it, with other vague statements that have already been made – with things you've read in the paper, or heard people say in the pub. No one you say them to expects them to represent very much more than that. In any case, even if you go out and research public opinion in Tunbridge Wells, or hang out with young people in Brixton, you're probably going to have to rely on what respondents *tell* you about their attitudes. So in either case the truth conditions for these statements have *something* in common with those for statements about the identity of the Queen's husband.

Different once again are those for the statement you make to the press about the style of the Queen's shoes after you have just been gazing at them as you knelt in front of her to be knighted; different yet again for the expression of your feelings of pride and honour as the sword touched your shoulders; and again for your conjecture about *her* feelings as she wielded the sword.

Now try shifting the tense of the verb. What are the truth conditions for 'The Queen will be the wife of the Duke of Edinburgh tomorrow'? Or '... the wife of the Manchester United goalkeeper in half-an-hour from now'? Here there doesn't seem to be any existing state of affairs out there for the words to represent, or fail to. The element of judgement seems to have taken over completely. In the case of the Duke of Edinburgh you feel you want to ease back from the word 'true' altogether, given the vicissitudes of even the healthiest and best-ordered lives, and opt for something more like 'highly probable'. But it's much more than highly probable that she won't be the wife of the United goalkeeper in the next thirty minutes – it's *true* that she won't! Isn't it? You hesitate – you don't want to make any rash claims here. Oh, come on! Do I have to explain to you about the law on divorce, the conditions for a Special Licence, the Royal Marriage Act . . . social class . . . trains to Manchester . . . ?

Now think about the truth conditions for 'If she were not married to the Duke of Edinburgh she *could* be married to a professional footballer', ' . . . *would* be married to a professional footballer'. But I won't revert to the old problems with counterfactuals.

*

You will recognise in all this the influence of Yeswell Sortov, the philosopher of vagueness, imprecision and ambiguity.[20]

Sortov was suspicious of all categories. Everything, he believed, was sort of vague and confused, a sort of seamless flux, though this is making his ideas sound much clearer than he himself found them. Everything, he taught, was sort of itself, and sort of not another thing, and yet at the same time was sort of inextricably tied up with everything else. Indeed, although everything was sort of itself, as soon as you tried to talk about it you realised that it had sort of turned out to be sort of something else altogether.

He believed that Everything is One, or at any rate probably somewhere between Nought and Two. He was greatly influenced by Heraclitus; but he went further, and considered that stepping in the same river even once was an unrealisable goal. He was a nominalist, of course; but he doubted whether there actually were any real resemblances between individuals, when you came right down to it, or whether individuals could be classified at all. He took a particular interest in quantum mechanics; but questioned whether measuring even one cognate variable of a pair was really much more plausible than measuring them both. He was not at all surprised that particles could be in more than one place at the same time, or that things changed as soon as you tried to look at them; he said that the contents of his study were a good macroscopic model of a quantum state. He developed an early prototype of multiple universe theory; but held that all the different universes were actually present in front of our eyes, piled on top of one another in inextricable confusion.

His contribution to human thought was immeasurable. Immeasurable, indescribable; a ray of darkness in the light; as broad as a razor and as sharp as the wide Sargasso Sea; here today and gone yesterday.

II: Actions

Why the Marmalade?

Intention and purpose

So whichever way you look at it you come back to the same thing. What gives the world around us form and substance is our contribution – the ways that we have developed for coming at it and dealing with it. We are active agents in the traffic, co-originators of it, autonomous members of the great cooperative. Is this what philosophers and scientists alike dismiss as anthropocentricity? Not really. We remain peripheral, a trifling irrelevance in the great singularity. In some ridiculous sense, though, it is our participation that brings it into existence. We are the partners in creation; the world began, in this sense, not on Day One but Day Six. If there is a word for what I'm trying to suggest I haven't come across it. 'Co-determinacy', perhaps. Or (a barbarous mixture of Greek and Latin) 'anthropocogenesis'.[1]

How do we maintain our side of the dialogue? How do we originate thoughts, actions, and perceptions?

Out of all the things in the world that we try to explain, human behaviour is surely the one that we work hardest at, and the one with which we find most difficulty. There is an obvious problem in the case of other people's behaviour, because so much of the evidence is unavailable to us. All we know of how they see the world, and what they feel about it, is what they are both able to articulate and choose to reveal, and what we can infer from their actions and appearances. But this inference, and even our understanding of what they tell us, depend upon relating the evidence to our own way of seeing the world, our own feelings, and our own behaviour.

At least we have unrestricted access to our own thought processes, and to the ways in which they have influenced our actions. Or so it seems while they are at the back of our minds, unexamined. But when we turn to look at them direct it's as if a millionaire had turned to look at the wealth whose warm glow he had always felt at his back – and found it all to be intangible, unrealisable, unlocatable, unappraisable.

Even the kind of fixtures and fittings like the apparatus of scientific laws and causality that we believed we could locate behind physical phenomena seems to be missing. We are in a world of pure fluidity, of unlimited relativity.

We really do need to know what's going on, though! Upon the explanations we offer for the behaviour of others depend all our efforts to allot blame and praise, to persuade and control, to help and cooperate, to love and nurture. Upon the explanations we offer for our own depends the whole course of our lives, our understanding of the past, our interpretation of the present, our plans for the future, our ability to govern ourselves. The problem is tantalising. It's not like quantum mechanics or astrophysics, where the observations require arcane technology, abstruse mathematics, and strange counter-intuitive concepts. Everything lies to hand.

And yet the actual workings, even in the case of our own behaviour, are even more elusive than the internal workings of the atom. There is a tangle of interconnected difficulties involved, some of which we shall come to later. High among them, though, is the fact that, while human behaviour involves all the forces of physical causality, it also involves another *kind* of causality.

We know what it is to behave like the red snooker ball when it's struck by white – to fall helplessly because we are pushed – but most of our conscious behaviour isn't like that. We respond to events not because of the physical force of them itself, but because we can *feel* the force; we respond not to the events themselves but to our perceptions of them. We move, unlike the red ball, because we *feel* ourselves being struck, or because we *perceive* ourselves to have been. We have some choice in our response, or seem to. We move because we think we are *about to be* struck, and don't want to be; or *remember* being struck; or can *imagine* being struck. We also move not just *away* from the white ball, but *towards* the black – because we see something that we want, or think we want. We move hastily, reluctantly, panic-stricken, after a pause for thought. We also sometimes move, or seem to, without perceiving either the repulsive power of white or the attractive power of black, but simply of our own accord. We initiate actions, or seem to, which are not responses at all, are not part of any existing causal chains, but which are new beginnings.

Even if we are deceived about our capacity to choose our response, and to initiate action – if even our most spontaneous-seeming act is as thoroughly determined as the move of the snooker ball – the machinery of perception and volition through which this takes effect is plainly much more complicated than the transfers of energy and fields of force through which physical causality works. The explanation of our behaviour involves the concepts of motive and intention, which may have a superficial resemblance to the forces involved in physical causality, but which become less and less like them the more you examine them.[2]

The question of whether Josef K murdered his aunt has been settled. The evidence is clear, and Josef K has pleaded guilty. The question for the court now, before a just sentence can be passed, is *why* he murdered her. What was his motive?

The prosecution says that he did it because he wanted to inherit her money. The police have established clearly that he knew he stood to benefit under her will, that he was being threatened with violence if he didn't settle his bookmaker's bills, and that as soon as she was dead he did in fact borrow money on the strength of his expectations to make the necessary repayment.

He himself says he did it because his aunt tyrannised his ageing mother.

A psychiatrist called by the defence, after long sessions with the accused discussing his childhood and his relations with his parents, says that he did it because his aunt represented a forbidden object of sexual temptation to him.

Each of these explanations has its strengths and weaknesses. The first has the great advantage of being objective. It is constructed from evidence which can be inspected by any observer, and which is publicly verifiable (or as publicly verifiable as historical statements ever are). But, plausible and persuasive though the conclusion is, we feel that it leaves out of account the one essential element – what was actually taking place in the mind of the accused. All the business with bills and wills is relevant only in so far as it suggests this. There have been plenty of cases, we know, where external appearances were misleading, and fatally unjust. This, of course, begs the question as to *how* we know this. Not from external evidence, obviously – or if so, only because one set of external evidence has come to outweigh another. From our own experience, then, when we felt from inside that our motives had been misjudged.

The strength of the second explanation is that it comes from the one observer who has access to his own thoughts and experiences. Its weakness is only too obvious – that this observer has good reason to be misreporting his experience to us: he hopes that it will present his crime in a better light, and lead to a less severe sentence. And of course it goes deeper than that – he has rather similar reasons to be misreporting his experience to *himself*. And deeper still, because even if he is examining his conscience with the most scrupulous honesty, we know that he may be mistaken in his account of what he finds there. Deeper yet, because even if he is right about his (conscious) intentions he may be wrong about his (unconscious) motives.

Again, though, *how* do we know? Only because we know of cases, parallel to the ones that made us hesitate over the external explanation, where someone's account of his own intentions has changed, or because his estimation of his motive fails to coincide with the account deriving from the other sources open to us. Not that this is in itself a necessary reason for preferring the alternative version – merely a reason for keeping an open mind about both.

The testimony of the psychiatrist, we hope, combines the advantages of the other two. It is based upon an examination of the one really pertinent issue – what was going on in the mind of the accused – and the examination is of a type that is, in theory, open on equal terms to any similarly qualified observer. But at the same time it combines the disadvantages of the other two methods. The proceedings in the mind of the accused have not been observed directly, but reconstructed from clues in much the same way as the motivation suggested by the police. And the clues have been supplied by the accused himself. However hard he has tried to be honest about his childhood experiences, his account to the psychiatrist is open to all the same objections as his account to the court – he may have misremembered, he may have misinterpreted.

He has also organised his thoughts in a certain way even to communicate them to the psychiatrist. Just as your inchoate physical symptoms are categorised to fit the grid of possibilities offered by the doctor: Are they stabbing pains, or is it a dull ache? Here? Or here? How often does the pain occur? – so, to put your thoughts and feelings into words, you find yourself shaping them towards a certain coherence, edging them towards a pattern. You are beginning to suggest one of a range of possible interpretations. Or at any rate you are beginning to place them, in the psychiatrist's mind, in a category of related cases, which may interlock with one or more of a range of formulated interpretations.

Is there any way at all of choosing between these three explanations? There *has* to be, in practice, and the choice is important, because the whole future life of the defendant depends upon it. We are trying to locate the crime in the world of cause and effect, and the standard analysis of cause and effect is regularity and predictivity. Does the judge, as he deliberates his sentence, make some covert appeal to this principle? Is he debating with himself whether, if provided with more gaming debts and another beneficent aunt, the accused would be likely to kill this one as well? Or whether he would murder again only if someone were found to terrorise his mother, or to incite him to incestuous longings?

It's difficult to imagine the judge's deliberations proceeding along these lines. Even with a judge, supposedly so much more resistant to emotional appeals than the laymen on a jury, the tides sweeping back and forth through his mind are likely to be tides of feeling. Of sympathy, or the opposite, with the accused as a person; with the psychiatrist as a person; with psychiatric explanations in general; with gambling and its perils; with bedridden mothers; with sexual susceptibility. You can imagine a judge who accepts *all* the evidence, who puts himself in the place of the accused and feels all the emotions alleged by all the witnesses – the terror induced by the threatening creditors, the boiling indignation on his mother's behalf, the ache for sexual possession. How is it possible to know which, or what combination of them, was dominant? Doesn't it in the end come down to an inclination in the judge's mind towards a more or less charitable view? Gradually a picture begins to take shape that seems *affectively* appropriate. And what determines this? Nothing, perhaps, that the judge consciously remembers. An expression that crossed the face of the accused at one particular point. The way the psychiatrist adjusted the sit of his jacket. Things buried deep in the judge's memory of his own fears, his own indignations, his own longings.

Of course, my investigation of the judge's mind and motivation is open to precisely the same range of objections as his original investigation of the mind and motivation of the accused. It's a plausible-sounding story, though. So is the version the judge fixes upon. We are both telling stories, once again. And I have the advantage over the judge here, because he is a fiction from first to last. He is my fiction, and as his creator I am allowed to assert what I like about him without fear of contradiction. He in his turn has to do much the same as I did – but with material over which he has no inherent rights.

One of the things that distorts our view here is the freedom offered by fiction, and our familiarity with it. We are so used to explaining and learning about human behaviour through stories that it makes everything in life seem just as simple. The storyteller and the novelist have, if they choose, unrestricted access to the insides of their characters' heads, together with a godlike ability to determine the thoughts and feelings occurring there, and the connections between them. Indeed, they have a kind of *obligation* to determine them, because these thoughts and feelings, and these connections, are precisely what a story *is*. If fiction has a purpose this is it – to make plain what cannot be made plain in reality. A story that simply listed events, without any attempt to establish the emotions that accompanied them and the associations that give them their significance, or to trace the causal connections – between event and event, between intention and event, between event and feeling – would not be a story at all; or would, at best, be the limiting case of the story, a philosophical exhibit that defined the limits of fiction.[3]

Let's turn to an act where the motivation is absolutely simple and clear-cut. A case where there is no possible turmoil of confused emotions, where there's no time or occasion for anything except one flash of volition and action. Why did the motorist suddenly brake so violently? – Because he saw the child step out in front of the car.

There are no holes here, surely. The child stepped out in front of the car. He braked. The first event is an entirely adequate and unambiguous explanation of the second. No other explanation is needed; none is possible.

But now suppose the case is just as simple, but slightly different. What if the motorist *hadn't* braked? What if the child had stepped out and he'd accelerated? The story now reads: 'Why did he accelerate? – Because he saw the child step out in front of the car.'

We feel very uneasy about this as a complete, open-and-shut explanation. Something else is needed, surely. 'He attempted to brake, but he put his foot on the wrong pedal.'

Or: 'He is a psychopath.' Though here there is a strong element of circularity. We are saying that he accelerated when the child stepped in front of the car because he is the sort of man who does things like . . . well, accelerating when children step out in front of his car. We are, at best, generalising his behaviour. We are not explaining it. The explanation has now moved back a layer. Why does he tend to behave like this? – Because he has an irregularity in his genes, or had certain experiences as a child, which other people who behave in the same way also seem to have, or have had. – Why do we think that these genes, or these experiences, cause people to behave in this way? – Because . . . people who have or have had them tend to behave in this way.

Are we any closer to an explanation than we were at the beginning?

And now the explanation for the first driver's behaviour in braking begins to look a little bald, a little in need of expansion. When we say he braked, we mean that he attempted to brake, and that he did in fact succeed in putting his foot on the appropriate pedal. We also mean that he attempted to brake because he is *not* a psychopath – not someone who does things like accelerating when children step out in front of his car, but who . . . well, tends to brake in such circumstances. Why does he tend to behave like this? Because of his genetic inheritance. Because he had good experiences as a child . . . And off we go again.

Two more drivers come pelting down the road. The first brakes because he sees a child step out in front of the car – even though he is an attested psychopath. The second, as he admits afterwards, stunned, accelerates because he sees the child step out – even though everyone who knows him agrees that he is the kindest and gentlest man alive.

All one can say is that, yes, people do sometimes – often – behave uncharacteristically. Where does this leave our explanation? Imagine that in one particular experiment, which has so far proved unrepeatable, we got the result $2H + O = HCl$. Do we shrug our shoulders, and say, well, hydrogen and oxygen are usually pretty reliable about producing water when they combine rather than hydrochloric acid, but they obviously act a little out of character from time to time? No, we dismiss the experiment, as we do with so many scientific experiments, because the results plainly cannot be right; the sequence of cause and effect is so well established that it overrides apparent exceptions. Now researchers in Osaka and Moscow begin to report other cases of hydrogen and oxygen apparently combining to form hydrochloric acid. At this point, perhaps, we have to begin to look at some of the fundamental assumptions of chemistry again.

Shrugging our shoulders over the vagaries of human behaviour is not the only reaction open to us, of course. For any human action, however uncharacteristic, an explanation can be found. This is a demand-led economy. You tell me what you want explained and I'll manufacture you an explanation for it. Rival explanation-mongers will manufacture you alternative explanations – as many as you like. One might propose a theorem to the effect that for any human action there is an indefinite number of explanations.

Am I trying to deny the existence of any link between the child's stepping out and the driver's braking? Of course not. This is why he braked – because he saw the child step out! All I am demonstrating is that his explanation is not as cut-and-dried as it looks; that it could be extended indefinitely into an examination of the driver's moral nature, his social relations, and his experience as a driver; and that the same machinery of explanation could be used to explain precisely the opposite outcome.

Some of our actions, of course, are completely involuntary. My foot jerks – and if you have just tapped my knee with a rubber hammer then this is a complete explanation of the event. I gag – and if this follows the obstruction of my throat then we need look no further for a cause. No doubt my braking when the child steps out is also largely an involuntary reaction; but in so far as the possibility of my having done otherwise existed, then there is some irreducible element of choice, and for that element, however small, a universe of explanations and counter-explanations opens up.

You want to say, yes, but all these attempts at explanation are coming at *something*. At what, though? At some entirely objective but somehow unsayable aspect of the universe? (The objective correlative expressed by the scientific law once again, the meaning expressed by the proposition, the proposition by the sentence.)

Do I thieve because I'm poor? Or am I poor because I thieve? Are you corrupt because you're rich? Or are you rich because you are corrupt? Some say one, some the other, some a little of each. But here 'because' seems inappropriate; cause and effect are scrambled together in a way that makes nonsense of both. The thieving and the poverty, the wealth and the corruption, are intertwined like the fibres of a rope. They are part of a common texture. Does the warp support the woof, or the woof the warp?

A particular explanation of why someone did something *feels right*, all things considered. This doesn't exclude a later conviction that some completely different explanation feels better. The explanation is an artefact. And this artefact is more like a piece of sculpture than a piece of machinery. You can't judge it by its function, as you do a machine. This would be to begin with the end. The explanation may serve a function – may support your theory of child-raising or economics. But it can serve that function only if there are independent grounds for accepting it. And these must be similar to aesthetic feelings. An explanation of someone's behaviour (even one's own), like the sculpture, feels somehow satisfying, or fails to; feels satisfying to me, but not to you; feels satisfying today, but tomorrow will feel clumsy, naive, inappropriate.

All our struggles to establish why Josef K killed his aunt began from one central assumption – that he *did* kill her. Nothing, on the face of it, could be plainer, nothing could be more cut-and-dried. There's no question of having the wrong culprit. Our man was seen by unimpeachable witnesses to pick up the revolver and shoot her; we have it on the security video; he admits he shot her; and as a result of his shooting her, the pathologists agree, she died.

But when we turn to look at this part of the story, with the caution we have developed about motivation, we might begin to feel a little uneasy once again. For a start because saying that *he* killed her suggests some kind of purposive action; it raises the question of intention, and once again we are back in the same cloudlands as we were with the question of motive. Once again, though, the question is crucial. If he was able to persuade us that he had picked up the gun with the friendliest of intentions – to give it to his aunt as a Christmas present, for instance – and if we had good evidence that he had stumbled, or had his elbow jogged, or that some rictus had involuntarily contracted the muscle of his trigger finger, we should surely be a little uneasy about saying he killed her. 'He killed her' does seem to imply that the action was undertaken by him as a complete and autonomous person, not merely as an involuntary conduit for the actions of others, or as the uninvolved location of some impersonal muscular event. To say that he killed her, if the action was entirely involuntary, conceals more than it reveals; it reduces this huge moral action to the level of 'He sneezed'.

You may object that rictus of the trigger finger, though conceivable, is so outlandishly improbable that it scarcely casts much of a shadow over the simplicity of our statement. People with guns in their hand *do* notoriously stumble and shoot someone, though. There is also, more interestingly, a whole range of possibilities where the intention is more subtly or confusingly clouded. Supposing the accused claims that he picked up the revolver, not with friendly intentions, certainly, but also not with the settled intention of using it to kill his aunt. He claims – and this would seem to be very characteristic of human thought and action – that he picked up the gun with intentions that he had never formulated to himself. We might think, from outside, from all the circumstantial evidence, that even if he had formulated them (supposing for the moment that intentions can ever be formulated) they would have been incoherent and inconclusive. We might be inclined to think that when he picked up the gun it was not because he had decided to kill her, but with the desire to feel that he had in his possession the means of killing her if he later decided to; that it was the formulation of the choice in front of him rather than the resolution of that choice.

You might say that a counsel who put this forward wouldn't have advanced his client's defence very far. After all, by one means or another the gun still did in the end go off! But we are not talking about practical matters of criminal responsibility. We are talking about whether, even in theory, a particular event occurred that is precisely described by the proposition 'He killed her.'

If he picked up the gun not because he had decided to kill her, but as a practical way of putting the question to himself, then might not the same be true of each of his subsequent movements? He turns towards his aunt – he releases the safety-catch – he raises the gun – he takes aim . . . With each action he brings the decision he will have to take more closely into focus. And even if it wasn't like this – even if he picked up the gun with what he felt to be the settled intention of killing his aunt – he might still have altered that intention at any point up to the actual squeezing of the trigger. This is not mere semantic nit-picking. People *do* revoke their decisions – and the consequences may be practical and vast – may be (as here) matters of life and death. So even in this case the *real* decision still has to be taken!

In the end, though, either his right index finger contracts the final millimetre or it doesn't. Yes, but this tiny movement now begins to take on an air of arbitrariness that surely mocks the concepts of intention and motive. Can so complicated a construction of thought as would be required to locate an intention or a motive be expressed in this scarcely perceptible muscular contraction?

When we say that he killed her, we mean more than 'The muscle of his right index finger contracted slightly.' We also mean more than 'He slightly contracted the muscle of his right index finger.'

There is a fundamental circularity involved here. We find that we need to know a lot about the intention before we can even specify the exact event, the intention behind which we are attempting to establish. Once again a whole narrative is required in which almost all the main events are ambiguous, open to interpretation, accessible on a different basis to the subject and his observers, questionable as to their objective existence.

There are many other human activities which seem to be open-and-shut, merely mechanical, with definite criteria for success and failure, but which turn out on examination to depend on this aesthetic *feeling* of rightness or wrongness. Imitation, for example. Think what it's like when you try to capture the spoken inflections of another language. It's not a mechanical process at all – it's not like recording. You listen and struggle to repeat, and it's hopeless. Then suddenly, out of nowhere, you find you're saying something in the other language, and it sounds somehow plausible. The new voice is coming in some way from inside yourself, not from outside. You're giving a performance. It's like the moment Wittgenstein describes of 'knowing how to go on' – of grasping in practice how a formula works, of being able to use it, without necessarily having any conscious understanding of what that formula is. It's akin to getting someone's likeness in a drawing – not something that anyone would be tempted to consider merely mechanical. But, less obviously, one might think of what the difference is between copying a painting by photographing it, and copying it by painting it. In the latter case you have to find the brush strokes of your own that achieve the same effect. Even if you are attempting to reproduce the master's brushwork you still have to find the brush strokes of your own that mimic his.

Then again, explanations of behaviour are not neutral and dispassionate. They are further blows struck in the battle, contributions to the traffic. An explanation of why you did what you did is a move to incriminate you or exculpate you; to support or discredit a more general theory; to demonstrate consistency with the explanations offered in other cases; to establish or challenge the rational basis of the world; to bring out or to deny a community of feeling between the person explained and the person explaining.

De gustibus . . . Are explanations of motive also not meet to be disputed? We *do* dispute them. But then of course we do dispute matters of taste, whether fittingly or not. And when you hear people actually discussing (in a jury room, for example) why someone did something, it does often seem as if they are making statements of their own tastes as much as anything else. 'Well, I always feel that . . .' How do you shift other people's opinions about a sculpture? By association. You remind them of other works by the same sculptor, of other people's views, of their own views of other works of art. You mention that the sculptor was influenced by Kierkegaard, or on the other hand was appreciated by leading Nazis. Isn't it the same with their views on motivation?

The law is an extreme example of the struggle to accommodate the chaos and indeterminacy of the world to a set of standardised shapes, as anyone who has ever served on a jury knows. Counsel attempt by their questioning to select strands from the tangle that might possibly be woven together to make one of the authorised patterns, but the more tightly witnesses are constrained in their responses ('Please answer yes or no!'), the greater the sense of a world beyond the walls of the courtroom where everything is inchoate, both odder and more ordinary, than anything the law allows for. And the requirement in criminal cases to fit the whole confusion into one of two categories, guilty or not guilty, often seems as insane as the expectations of religious fundamentalists to see mankind divided at the last into the saved and the damned.

On the basis of mere feelings, then, and on an over-simplification of reckless irrationality, a man is going to spend ten years in jail, or else go home and have supper with his wife. Yes, but then on the basis of mere feelings, and on another leap into the more or less arbitrary, this sculptor is going to lose his grant, and return to industrial welding. On the basis of mere feelings, and on the need either to do something or not do it, we are going to halt the trade in live snails, and so destroy the economy of Burgundy, or go to war with Lower Pannonia.

Isn't there, at the heart of all my choices, another, inner element that resists all explanation, not only the hypothetically objective explanation of the determinist, but even my subjective one? Isn't there, once again, as in the mutations that occasion the choices of natural selection, an element of the completely random? *Pour un oui ou pour un non* we brake or we accelerate. The synapse fires, or fails to fire, and neither human choice nor the great causal chain determines the outcome.

You brush this aside impatiently. If the synapse fires, then it's because of this and that. If it doesn't fire, it's because of that and this. But why shouldn't it turn out to depend, when you come right down to it, on the indeterminate behaviour of single particles? And even if the internal causality of the brain is entirely deterministic, why shouldn't the collision between it and the separate causality of the child's behaviour – entirely indeterminate at least within the context of our own mental processes – produce a random outcome, just as the same sort of crossover does in Valya's machine?

The apparent simplicity of the choice between determinism and untrammelled free will as an explanation depends in any case entirely upon the simplicity of the choice we are attempting to explain. On/off, tea/coffee, guilty/not guilty, shoot myself/not shoot myself – yes, let the debate between Government and Opposition rage, because we end up by either passing the motion or failing to. But now look at the kinds of things we are trying to explain to ourselves most of the time. The way in which the painter has laid the white on the blue just *so*; the way in which the pianist has distorted the time in each bar by just so much; the way in which I have constructed this sentence with two semi-colons. At once the idea of both pure choice and an objective causal origin seems inappropriate, strained, absurd. Something inherently more complex is going on. The colours, the hesitations, the semi-colons are offering themselves; they're dancing partners with the painter, the pianist, and me, in a dance which cannot be broken down into units of conscious decision. We are all in the story together. Is the painter leading the white paint out over the blue, or the paint the painter? Am I telling the story, or is the story telling me?

And in all this, notice, we are talking not about the difficulty of *prediction*, which is so notorious in human affairs. We are talking about the past, with every possible advantage of hindsight.

Nor are we in the realms of probability here, in choosing between these different versions. Not, at any rate, in the scientific sense. The judge may well say to himself, 'Probably he did it for the money.' But this is merely to express a cautious preference for this version. There is no way in which the statement can be given a statistical basis, as statements of scientific probability are. The judge does not mean, could not mean, that in over fifty per cent of cases examined by researchers, exactly similar people accused of exactly similar crimes were shown to have done it for the money – not only because of the impossibility of finding exactly similar people or exactly similar crimes, but because, even if you could find them, precisely the same impossibility of establishing the probability of the explanation would arise with each of them, so that the formulation is circular.

Or is there in theory an objective procedure that would enable us to sidestep probability in our choice between explanations? Some philosophers have tried to see human actions in a briskly scientific and causal light by declaring that they result from physically determinable chains of brain states and nerve states, etc. Well, yes, in the sense that the Russian Revolution was caused by certain soundwaves generated by various larynxes, and by certain arrangements of printer's ink on pieces of paper. What does this account *explain*, though? You could have a total description of all the soundwaves generated by orators and shouting crowds, of all the arrangements of molecules of ink on molecules of paper, and it still wouldn't cast any light whatsoever on why the mob stormed the Winter Palace. That requires understanding the *meaning* conveyed by the soundwaves and the ink, and the way in which this meaning was understood by hearers and readers; and this is not a mechanistic explanation at all.

So, if we managed to detect the brain states in Josef K that preceded and accompanied his act, could we then find an objective semantic *interpretation* of them? Well, if the accused could be honestly deceived about his motivation, so could we, faced with the same material. Ah, but we're looking at it from outside. The flash of something that he felt, and which he construed – honestly, dishonestly, or with some mixture of the two – as a flash of altruistic indignation, we can see has the characteristic wave pattern of self-regarding panic. As we know from all the other similar wave forms we have examined in our studies, which turned out to be self-regarding panic – according to some other, independent way of assessing them, of which we thought highly at the time . . .

Yes, but some specific events must have been occurring in his brain in the moments leading up to the shot! No doubt. Perhaps even the proposition 'I am doing this because she tyrannises my mother.' But if this sentence came up, simple and unambiguous, on our printout of his mental events, would it for a moment begin to solve any of the problems? We should have to go through all the usual tests to decide whether we believed it or not. We might decide that it constituted a piece of circumstantial evidence in favour of what it claimed. Or we might decide that it was as self-deluding as the explanation he offered the world later. We might just as well not have discovered it, for all the use it would be to us.

Or perhaps what we should find was wholly occupying his consciousness at that moment was a splitting headache. Should we be tempted to say that he did it because of the headache? We might just as well say that he did it in spite of the headache. Or supposing we found a passing fancy for a glass of beer, or the recollection that his car needed servicing? They might suggest something – a certain cold-blooded detachment, perhaps. Or they might not. We all know, warm-blooded and spontaneous as we are, that disconcertingly irrelevant thoughts come into our heads at the oddest moments.

Anyway, we learnt long ago, from reading Freud, that the source of many of our actions is not accessible to our conscious minds at all. It lies in the unconscious, a vast underground warehouse inside each of us to which we have lost the keys, but from which all kinds of noxious fumes and radiations emanate. In practice this usually turns out to be merely another way of saying that we don't know why we do what we do, but in theory, at any rate, there are techniques (hypnosis, association, the analysis of dreams) for revealing the contents of this hidden world. With their help psychoanalysts unearth a range of suppressed memories, unacknowledged terrors, and unacknowledgeable desires. The causal connection between this festering hoard and our behaviour, though, remains as elusive and as subjective as with conscious thoughts, because we still have no independent objective criterion for determining what causes what.

The justification for an explanation of this sort is simply that it feels plausible (if it does). This is what it comes down to once again. It has a convincing ring, like a good story well told; it is aesthetically pleasing; we are inclined to accept it; Josef K himself feels that it has given him insight into his own behaviour. Even where the analyst can't find anything beneath the floorboards he might remain confident that it must be there somewhere, sending out the occult waves that incline us to pick up the knife and raise it, or to turn and cross the road 15.7 metres from the corner instead of 15.3. But now we are out of aesthetics and into metaphysics; the existence of the hidden sources is validated only by the principle – a principle that could be validated in its turn only by the existence of those very sources.

Let's abandon our exploration of the opaque mental processes of Josef K for a moment, though. Let's make the situation as simple and basic as we can. Let's think about our own. Or rather, you think about yours, and I'll think about mine. It's me in the dock (I tell myself, you tell yourself), pleading guilty to murdering my aunt (you to yours). The judge has retired to his chambers, considering sentence, thinking about my motivation. I, back in my cell, begin to think about it myself. And now I find myself wondering if I really did do it because my aunt tyrannised my mother, as I have maintained all along, and maintained for the simple reason that it seemed to me to be true. Perhaps I did it to put myself here in the dock – to be the centre of attention, to have articles written about me, psychiatrists and judges puzzling over me, philosophers philosophising. Now that it's occurred to me it seems obvious. So what new evidence am I considering? What have I dredged up out of my memory? Well, maybe I've remembered some detail that seemed irrelevant before, some moment of jealousy of a friend who was the centre of attention. Or maybe I haven't remembered anything at all. I have simply begun to see things from a different perspective, from the ruins of my life, during the first moment I have had for calm reflection since the fatal shot was fired, with the afternoon sunlight coming through the window of the cell at just the angle it is now . . .

It may well be as hard for me to explain myself, even to myself, as it is for everybody else. Would the difficulty vanish if we got closer in time? Far from vanishing, it might increase. If I could somehow have collected myself sufficiently, as my trigger-finger tightened, to ask myself why I was doing what I was doing, I very much doubt if any words would have formed at all. Some great rush of something would have taken over, and only afterwards would I begin to ask myself whether it was fear or rage, and whether there was pleasure mixed in with it.

Tom, at the age of six, having knocked over a glass of water at table: 'Did I do that on purpose?' The only thing that's odd about this question is its being asked aloud, of others. It's very much the kind of question that one might ask oneself, in the small hours, with the most painful seriousness. And the answer, as with all questions about one's own motivation, might be highly elusive, and remarkably arbitrary.

You might reasonably object that the murder of my aunt is a rather extreme case, and that it's not surprising if I, even more than psychiatrists and lawyers who spend their lives dealing with outlandish behaviour, find myself a little out of my depth in trying to reconstruct what I was up to. Most of the time, you might say, with most of the ordinary everyday purposive acts that I perform, my intentions are simple, and I know perfectly well what they are. Elizabeth Anscombe, in her meticulous examination of the subject,[4] concentrates upon these ordinary everyday acts – my going upstairs to fetch my camera, for example – and finds that the class of intentional actions, and indeed of mental causality in general, is a sub-class of the class of things (such as the position of one's limbs) that are known without observation.

It's true that most of my intentional acts are in practice unproblematic. I want to take a photograph, and even without consciously reviewing the question of where I keep my camera, or making any conscious decision to fetch it, I find myself going upstairs. In what sense do I know what my intention is? Well, I might articulate it to myself (and as one gets older one often has to, because otherwise one *doesn't* always continue to know what one's intention is). Most likely, though, I know it in the same way that I know what the capital of Norway is: I could tell you if you asked me, or if I asked myself (and indeed, as Anscombe says, do it without any need to observe my own behaviour).

On the other hand, even with this simple case, I might (quite innocently) have more at the back of my mind than I could easily articulate to you – more than I could fully articulate to myself – more even than I fully know in the pre-articulate form, if this can be said at all. The possibility of picking up a sweater while I'm upstairs, as well as the camera, might be lurking somewhere; the not yet quite consciously sensed need to have a pee; a slight sense of relief, source unidentified, but which you might feel, if you knew about it, came from finding the opportunity to escape for a moment from the very extensive and detailed account you are giving of your recent holiday.

Anscombe also considers more complex cases – some of them of a most ingenious baroque convolutedness – and suggests means of discriminating by external observation and interrogation what the real intention is. By the intention, however, she still seems to mean something of which the person concerned is aware (as distinct from his motives, about which he may be deceived or of which he may be unconscious). And, in the case of my little expedition upstairs, you might be able to reveal some kind of residual consciousness of the penumbra of my intention by these means. I might be a little slower to return to the conversation than I need be. You might raise an ironic eyebrow, and get a slightly shifty smile in return. But *did* I know these things until you revealed them to me? Isn't my intention, even in this simple case, even in my own mind, an indeterminate concept, as open to challenge and interpretation as my motives for murder?

Let's lower our sights a little, and make a more modest claim: I know at any rate whether I did something because I *wanted* to do it!

Do I? Even leaving the 'because' to one side for a moment, often I don't even know whether I want to do what I'm doing or not! And even if I'm sure I want to do it, I often can't help asking myself if I'm accommodating my tastes to necessity – if, for instance, I'm justifying to myself some choice I've made by discovering a taste for its hitherto unforeseen consequences. (Can't afford a taxi – discover that travelling by Tube enables me to further my studies of human behaviour. Too lazy to take the Tube – discover that travelling by taxi enables me to think more profoundly about the behaviour I have observed.)

All right, let's say that the sanction that constrained me disappears, and I continue to behave in the same way. So *now*, at any rate, I must be doing it because I want to do it . . .

Not necessarily. I might simply dislike change. And if I now stop doing it, on the other hand, it might be merely out of lethargy. You might feel when you look at me – I might feel myself – that I am less happy doing what I apparently want, nothing, than I was previously when labouring under the constraint to make some effort.

Marmalade and honey on the breakfast table this morning, and I chose the marmalade. Why? I like them equally well. They were the same distance from my hovering hand. There was no hidden agenda – no one to impress by the social correctness of my choice, no fair-trade orange-growers or beekeepers to demonstrate support for. I could simply take whichever I wanted. So I did, and there can only be one reason: precisely because, at that particular moment on this particular morning, the marmalade *was* the one I wanted to take. An open-and-shut case of motivation, if ever there was one.

So how did I *know* that I wanted the marmalade? Did I have to examine my feelings before my hand moved? Did I have to ask myself, as if I were doing a survey of private opinion, and see what my answer was? Well, all sorts of things may have gone through my mind.[5] I don't need to make any great study of them, though, because there's a much simpler way of answering the question – one that cuts straight through all the complications, one that you can verify just as well as I can. I knew that the marmalade was the one I wanted from the fact that I took it. The question of whether the marmalade was the one I wanted was incontrovertibly settled when my hand moved towards it.

So what am I telling you? That I took it because I took it?

You smile indulgently. You have a more profound answer. Examining my behaviour with your usual dispassionate care you announce profoundly: 'You did it for the same reason as we all do everything in life – out of self-interest.'

This sounds solid enough, certainly. You have surely located the great mainspring that drives the clockwork inside every human psyche. And what you're alleging has a function – it seems to be suggesting that at any rate I didn't take the marmalade for altruistic reasons. Does it have any substance apart from that? Does it have any *more* substance than my explanation? Any more than the shadowy events that it is supposed to explain?

Well, it has *generality*. It places my behaviour in an established tradition. But supposing something about the look on my face as I reached for the dish suggested to my breakfast companion, a kindly aunt who always tries to see the best in everyone, that really I did it because, without even being conscious of my benevolence, I wanted to leave the rest of the honey for her.

Your heart is not moved for a moment. Self-interest once again, you announce, because even more than marmalade or honey, evidently, I wanted to win my aunt's approval, or some kind of moral advantage over her, or simply to get pleasure from messing up sensible people's views of motivation. And if I managed to do it without even being aware of my own honourable intentions, then it was still self-interest, because I wanted to enjoy the virtuous feeling of not feeling virtuous.

This sounds a rather more interesting judgement, because it really did look for a moment as if a bit of altruism was creeping in here. The problem is that the generality of the explanation is turning out to be so all-embracing that it explains *everything* – even altruism, the apparent opposite of self-interest. The monotonous howl of the uboama is heard in the breakfast-room.

So there is this fatal indeterminability at the source of our actions. None of the explanations that we offer for our behaviour seems to have much objective reality. Over some large and important area of our experience we are inventing our account as we go along. We are weaving a story, not writing a report. Yet it would be absurd if we talked ourselves into feeling that we couldn't offer any explanation at all for what we and others did. Of course we can, and do, and always have, and always will. Just as we can tell ourselves stories, and do, and always have, and always will. And these stories that we tell ourselves in the case of our motives and intentions are important ones, about historical events and historical characters.

Now we turn to the history in serious history books, and it seems more sewn-up and grown-up – much more *historical* – than the halting and muddled accounts we've given ourselves of the things in our own experience. The writers have made scrupulous use of evidence – of written records and contracts, of depositions and eye-witness accounts, of wills and letters, of trade and tax returns. The same indeterminacy remains, though, only here it has entered the system at an earlier stage – in the formulation of the evidence. Those letters and eye-witness accounts are plainly on a par with our own amateur efforts. You think that the contracts and tax returns are much better? Thirty-seven cartloads of oats were contracted to be delivered on 3 November, for 2,380 rubles. What you don't know is that the oats harvest failed, and the parties agreed informally to substitute barley. An unknown amount of the barley was eaten by rats in transit, and the recipient defaulted on payment. Do you think that *any* of the contracts actually got fulfilled to the letter, and not varied to meet changing circumstances? As for the tax returns . . . Do you seriously imagine that even the *tax inspector* thought they bore much relation to reality?

Our assessment of the causes of the Thirty Years War is no more determinate for having been precisely determined by the documentation than the decision of the judge about the motivation of the accused is determinate for being based upon a painstakingly exact note of the proceedings in court.

What we need to examine is not the motivation for our decisions, but the decisions themselves. These must occur, and must often be unambiguous and in clearly statable form, because they so often have unambiguous and clearly statable consequences in the outside world.

Mustn't they? Don't they?

How the Marmalade?

The act of deciding

We spend our entire life making decisions. It begins as soon as we open our eyes in the morning, while we are still in bed, because at once we have to decide whether to get *out* of bed. Or to be precise, since we certainly don't intend to spend the day malingering, whether to get out of bed *now*. Or whether, instead, to wait a moment or two, and get up *now*. Or *now* . . . [1]

The choices are entirely mine, however powerful the reasons are for getting up, however compelling the urge to postpone it. From them, as from any other event, consequences will flow – some trivial (a hasty breakfast or a leisurely one), some perhaps more significant (the train missed – the meeting not attended – the committee's policy changed in my absence). Many of the ensuing day's long series of decisions will be so minor that they will be taken almost before they are formulated. Start brushing teeth on left or right, top or bottom? Many will require a moment's thought: white shirt or blue? Streams of decisions will apparently take themselves without my even noticing. Lift coffee cup as I read the paper. Sip. Put coffee cup back. My feet, walking themselves along without any conscious contribution from the management, will turn to carry me across the street twenty metres beyond the corner, or twenty-one, or 21.13, or fifty-six. Perhaps, when I become belatedly aware of what's been happening on the shop-floor, I might consciously have to decide to cross back again, because management doesn't want to be on the other side of the street at all this particular morning.

Many other decisions, as the working day develops, will be hard but not impossible: which word to choose next, since this is my job, which construction to employ. You will be making a lot of decisions like this yourself, even if you're not a professional writer, even if you're going to be spending the day gossiping with your friends. Every now and then I'm going to find my mind reverting to the hopeless consideration of some major question which I can see no way of ever resolving. So, I imagine, will you find yours. Should you marry, separate, put your mother in a home, move house, enter a monastery . . .? Meanwhile the coffee cup will have been lifted and replaced an indeterminate number of times, the pages of the paper turned, the street crossed . . . Even some of the larger decisions will have materialised no less mysteriously. The elusive words will have appeared on the page, your house will have been put on the market, and the estate agent will be coming to look at your mother's house as well.

All these decisions, great and small, are going to create the world we live in.[2] The more leverage our technology and our powers of social organisation give us, the more far-reaching the consequences that the decisions will have. This is the export side of our traffic with our surroundings.

So here I am again, back where I was in the last chapter, with the marmalade and honey on the breakfast table in front of me. Not this time, though, in the hope of discovering *why* I decide what I decide, but simply *how* I do. Whatever the history that led up to it, how does the decision emerge? What sort of entity is it? How does it take effect?

This is one of the two specific problems to which I thought philosophy might give me an answer when I first began to study it as a young man. Why was it that I could never actually catch myself taking a decision? Was it because I was so indecisive that I never really took one? I seemed to go through all the proper preliminaries often enough, all the appropriate anguish. My complete attention would be engaged by a serious practical question about, for example, whether or not I should boldly ask the girl I'd just met to come out with me. (This was the kind of decision I was really worrying about at the time.) I would picture to myself the possible consequences of the two alternatives. I could already hear the wounding transparency of the girl's excuse as she refused. I could already see the difficulties I should have in getting out of the room and coming face to face with her on subsequent occasions. On the other hand I could envisage the disastrously embarrassing evening that was going to ensue if she reluctantly agreed simply because she couldn't think of an excuse fast enough. I could even, occasionally, glimpse the shimmering nimbus of indeterminate delights that would hover over the evening if everything went well. I could certainly feel the fury and despair at myself that would descend upon me if I didn't at any rate give it a try. What I couldn't see was any possibility of ever finding a way to bring all these various and disparate considerations into relation with each other, and of finding some resolution to the anguish they caused.

And then it was past. There I was, asking her. Or there I was, slinking shamefully out of the room with the words still in my throat.

Sometimes, between the turmoil of thought and my subsequent action (or lack of it), it seemed to me that there *was* a conscious decision inside my head. The words 'I will ask her tomorrow without fail' were imprinted on my consciousness in a typeface so clear that it could only be the Memorandum of Agreement at the end of the negotiations, a document that embodied a commitment in the way that a legal declaration does – though quite how this document had emerged from the negotiations was no clearer than if it had dropped from heaven. But when tomorrow came I would find that a different text – something more like 'I may well ask her next week, if circumstances are right' – had taken the place of the former one inside my head as mysteriously as the rewritten constitution of Animal Farm.

And whatever text I had most recently found inscribed on my mind, and however it had got there, its connection with any subsequent action was entirely haphazard. Sometimes the decision resulted in the appropriate action, sometimes not. Quite frequently what emerged was some other action altogether, not consciously contemplated before. What I could never manage to observe were the actual moments when consideration became commitment, and (even more to the point) when commitment became action.

Or when *anything* became action, with or without consideration, whether there seemed to be a commitment to it or not.

I came to see as I began to study philosophy that the difficulty of catching myself deciding was in the first place simply one particular instance of the general problem of introspection and of attention. The mind can be undertaking only one major conscious operation at a time, so that if it's making a decision it can't be looking to see whether it's making a decision or not, and that conversely if it's looking to see whether a decision is getting made then there's no capacity left over for actually making the decision.[3]

The elusiveness of deciding, though, seemed even more acute and more systemic than the elusiveness of everything else that goes on inside one's own head. And this, as it turned out, was something upon which philosophy – and particularly the linguistic philosophy being taught at the time – could indeed cast a certain amount of light. I began to understand that deciding, like winning a race, is not the name of an activity – it is an 'achievement verb'. The activity is the running of the race, the turning over of arguments and alternatives. You can be aware, painfully and continuously aware, that you are running a race, and no less painfully aware that you are thinking (though perhaps rather more intermittently, given the difficulties of giving your attention to an activity that already takes up a large part of that attention). Winning the race, and deciding the outcome of this internal debate, constitute not activities in themselves but the successful completion of the activities. You can't win without having gone through the business of running the race, but you then, if you cross the finishing line first, win in that very instant. And the finishing line, in the case of a decision, is the initiation of the appropriate action. In a sense you can't win a race – you can only move, seamlessly, from being about to win to having won. In just the same way there is no 'moment of decision' in the sense of a moment when you can be conscious of deciding – there is only the moment of recognising that the decision has been taken.[4]

This also makes clearer what's happening in cases where we feel that a decision has been taken, but discover that it fails to issue in the appropriate action. There was a striking example of this in the paper recently. A man recounted how he had decided, after long and anguished consideration, to call on his wife, from whom he was separated, and to ask her to consult a solicitor and begin divorce proceedings. He duly called on her – and found himself asking her to take him back. Well, decisions often get changed – there's no mystery about that. We decide to go to Ithaca, but on consideration decide to go to Cephalonia instead. The man in the story, though, would probably say that he *didn't* change his decision – that no further process of consideration occurred between his knocking at the front door with his decision in mind and hearing the opposite of what he had decided to say come out of his mouth. If we return to the analogy of the race then we see that the man in the story is a little like a runner who is winning all the way through the race – until he is pipped at the post. We recognise with hindsight that our use in these cases of 'winning' and 'decided' was predictive, and that the prediction, though it stands historically as a prediction, fails historically as history.

There are parallels here with perception, though in perception the process is usually so speeded up as to seem instantaneous. The mind darts back and forth over a selection of inchoate visual information; and now (without any examinable procedure having occurred to explain how it has happened) we are seeing a bus or a porcupine. There are parallels, too, with narrative. Any narrative touches only here and there upon the great sea of events, like a dragonfly skimming over a pond, or a particle leaving its trail of discrete collisions through the cloud-chamber. It takes time to unfold; there is no one single point of time at which the story occurs; its completion is at an arbitrary point in time.

So I began to understand why I couldn't catch myself in the act of taking a decision – because there was no such act. What remained as mysterious as ever, though, was the conduct of the events whose outcome would constitute the decision. *Something* was going on in my head as I struggled to decide – some activity, certainly, almost as strenuous and as noticeable (to me, at any rate, and quite possibly to the girls concerned as well) as running a race. But what? Well, I was weighing the various arguments and alternatives – which makes it sound as formal as the proceedings at the local Debating Society; and I think it did sometimes take on this kind of air. The actual words of a proposition would form inside my head: 'One argument for asking her out is that I shall feel ashamed if I don't.' Then the actual words of another proposition would frame themselves, as if the Opposer of the Motion had risen to speak: 'One argument for not asking her out is that I shall feel humiliated if I do and she says no.' But how did I . . . *feel the force* of these propositions? How did I weigh one against the other? There were no other debaters to convince, no vote to be taken. The audience was also the Proposer and Opposer, and they had only one vote between them.

Well, whether I had articulated the arguments like this or not, I would imagine myself having the feelings that they described. So now I'm feeling the shame. Or, at any rate, I'm feeling a representation of the shame. It's as painful, in its way, as the feeling of actual shame . . . I'm not sure, in fact, that it's possible to make a distinction in the reality of the *feeling*, whether it's of actual shame or its representation . . . And if the feeling is indistinguishable, I'm not sure whether it's possible to distinguish between even the actual shame itself and its representation . . . I suppose that in the case of the latter I am always conscious that I am *experimenting* with it – that I can stop feeling it whenever I choose . . . Except that it's so vivid that it trails on, like a bad dream into the waking world . . .

Now I'm feeling the humiliation. Or a representation of the humiliation . . . And of course all the same considerations apply . . .

So which is worse, the shame or the humiliation? I try the shame again . . . Yes, the shame, certainly! Now the humiliation . . . No, the humiliation is even less endurable!

The shame . . . The humiliation . . .

How do I get the two imaginings into the ring with each other? How *did* I manage it, as I must actually have done so many times?

Probably decisions are often made in the same way as Valya's machine made them. You can't keep your finger on the button for ever, even if you wanted to. So you take it off, and if it happens that you're in the positive phase of the cycle rather than the negative at that particular moment, the phase of feeling the shame of inaction rather than the phase of feeling the humiliation attendant upon action, then that's the lamp that lights up.

But the difference from Valya's machine is that you know which cycle you're in. So even if you bring your consideration to an end simply because it can't go on for ever, the choice of whether you bring it to an end in this phase or in that remains in your hands. You still have to decide whether it's the shame or the humiliation that is to be outweighed by the impatience and weariness.

The decisiveness of decisions is as elusive as the decisions themselves. It recedes like the intentionality of intention.

Not even a death sentence is final until the trap has been sprung, and the ground gone from beneath the hanged man's feet. Up to that point the apparent finality of the decision is a matter of convention. The two brutal syllables of the jury's verdict echo unanswerably, like a prison door clanging shut. But the shut door may reopen if the judge rejects the verdict. It is the judge's sentence that turns the key inexorably in the lock. The key may be turned back, though, if the sentence is rescinded by the court of appeal. And if the court of appeal upholds the sentence it may still be annulled by the clemency of the minister; the minister's clemency by the intransigence of the Cabinet; the intransigience of the Cabinet by the pardon of the sovereign; the pardon of the sovereign by its delay in reaching the hangman; the action of the hangman by the failure of the trap.

But now the trap has opened and gravity has taken over . . . only to be frustrated by the miraculous intervention of the appropriate saint.[5] Call no man happy until he is dead – and no man dead until rigor mortis has set in. Call nothing decided until it is done.

Often, of course, we get up in the morning without taking any conscious decision at all, just as the coffee cup will be lifted while our mind is on the newspaper. The alarm clock rings, and we put our feet on the floor. Or we open our eyes, remember all the things that have to be done – and we are up. Up, teeth cleaned right to left, coffee drunk . . . We do what we do, and that's all there is to be said about it. The decisions that produced the actions, if decisions there were, seem to have taken themselves. We cannot locate them even with the combined benefits of hindsight and philosophical analysis.

On the other hand we are sometimes haunted, even as we go through the performance of weighing the evidence and agonising about the choice, by the sense that somewhere inside ourselves the decision has already been privately taken, perhaps in our unconscious, so that the inquiry we are conducting is a sham, undertaken merely to satisfy appearances – as if we were a government setting up a Royal Commission so as to give a public impression of concern and responsiveness even though the policy in question has already been settled. We should be inclined to abandon the whole charade, except that . . . we still don't know what this private decision is! Somewere in the dark depths of ourselves, in the Cabinet Room, behind locked doors, it has been taken; but so deep are the depths, so secure are the doors, that we remain as ignorant about the outcome as the man in the street outside.

At last the moment comes when we announce the secret decision to ourselves . . . Now we are haunted by the feeling that this in its turn may be a blind, yet another subterfuge of government which at the last moment will be jettisoned to reveal its opposite. And yet we know that we remain personally responsible for all our decisions, even if they are taken without conscious reflection, or at the prompting of unconscious urges. The decision to fly on autopilot, and to accept the readings on the cockpit computer, remains with the captain of the plane. At every moment he can choose to resume control. At every moment he can choose to override the computer. So that at every moment when he doesn't he is choosing not to.

Some decisions take themselves not while we are asleep, or thinking about something else, or under the influence of unconscious forces, but in the full floodlight of conscious attention. This happens, for instance, in cases where a decision has to be taken too fast to allow time for consideration. You're driving along the motorway, and suddenly something is falling off the back of the lorry in front. You have to swerve to the left or you have to swerve to the right. There's just a fraction of a second to register the choice, no time at all to examine the merits of the two alternatives. If it were Buridan's Ass in the driving seat instead of you the poor beast would die just as surely as he would between the two heaps of hay, and a lot quicker. But no – you've swerved, one way or the other. In the full glare of your most intense and urgent attention a decision has formed and executed itself – and you have not the faintest notion how.

Doesn't much the same thing often happen, though, even when there's no urgency at all about the decision? Marmalade or honey – it's not a matter of life or death, and in any case it's Sunday, so I have the whole morning to make it. But I don't take the whole morning. I simply reach out my hand and . . . pick up the marmalade.

Never mind this time *why* I did it. *How* did I do it?

I think that what happened in this particular case was this. I noticed the marmalade, and it somehow caught my imagination. I considered the honey carefully as well, of course, or at any rate I went through a brief performance of considering it, just to meet the requirements of Equal Opportunities legislation, just to spare the feelings of the honey, or at any rate to make myself feel that I was really exercising a choice – but I knew all the time that it was the marmalade I was going to take in the end. The marmalade was my first-born. It had acquired squatter's rights in my mind, and a wish for it crystallised around it *post hoc*. So, although I made a conscious choice, it was in fact a random one, because there was no clear causality that brought the marmalade into my eyeline first.

A week goes by, and next Sunday I go through the same procedure as before. This time, though, for some reason, it works the other way round. Even though I see the marmalade first the memory is somehow overlaid by the sight of the honey. The attractions of the marmalade are supplanted by those of my new love.

Before I lifted my eyes from the newspaper I didn't know which of the two pots they were going to fall on first. But then nor did I know which of the two *principles* – the primacy of the first seen or the primacy of the more recently seen – was going to apply on this particular Sunday morning. The choice of principles took place only simultaneously with the event. And there seems to be no objective determinant of it – nothing that is explained by the barometric pressure or the phases of the moon, or even by the shortcomings of my parents. Was it as random as the direction of my glance? Or was I making a personal choice of principles, even if not of spreads – *allowing* primacy to this or that principle?

In the end, though, I reach out and take . . . not the honey, after all, but the marmalade again – precisely because it's the one I *don't* want, just to demonstrate to myself that I'm not fettered by random circumstance.

So why do I sometimes allow myself to be fettered and sometimes not? Isn't my choice not to be fettered just as capricious, just as out of nowhere, as my taking the first choice that my eye happens to fall upon?

In actual practice my choice is probably determined by some quite other principle – a personal principle which I have just this morning adopted precisely in order to save myself from that brief vertiginous moment of uncertainty. I take the marmalade because from now on I am always going to take the marmalade on Sundays. Marmalade on Sundays and Wednesdays, this is going to be the rule, and honey on all the other days – except when the weather is depressing, or the honey pot is sticky. So I have chosen the rule. And I have also chosen to apply it on this particular Sunday morning, when I could have chosen to suspend it for once.

In all these cases there is at any rate one objective determinant to my choice – the range of spreads on offer. But then my acceptance of this limitation is itself a further decision on my part. Faced with the choice of marmalade or honey I might well have decided, on this particular Sunday morning, that what I wanted was Marmite.

A pure and undetermined choice it would have been this week, then, if I had. Or would it? I suspect that no thought of Marmite had entered my head until I was offered something else. My wish for Marmite was partially determined by my not being presented with it. I was constrained once again – to a much wider, indeed indeterminate, field, but constrained none the less. But, then again, I had chosen to accept this different constraint . . .

So many decisions that have to be taken, consciously or unconsciously, before I can sink my teeth into the toast! It wouldn't be surprising if I made just one single decision instead – to go back to bed and make no decisions about anything at all. Except that making no decisions would be a very active and difficult programme, which would require at every moment, as new situations that seemed to require decisions presented themselves, a constant stream of decisions not to take them.

A hardline determinist might insist that all this fuss about decisions, never mind the motivation behind them, is beside the point. The process of deciding is illusory. The output follows from the input (even if we don't know what it is) as remorselessly as the striking of red on black follows from the striking of white on red. Not that this will save them from one single moment of effort, though, even as they struggle to explain their thesis, about what words to choose to express the thought, or from one single moment of anguish, even as they do so, about what to do with their aged mother. So presumably they have to accept that this effort and anguish, this false perception of an open choice in front of them and of a personal responsibility in closing it, are themselves inescapable elements of the causal process, even if the result is a foregone conclusion, theoretically predictable by anyone, except, apparently, themselves.

Just like the rest of us, too, a determinist would struggle to rationalise at any rate some of the more important decisions that he seems to make. He would try to ensure that he gave fair consideration to all the alternatives, to observe some explainable selection procedure, and to abide by the results. He would understand just as well as the rest of us the further set of external difficulties that arise when the decisions of separate individuals have to be brought into relationship with each other to produce corporate decisions.

The difficulties are compounded for all of us because, even where decisions are rationally based, they are not the product of any objective state of affairs but of our *perception* of those states of affairs. Different people's perceptions of the same state of affairs are notoriously varied – as varied as the decisions they then proceed to make. So even for the determinist, the determining cause is an inward experience, which, in the case of others, is not directly accessible to empirical examination, and in his own case subject to all the familiar difficulties of introspection and self-knowledge. His views on the subject may be a matter of reason, or more probably of faith, but they are certainly not based on observation or self-observation.[6]

It might seem rather curious, given all this, that determinism lasted slightly longer in human affairs than it did in physics. We gave up the classical clockwork of the atom in 1926; the classical clockwork of Marxism, with its picture of human beings operated like the figures on a clock by the great mainspring of history, lingered on for another sixty years. 'Historical determinism' was always more ambiguous than it looked, though. Even though the transition from capitalism to socialism was seen as inevitable, it was usually understood that the historical mechanics could only operate through the indeterminate media of human perception, feeling, and choice: through what people perceived to be happening, what they felt about it, and what they chose to do about it. Communists saw themselves as being among the determining agents – of choosing, out of their own free will, to participate in what was always characterised as a struggle. Their role in this struggle was not mechanistic – it involved great efforts of persuasion and coercion. Like fascism, communism was an attempt to change the world by the exercise of the human will, and the metaphysical belief that these all-too-human efforts were somehow part of an inevitable process was a psychological tonic, an antidote to the sense of helpless arbitrariness that threatens to engulf us all from time to time.

The same applies to the determinism of the market, the supposedly iron laws which also take effect through the same distinctly non-ferrous procedures of perception, feeling, and choice – and which indeed have to be reconciled with the freedom of the individual. What the laws of the market determine, though, is the form and framework of economic competition. You might say that the market is the continuation of evolution by other means. You can imagine someone arguing that evolution was a deterministic theory – that changes emerged absolutely precisely from the various forces involved. But in the first place there is a fundamental indeterminacy at the heart of evolution – the random genetic changes upon which selection operates. There are many other random inputs (events that are random as far as genetic reproduction is concerned) – floods, droughts, plagues, volcanic eruptions, and the shift of continents. In the market, likewise, quite apart from human vagaries, there are the same random inputs, and more still – uncovenanted inventions and chance discoveries, the opening up and drying up of natural resources – upon which competition operates.

What neither the laws of genetic reproduction nor the laws of the market predict is the outcome of the struggles they give shape to, any more than the rules of football predict which team will win.[7]

Some people never quite shake off their longing for the reassuring objectivity of determinism. In religious history the idea of the human being as sovereign, of his having the freewill to accept God's grace or not, has been repeatedly rejected by Calvinists on the one hand and Jansenists on the other who cannot reconcile themselves to such a disorderly – such an incipiently humanistic – procedure. The same battles are fought now, not only in politics and economics, but in neurology, where many of its practitioners want to see not only the human brain but the human mind as a machine.

In the end, though, a machine is exactly what it's not. Or not, at any rate, a machine that can conceive of *itself* as a machine. And it's only what we can by one means or another conceive of that can take its place among the constituents of the universe.

Your life has been shaped by decisions – by all the dozens of conscious major decisions you have made yourself, by all the millions of minor ones, and by all the billions and billions of entirely unconsidered ones.[8] It has also been shaped by the decisions of others, as you may be gratefully or only too painfully aware. As the human tribe has become ever more widely and densely settled across the face of the earth, and as we have gained ever greater control over the natural forces that affect us, so our environment has become ever more man-made – and therefore ever more the product of the decisions that we take. Once, when a river burst its banks or the crops failed, we wondered helplessly at the inscrutable decisions of God. Now the media investigate why the constituted human authority took a decision to save money on flood barriers or irrigation canals. Or why it failed to take a decision on these matters at all – because the failure of foresight and imagination, too, is a human responsibility. Human choice (whether determined or not) accounts for more and more links in the great causal chain that we perceive.

But then the world that we inherited, and which we have all done our bit to modify, was already built upon the decisions of our forebears. The laws, civil and scientific; the cities and the first habitations around which those cities clustered; the philosophies and outlooks; the languages we speak in, the miles and milligrams and the degrees

Celsius and Kelvin that we measure in. Just as you and I, in order to keep up our conversation, make a decision with every word, so we and our forebears, in order to maintain our endless dialogue with the universe, have made a stream of decisions about how to frame our concepts. In framing those concepts we have given our relationship with the universe the form it possesses. It is the human mind that has decided how to measure the distance of the galaxies, and how to picture the relationship of gravity and light. The universe we live in is a coral reef, formed from the bones of countless decisions left behind after their authors' deaths.

God's creation of the world by a pure act of will – by a decision – is the mythological summation of this incomprehensibly complex skein of decisions through which human beings have so lengthily and laboriously performed the same glorious feat.

Actually the world was being shaped by decisions long before human beings even existed, or God had been invented, because we are surely not the only species that is capable of making choices between different alternatives. How, after all, did man evolve? – Through the decisions of his animal forebears. Individual creatures, and communities of individual creatures, have prevailed over each other or been eliminated through the fundamental choices they have made, in encounter after encounter, between fight and flight. Is it possible to regard these choices as purely deterministic, simply because the creatures involved have no language to explain themselves to themselves, or to others? Look at two cats confronting each other. Each animal is plainly *assessing* the other – estimating its resources and judging its intentions, probing and feinting, then observing the reaction as carefully as any military commander. A high level of perceptual skills is involved, and the interplay of a great many considerations of quite different sorts. How could such complex and disparate material ever be assembled mechanically to provide a mechanical stimulus? A judgement is plainly being made, even if the only expression it can be given is in the physical action that results (which is, after all, how even we, articulate as we are, give expression most of the time to our judgements). The behaviour of two animals confronting each other like this often involves a *display* – the adoption of a posture or behaviour to suggest size or power or aggressivity. What purpose could such a display serve except to produce an *impression* upon the rival? What could the point of producing an impression be except to *persuade* the rival to back down? How could such persuasion take effect except by the rival's consenting or choosing – *deciding* – to do so?

Fight-or-flight is not the only decision you can see animals taking. Watch two cats playing. What account could be given through permutations of stimulus-and-reaction to explain why they yowl and roll just so, why they suddenly roll apart and twitch their ears and feign indifference, why they resume the game, and do it just precisely now?

Of course we are talking about higher-order animals here. Move down the scale a little and you don't know quite what to say. Watch a lizard and an ant and you might feel that the ant-shaped image on the lizard's retina activates the flick of the tongue purely mechanically. Or you might not. Why does the lizard wait . . . wait . . . and *then* lash out? And when you get down to the level of the ant itself and its random scurryings then . . . you're forced to hesitate yet again.[9]

Entomologists did a famous series of experiments some years ago with ants, in which an ants' nest was located between two perpetually replenished sources of food.[10] Since no ant could detect either source of food until it had left the nest to prospect there was no temptation for it to sit helpless in the middle, like Buridan's Ass, and die of hunger. Individual ants foraged randomly and stumbled upon one source or the other. Having discovered a source the ant would then return to the nest and stimulate another ant, or ants, by means of a chemical secretion (an effect which we can presume to have been produced quite involuntarily), to follow it back to the source. So a stable pattern of affiliations to one source or the other should have developed, and, since the affiliations had originated from a random distribution and the number of ants involved was large, the probability should have been that those affiliations were reasonably stable and equally divided.

But the affiliations turned out to be neither stable nor equally divided. They were subject to continuing wild and apparently random fluctations. This has suggested to some economists important analogies in the human sphere, in situations where the behaviour of individuals can be influenced by the behaviour of other individuals. It also makes one think again about the possibility of ants being in some sense able to choose their actions. What the results imply is that a relatively large number of individual ants, having located a reliable source of food, and well able to find their way back to it, actually opted to turn their backs upon it and go off in another direction altogether. I suppose that one could construct some device to explain this mechanically, without recourse to any notion of choice, by saying that ants have a tendency (not unlike human beings) to random variations in their behaviour, variations which are so strong as to override the normal learned patterns of reward and punishment. If random, though, wouldn't the variations in the behaviour of the various individuals tend to cancel each other out, so that the numbers abandoning source A would be balanced out by the numbers abandoning source B, leaving the overall numbers visiting each pile roughly constant? Something much odder and more elusive is surely going on here. I can't help noticing that the economist who refers to these experiments says quite simply and straightforwardly that the individual ant 'of its own volition . . . decides to try the other pile itself'.

So here is some respectable academic talking about 'volition' and 'decision' in *ants*! And for good reason: because it gives the simplest, shortest, and most immediately comprehensible picture of what happens.

And what about all those ideas that seem to present themselves out of nowhere – for the gift to a friend, for the plot of the book? Do they dump themselves down upon me like troops being billeted by a quartermaster against whom I have no appeal? Once they've turned up on the doorstep I'm surely free to admit them or not. In the end there is a decision to be taken, here as with everything else, and the decision is once again with me. I may hire a speech-writer to write the speech for me, but I'm the one who has to make it or not make it. I'm the one who has to take responsibility for it if I do.

And yet . . . how do I take *this* decision, any more than any other? One moment the idea is on the doorstep – the next either it's inside, or the door has been closed against it. Or else the door is swinging uncertainly back and forth in my hand. Hasn't the decision to admit or not to admit, or the decision not to take a decision, come from just the same mysterious inner world as the idea itself?

So there are two stages, is that it? First the thinking, which is not under central control, and then the expression of that thinking (in words, in actions), which is. Well, it's true that we often do distinguish the two elements. There are many occasions when I feel that I've spoken or acted *without* thinking, for better or worse; and perhaps some when I feel the converse – when I emerge from what seems to have been contemplation, of absorption in what seems to have been thought, without any words or actions to show for it.

If I *do* express something, though, even if I'm not conscious of any serious reflection first, *something* must have happened, surely, for that something to be expressed! Yet how can I possibly have any knowledge of what that something is, if it's unexpressed? Well, you might investigate the workings of my brain, and trace a train of electro-chemical events. But what significance would those events have, what interpretation could be given to them, except in terms of any expression that they eventually take? And so I'm driven to the contrary conclusion: that the expression *is* the thought, and that's all there is to say about it.

You might object that the entire thinking process can also be completed internally, that it can terminate in the preparation of the words (images, actions) inside one's head, so that the actual expression of them, in the sense of their being pushed out into the external world, is a merely mechanical follow-up, like the printing up of a document from a computer file. (Mandelstam, for instance, is said to have completely composed all his poems in his head before he put a word on paper.) But in this case I can't help feeling that it was this internal assembly of the words that constituted the act of expression – that the thought was pushed out, if not into the outside world, then into an ante-chamber of that world. It's as if the document to be printed had been assembled first in a buffer inside the computer, ready to be downloaded.

We all use this ante-chamber constantly, even if we don't do it as thoroughly or as systematically as Mandelstam. We all try out thoughts, or parts of thoughts, inside our head before we launch them into the world; we hold private rehearsals before the public ceremony takes place, then modify the text, perhaps abandon the public performance altogether. But we know that, once again, the script of the show in the first place, and the edict authorising it, have emerged from inner chambers of the palace to which not even we ourselves have been granted access.

'To be free is to be able to make an unforced choice, and choice entails competing possibilities at the very least of two open, unimpeded alternatives . . .' (Isaiah Berlin)[11]

Questions of choice always seem to be presented like this, in clear-cut, digital terms – and usually binary ones. Some of our choices really are of this sort. Do I pull the trigger or not? Shall I ask her out or shan't I? Do I agree to sign or refuse to? Which of the seven candidates do I give my vote to? In most cases, though, the choices I am confronted with are not like these neat paradigms. They are more likely to be graduated – and often so infinitely graduated that I may well fail to recognise them as choices at all. Do I cross the street here . . . or here . . . or three metres further on? Do I cross at this angle . . . at this . . . or 3.7 degrees to the right of that? How eager am I to go to Trebizond? I might say fairly, or very. But *exactly* how eager am I? The question is ridiculous, because there is no way of *measuring* eagerness – but even so, some answer has to be found, which will be exactly expressed in whether I actually do go to Trebizond or not. I draw a line, curving it slightly to the left – now round in a loop – now straight, now zig-zag . . . Am I making a series of discrete choices? A series as infinitely numerous as the series of mathematical points along the line? Well, notionally, I suppose, in that anywhere along the line I could do, or could have done, something else, just as the captain could at any moment choose to switch off the autopilot. But the choices are not part of my conscious experience as events separable from the results. My hand is moving this way – now it's moving that way . . .

In any case the range of choice is often not pre-defined – not limited to situations like crossing a certain stretch of road, or to what might be drawn with a pencil and paper. I have to improvise out of an indefinite range of materials. Night after night I lie awake, wondering what to do about a recalcitrant colleague or a difficult neighbour. Perhaps, as with the inspiration for the gift or the plot, a solution presents itself. Or perhaps no clear expression results, either externally or internally. I rehearse angry speeches – I rehearse conciliatory ones. I experiment with feelings of resentment, condescension, helplessness . . .

And then whatever emerges from this process becomes modified even as I try to express it. When I come face to face with the person I have thought so hard about, neither the angry speeches nor the conciliatory ones get spoken. The result of all that thinking seems to be merely that my irritation is a little more marked, or a little more controlled. Perhaps some of the phrases I tried over do actually emerge, but curiously garbled, in the wrong order, lacking the cogency that recommended them. Whatever I manage to express, in any case, leads to fresh and unforeseen recriminations and obduracies from the other side, or to unexpected concessions, all of which have to be dealt with in the light of some cloudy general policy that I have apparently adopted. A policy which seems to be infinitely flexible. A policy which is certainly unexpressed and almost certainly inexpressible. Or one that is expressed and expressible only in terms of the behaviour that seems to be emerging subsequently.

Even where the range of choice is mercifully limited (as in crossing the road or drawing a line on paper) what is it that limits it? My own choice! I don't have to cross the road at all if I don't want to – I don't have to use the paper and pencil to draw a line. I can stay on the pavement – or lie down in the guttter. I can set fire to the paper, and stick the pencil into someone's backside.

Those apparently simple binary choices between clear-cut digital alternatives, for that matter, usually leave us with plenty of opportunity to blur their clarity. Pull the trigger or not – or perhaps apply just a little pressure on it, like this, and let the gun itself decide whether to go off or not. Ask her out or don't ask her out – but privately keep open the possibility of just happening to run into her without any pre-arrangement. Agree to sign or refuse to – or sign with a codicil.

Even the starkest, most immediate decisions, where there is no time for this kind of organised subversion, may well conceal the same kind of complexities. Swerve left or swerve right, certainly, as the load comes tumbling off the lorry – but exactly how far to the left or the right? Only *so* far, and we still get sideswiped by the falling load. As far as *this*, and the car spins out of control. Once again the policy is infinitely subject to revision in practice, as the tumbling girders bounce to this side or to that, as the car begins to break away. Once again, what we turn out to have decided is not simply left or right, but a general course of action expressed and expressible only in terms of what we in fact subsequently find ourselves doing.

The traffic continues. We think and act on the move. On the move ourselves, in a world that moves around us. By the time the thought is complete the situation has changed. Thoughts are like still pictures, cross-sections of the world from which depth and time are missing. Consciously articulated decisions are like architects' plans, flat surfaces to be projected on to a world that also extends away from us, beyond the surface of the paper, unseizable.

Most of our everyday decisions are taken so fast and unconsciously that they merge into the perceptions from which they arise, and share some of their characteristics. We alter course to avoid a streetlamp as we walk, without consciously noticing either the streetlamp or our decision to avoid it. Just as the concept of the streetlamp must be held in store inside us, with all its hardness and all the undesirability of walking into it already etched upon it, waiting to be evoked by the briefest flicker of passing information in the middle of the visual field, so too the avoidance procedure and the decision that triggers it must also be prepared and ready to be released.

In fact you can't help feeling that even the most consciously considered decisions have something of this pre-packed quality. The mind moves back and forth indeterminately over the various considerations, with no procedure to define what constitutes a complete examination, or how the process is to be concluded. I look up into a tree, trying to decide what, if anything, to do about a child's kite that's tangled in its branches. Should I climb the tree? Find someone else to climb it? Look for some kind of pole . . .? The range of choices is open-ended, the considerations as tangled as the kite-string. The fragility of aged bones . . . the child's distress . . . the kinds of thing that would serve as a pole . . . the child's smile as I hand the kite back to it . . . the ethics of asking someone else to risk their limbs . . . the tingling feeling of the drop beneath me . . . the effect of throwing some heavy object . . . What protocol is there for bringing all this to a conclusion? Sometimes an arbitrary cut-off is imposed by an external deadline – I've got to catch a train in ten minutes' time. Otherwise I suppose I have to wait for . . . the *feeling* that a conclusion has emerged. And now it seems rather as if it had been waiting inside me all along, and had simply been *evoked*, as a perception is, by the ragbag bits and pieces that have passed through my mind.

Neurologists have established that, even in the case of a purely arbitrary and autonomous decision (to press a switch at any time the subject chooses) the act is *preceded* by a change in neural activity. What this appears to suggest is that the decision has already been taken internally, as one sometimes suspects from one's own intuition, before one is conscious of it. Some neurologists feel that there is an apparent contradiction here with the notion of free will. Isn't it once again, though, just another example of the same process that produces a reasoned decision: an unconscious activity of which the conscious decision is simply the successful outcome?

In any case, an explanation for the initiation of the unconscious, and apparently uncaused, neural event still has to be found. But then so it has for the beginning of the conscious, or partly conscious, debate about whether to ask the girl I've just met to go out with me. Did I *decide* to start the decision process? It surely wasn't random, nor yet a conditioned reflex, because I didn't go through this agony with *every* girl I happened to set eyes upon (or at any rate not quite every one) – but was responding to this one particular girl to whom my reflexes can scarcely have been conditioned because until this moment she hadn't been there to condition them. In both cases, it seems to me, my own (unconscious) act, in some inner chamber of the mind, is the least incomprehensible explanation. And that inner chamber is still a part of myself.

We are traditionally offered a pair of accounts of human decision-making to choose between: free will and determinism. It doesn't seem much of a choice (if, of course, we *have* any in this matter, any more than in any other). The idea that my crossing the road at an angle of twenty-nine degrees to the kerb, my tweaking the end of the line I am drawing just so, and the dog's urinating on the right-hand side of the third lamp-post, not to mention my inclining to whichever view of all this takes my fancy, are all parts of an inexorably linked chain that goes back to the moment of creation, seems simply . . . *inapt*. Inapt because it can't be used for anything, either to explain the procedure or to differentiate it from anything else. In what language, however laborious, in what figures, however copious, could such a chain of events ever be traced? At least the idea of free will – 'The explanation is that there's no explanation; I just do it' – differentiates between my own watering the garden and the rain's doing it, between the squiggle I drew by myself and the bit at the end where my elbow was jogged.

There *is* a genuine choice of accounts, but a different one: between, on the one hand, decision as the act of an agent operating as a complete entity and, on the other, the act of some sub-system within the agent. As with the choice of explanations in physics, between the classical and the quantum-mechanical, each of them is appropriate in a different context, on a different scale. The classical account, decision as the act of the agent, may well be adequate to describe the macroscopic world of human behaviour viewed from outside. But when that agent attempts to examine the decision-making process from the inside, from his own point of view, he has crossed a threshold beyond which classical concepts no longer seem to apply.

I can follow the general outline of someone's thinking, even my own, from the conclusions and actions publicly manifested (or, in the case of my own thoughts, assembled in the ante-room ready to be manifested). I can identify in broad general terms the reasons and emotions that seem to prompt my behaviour. But when I try to see precisely what's going on it eludes categorisation and exposition. I end up convinced by an argument; but where exactly am I at each moment while the process of being convinced is going on? By what means do I balance this argument against that argument, this emotion against that emotion, this emotion against that argument? I am perhaps conscious of intermediate milestones in the process. Having considered the arguments for a course of action I feel that I'm convinced by them. Then I plunge out of sight of myself again as I consider the arguments against, and remain out of sight until I emerge with the feeling that I have now been convinced in the opposite direction. More inaccessible consideration follows, more surfacings with temporary (and perhaps mutually contradictory) convictions in my hands.

In between whiles I remain in two contradictory states of conviction simultaneously – like Schrödinger's cat, alive and dead until the chamber is opened. This superposition cannot be examined, any more than the state of the cat can be, or the passage of the single photon through the two different slits, because observation precipitates the collapse of the superposition into one definite (classical) state of affairs. As soon as I turn my attention to what I am thinking, the thinking freezes in one position or the other.[12]

Actually I have the feeling that the situation is often even more elusive than that. What I observe, when I try to look at what's going on in my head, at once obligingly accommodates itself to what I'm hoping to observe – or at once disobligingly rebels against it. It's as if it knows which way I want the experiment to come out – which can only mean that somewhere, in the dark inner chambers, the decision on the result has really been taken already.

Of course, the indeterminacy of human thought and the indeterminacy of physical objects arise for very different reasons (human thoughts are not located in the way that particles are, in so far as particles *are* located, or at any rate not by the kinds of coordinates that locate particles). You might generalise the indeterminacy principle, though, so that it brought the two different forms of uncertainty into relation with each other. You might say something like: the more precisely you try to determine a particular situation at a particular time, the less you can know of the way in which that situation is changing; and, conversely, the more precisely you try to determine how a situation is changing, the less you can know about the situation. Or: the more precisely you try to determine the processes that constitute the universe, the less distinct they become. It is as if we had an inbuilt presbyopia, which no spectacles can ever alleviate, making what is clear and distinct when we view it from a distance become, as we approach, not clearer and more distinct but ever less so. The world is like a picture that yields not more information but less as we move closer to it, because it breaks up into brush strokes, or a screen of dots, or a chemical grain, so that, as we find out ever more about the brush strokes, dots, and grain, we know ever less about the picture, because the discrete elements carry pictorial information only in combination.[13]

All this suggests processes remarkably unlike the kind of organised thinking that is formalised by the laws of logic. But then you might see the Law of Excluded Middle (for instance) as a law in the legal sense – an edict promulgated to discourage the natural tendency of the mind to indeterministic misdemeanour. This is what makes a computer such a good tool for arriving at decisions – because we have built this law into its architecture. It is bound absolutely by it, and by the other laws of logic, as we are not – because this is how it has been designed. We have made its gate strait, and we have given it techniques for reducing the complexity of the world to a flow of digits that can pass through that strait gate only one by one. This is after all what a computer began as – a thought experiment for distinguishing between the decidable and the undecidable. The decidable was what the imaginary machine could, in a finite number of moves, actually decide. The computer that has developed from Turing's machine is able to decide because we ourselves have decided that it should be able to, and because it obliges us to submit material to it in decidable form. All the decisions, both strategic and tactical, have really already been made by human minds.

Could we ever design machines capable of imitating the characteristic indeterminacy of human thought? We can (and do) build in the element of the random that we have detected in some of our decision procedures. But to make them truly human we should have to supplement their logic with feelings, with the ability to have hunches and jump to conclusions, to find ways of outwitting the Law of Excluded Middle. Until then we shall have to go on coping unaided with all the aspects of the world that cannot be reduced to digits, and for which no decision procedure has been settled – but from which a course of action has to emerge.

Deciding the undecidable: the last human skill to be surrendered to the machines.

III: Stories

A Cast of Characters

Thinking of thoughts, speaking of things

The causality of human motivation and intention seems to be even more elusive than the causality of the physical world. But then it's not just our decisions that are so difficult to locate and examine. So is *everything* that involves human thought and feeling, regardless of what effect it has on our behaviour or the world around us.

Strange, when there's so much of it to study. Apart from breathing and digestion, and the various unconsidered involuntary processes that keep us physically alive, thinking and feeling are by far the commonest human activities. Even if you leave deciding out of it for the moment, we are thinking and feeling in some kind of way at every single instant of the livelong day. Five or six thousand million of us, from sunrise to sunset, from birth to death, observing, inventing, reflecting, daydreaming, lying, praying, promising, framing our words and directing our steps, resenting, loving, and worrying.

Every external action that we perform is accompanied inwardly by thought and feeling. We exult in what we are doing, regret it, imagine how we will recount it afterwards, discover ways of doing it better, remember something else that we should be doing. We shut the machine down for the night, and in the darkness it starts up again of its own accord, forcing dislocated snatches of narrative upon us, consoling and horrifying images, surges of longing and pain and happiness. All around us this universal activity boils on – and all of it scarcely more accessible to us than the neutrinos that also stream undetectably around us. All of it as elusive as the sources of action that lurk among it, known only by hearsay, only by projecting backwards from the actions themselves, from the words uttered, the expressions assumed, the gestures made.

All of it, that is, except the thinking and feeling going on inside one single head among the five thousand million. To some of the events in this one single head on my own shoulders, even if not to the occult sources of my actions, I surely have unrestricted access. And yet once again all the same problems of introspection arise. As soon as I direct my attention towards this internal universe it vanishes like the millionaire's wealth. Or rather, runs round behind me like the back of my head when I turn. Is transformed into the act of looking for it.[1]

Our feelings are not as volatile as our thoughts; they don't vanish as soon as we look at them. Even our feelings, though, seem to lose their determinate outlines when we actually try to examine them. Don't we even know that we are angry or that we are calm, that we are wretched or that we are happy? Of course we do! We know *something*, certainly, but the haziness of what exactly it is becomes apparent if you imagine trying to be scientific about it. Think of what it would be like to make a taxonomy of moods! To establish the species and sub-species and phyllae! What would the identifying characteristics be? Even if you abandoned all attempt to categorise other people's moods, and restricted your efforts to your own?

And yet we talk constantly about what others think and feel, and what we think and feel ourselves. We *think* about what others think, and what we think ourselves. We have feelings about their feelings, and our own. You can scarcely have failed to notice that I've been telling you about your thinking processes, and my own, even as I explained that it was impossible.

It all seems plain enough in fiction. Novelists often know exactly what their characters are thinking, just as they know exactly why they do what they do. It's very corrupting. We are tempted to believe that the great psychologistic novelists are writing *naturalistically*. We follow Musil's account of Arnheim's thinking in *The Man Without Qualities* quite unquestioningly:

> The realisation that his impotent annoyance about Ulrich at some profounder depth resembled the hostile encounter of two brothers who had not recognised each other gave him a very intense and at the same time comforting sensation. With eager interest Arnheim considered both their personalities from this point of view. The crude acquisitive instinct, the urge to get what advantage one could out of life, was something Ulrich had even less of than he had himself; and the sublime acquisitive urge, the wish to make the dignities and distinctions of this life one's own, was something he lacked in a

positively infuriating way. This man was without any craving for the weight and substance of life . . . Arnheim would positively have felt reminded of his employees if the selflessness that in their case amounted to being 'keen on their work' had not, as exemplifed in Ulrich, taken on a character that was uncommonly supercilious. It might be truer to say that here was a man possessed who did not wish to have possessions. Or what was evoked, perhaps, was the thought of one fighting for a cause in voluntary poverty. It seemed possible to express it, too, by saying he was an out-and-out theoretical being; only that again did not quite fit the case, because one really could not call him a 'theoretical man' at all . . .

So Arnheim's thoughts went this way and that . . .

Think about it for a moment (don't try to give me an account of what this thinking involves!) and you realise that this is like one of the *New Yorker's* old 'Shouts We Doubt Ever Got Shouted'. ('Death to Deputy-Chairman Wu and his quasi-objectivist anti-Party murmurings!') How did those implausibly complex slogans get reported as verbatim quotes from the demonstrating crowd? By the efforts of the reporter to find a coherent expression for the thoughts which he supposed – whether disingenuously or not – to be animating the confusion of sounds and actions before him. So, too, Musil is inventing an externally reportable correlative to the shifting currents inside our heads which we find it so difficult to observe, so difficult to separate into identifiable elements, so difficult to give any coherent account of. But the result is as non-naturalistic, as stylised (and as expressive and memorable), as Giacometti's reduction of the human form to pure height, as Seurat's reinvention of the visible world as spots of pure colour.

In spite of all the difficulties of introspection we do manage to catch glimpses of our own thinking. How is this possible? Well, how do people carry out more than one conscious task of any sort at the same time, as they do? (The common situation of women in particular, who habitually find themselves discharging full-time professional obligations at the same time as they plan the shopping, the cooking, and the care and education of their children.) Isn't it rather like the multi-tasking of a computer, which performs several functions in parallel, even though it has only one gateway, by switching continuously back and forth between them? In a computer this tends to produce a characteristic and disconcerting sluggishness in the behaviour of the machine – and in human behaviour often something rather similar. Waiting to cross the road, I watched apprehensively as a passing motorist simultaneously turned the corner, spoke on his mobile phone, and lit a cigarette; and if the standard of performance he was achieving on the corner was any indication, the cumulative effect of all this seemed likely to be not only dead pedestrians but garbled communication and burnt nostrils. What happens when you step back and catch yourself out with a thought still present (so far as I can tell by these same hit-and-miss procedures) is that you catch its afterglow, like the image that sometimes lingers on the retina after you've closed your eyes. But even if the words you were thinking, or the picture you had, is still there, the formation of the words, the assembling of the picture – the actual process of thought (whatever that process was) has ceased.

You also have moments of detachment from yourself, when you feel that you really are observing your own outward behaviour. Perhaps this happens most frequently when you're giving a performance – making a speech, acting in one sense or another (if only for an audience consisting of yourself alone) – doing something to produce an external effect.[2] Maintaining this performance must require thought, but the thought seems to be coming from some kind of auto-pilot inside yourself, and you have a certain freedom to sit back, like the captain of the aircraft sipping his mid-flight coffee, and observe what the plane's doing. Or perhaps it's more like being the director of a play, watching the production critically from backstage, knowing just how it relates to what was rehearsed, just where it's going better than usual, just where the evening is beginning to go wrong and turn into a disaster (and sometimes still, like the director, unable to intervene effectively).

But then the audience out front are doing much the same thing. There are some aspects of the event that they don't know about, there are some comparisons between intention and effect that they can't make, but the greater part of what the director sees and makes his judgement on is what the audience, too, are seeing and making their judgement on.

Even in my moments of self-observation, of course, my faculties may be affected by the very states they are focused upon. They may be dimmed by the depression, unbalanced by the euphoria, hustled by the eagerness, paralysed by the fear. The indeterminacy that clouds what I see in myself, though, goes deeper than this. I often don't really know what I feel, if it's anything less clear-cut than rage or exhilaration, even with things that affect me closely. Sometimes I'm not even sure whether I feel anything or not. A vague downness, perhaps, an unlocatable upness . . . Or both at once . . .

I cast around in my mind for some possible source. Now I remember the events of the previous evening, when I found myself treated more coolly than I had expected. Or I recall that I have a journey to make today which has some kind of promise in it. The nature of the event suggests names for the feelings. Resentment, yes. Anticipation. Now that I have the words I can begin to talk about the feelings, if only to myself. And once I can talk to myself about them . . . yes – I begin to feel them more strongly. There's a specimen on the slide distinct enough for me to focus the microscope upon. Now I can become positively botanical in the nicety of my discrimination. What I'm feeling, I discover, is not exactly resentment but a mixture of wounded pride and disappointment, not just vague anticipation but curiosity tinged with apprehension . . . But is it really my feelings that I'm examining? Or is it the events that aroused them? What I'm doing, surely, is constructing a story for myself, and inventing the feelings that must exist to make it work. And what I'm analysing so finely – isn't it almost the words themselves?

I reconstruct my thinking and feeling after the event. I assemble the clay into a shape. Just as I do with your thinking, and you do with mine. Just as Musil does with Arnheim's thinking. I compose a plausible and coherent fiction of emotions and occasions which seems to explain the indefinite colours of my inner experience by relationship to the external circumstances.

The act of naming is crucial. This *is* what I feel because I *say* that this is what I feel. The name I choose may tell you more about my general character, about my experience and view of the world, my scale of values, and my inclinations and desires, than it does about the inchoate something to which I'm attempting to give form. It may also be shaped by the perspective in which I view it. You, from your different perspective, may name it quite differently. The kind of feelings that I identify in myself as righteous indignation you may entitle sour resentment. Once again, as with so many things, the language seems to determine the object, rather than the other way round. It's as if language were a world of its own, with its private consistencies and conventions, and only fitful connections to the unspoken world.

The generative powers of language have often been noted.3 The Hungarian writer Frigyes Karinthy, in his classic account of suffering a brain tumour, and of undergoing one of the first operations possible for its removal (in the 1930s), describes the familiar fear that we have called something terrible into being merely by naming it, by entertaining the idea of it:

> Vienna again . . . At ten o'clock next morning I called at the Wagner-Jauregg clinic.
>
> Not until I found myself unwillingly plodding up the steps at last . . . did I realise with a profound sense of discouragement what it was that I found so repellent there. On that very staircase I had stood three weeks ago – only three weeks, yet they seemed as long as my whole lifetime! It was on that spot that I had flung out, as frivolously as if they had been some groteseque or flippant joke, the words, 'I've got a tumour on the brain . . .'
>
> I knew that the idea in my mind was an insane delusion and my education and scientific instinct rose up to protest against it, but superstitions of this kind have haunted me all my life. As I stood there I could not rid myself of the notion that my trouble began only when I spoke about it. Not only was it born at that moment, but *as a direct result* of my having given it a name. I felt that things happened because we gave names to them, and thus came to look upon them as possibilities. Everything we regard as possible comes to pass. Reality is the child of man's imagination.4

I suppose that the horror felt by believers that disrespectful thoughts should be expressed – even, sometimes, that certain things (God, the Furies) should be directly named – is a tribute to the power of words. Our sense of the power of blasphemy and obscenity goes very deep. The words we fear seem to be kept locked away like violent offenders in a separate chamber of the brain from the rest of language – from which they are occasionally released by neurological accident, as in the case of Tourette's Disease, to come tumbling in foul torrents, without either intention or the possibility of control, from the lips of the helpless sufferers.

*

Yes, when I tell you that I am happy, I am making some kind of assertion about how things are, some kind of report on my state of mind. But I am also doing something else: I am *enacting* a state of happiness, in rather the same way that a government, observing some situation of unrest or confusion, might declare a state of emergency. I am declaring myself happy in somewhat the same way that I might declare that a bridge is open, or that men have certain inalienable rights. I'm not saying the bridge is open because it is; it's open because I say it is. I have not discovered certain inalienable rights lodged irremovably in the cerebral cortex or the human immune system; men have rights because I say they have, and those rights are inalienable (in spite of being so often and so notoriously alienated) because I am now declaring them to be so.

In all these cases, of course, there is the assertory as well as the performative element. I'm declaring the bridge open because it is in fact in a condition – open – in which it's meaningful for me to declare it so. If the assertory element is false – if contractors' plant is still blocking access – my declaration fails of its effect, and becomes no more than a formal fiction. My declaration of men's inalienable rights has some plausibility because the possession of those rights is more or less the case in some societies, and at any rate a vaguely imaginable possibility in others; whereas my declaration that men have inherent powers of flight is plainly vacuous. I may declare myself to be perfectly happy in a voice audibly oppressed by misery. In so far as all these pronouncements are enactments, I am sovereign; in so far as they are assertions, you may well be better placed than I am, with my back to the building operations on the bridge, my study door shut against the injustices of the world, and a defence to make of my own way of life, to form a judgement on their correctness.

You are also free to challenge the validity of the enactment itself. You might point out that it is the Queen, not me, who has been invited by the County Council to declare the bridge open, so that, even though the tape has been cut and the contractors have removed their equipment, it remains closed. The chairman declares the committee to be in session; the secretary declares it to be inquorate. I tell the psychiatrist I'm happy; he smiles to himself, because he has a licence granted him by a recognised professional body to declare that I am in denial of my true feelings, and that my apparent exuberance is merely a mask for my underlying depression.

In practice all these ringing declarations become indeterminate. Traffic uses the bridge for weeks before the official opening – and is halted again when cracks appear in the structure immediately afterwards. No one was such a fool (in 1776) as to think that the men with the inalienable rights included slaves, or (until the twentieth century) women, or (even now) children. Once, when I announced the opening of a committee meeting I was chairing, the secretary pointed out that, though we were quorate, one of the most important members had left the room. I didn't declare the meeting closed again; I didn't insist that it was open but temporarily suspended. We all simply chatted about other matters while we waited for the missing member to come back. Was the meeting open or wasn't it? It didn't matter, because we weren't in dispute about the issue, there was goodwill, we could be relaxed and informal. We often observe a similar helpful indeterminacy with regard to our feelings. Am I happy or am I miserable? Well . . . what's for lunch?

Many of the same difficulties of determinacy occur when everything is out in the light and air – when we are describing not thoughts and feelings but the external events and situations in front of our eyes. I look out of the window at the end of a fine, cold winter's afternoon. Everything is very peaceful and still – I have plenty of time to examine it. I tell you that the sun is setting, the bare branches of the trees are motionless, the sky is pale lemon. And of course there is a strong assertory element in this. But, even here, in this simple factual report of what is before my eyes, of what is or could be before your eyes, there is also a performative element. I am *deciding* that the sun is setting, even though I can't see precisely where it is in relation to the horizon, which is obscured by houses and trees, even though we have no agreement on what precise relationship between sun and horizon constitutes the sun's setting. I am *proclaiming* the stillness of the branches and the colour of the sky. I haven't checked every branch of every tree; I can't exclude some motion. The colour of the sky is extremely difficult to put a name to, and it varies continuously from horizon to zenith. I am imposing the benefit of a simple and unitary code upon a lawless world. I am *declaring* sunset, motionlessness, yellowness.

All narration and description, even leaving its declarative aspect aside, is indissolubly subjective because it involves selection. The information transmitted about any situation can never be more than a fraction of what is available. There is no way of representing the totality of a situation except by the situation itself. And the *point* of representing it, after all, of turning it into transmittable information, is to make it amenable to examination and manipulation, to bring out what we wish to bring out for the purposes we have in mind. Even if, instead of describing the sunset outside the window, I silently point to it, I am distinguishing it from the room, and from the north, east, and south aspects of the view. I am implying by my gesture an attitude to it (interest, at the very least) and suggesting that you in your turn might have a similar response.

A lot of our information about the world – perhaps most of it – comes to us not through direct experience, but pre-processed, told to us in words, or shown to us in pictures. When we think and talk about information that we have acquired in this way we merely repeat or para-phrase the assertions that others have already made (and then repeat our own repetitions). The problems of representation have already been confronted, the decisions and enactments have all been made.[5]

In the 1950s a school of French novelists made a serious attempt to detach description from the ubiquitous human presence. The propo-nents of the *nouveau roman* tried to demonstrate that imaginative lit-erature could function as an exact and exhaustive description of certain aspects of the world, without the intervention of a human sto-ryteller. Alain Robbe-Grillet (the most interesting of the group, and the best novelist) says in his theoretical writing that he sees the task of the novelist as being to confront the world of objects – to establish their 'exteriority', their 'independence'. He wants to cleanse language of analogies – particularly anthropocentric ones – to allow the world of objects to be described for its own sake, without the superadded meanings with which we normally invest it. 'So that,' he writes,

. . . the first impact of objects and gestures should be that of their presence, and so that this presence should then continue to domi-nate, taking precedence over any explanatory theory which would attempt to imprison them in some system of reference, whether it be sentimental, sociological, Freudian, metaphysical, or any other . . . It is clear that those people who are the most aware are daily expe-riencing a growing repugnance towards words of a visceral, analog-ical or incantatory character. Whereas the optical, descriptive adjective, the adjective that is content to measure, to situate, to limit, to define, is probably showing us the difficult way to a new art of the novel.[6]

A heroic programme, but does it have any more chance of being realised than a manifesto calling for everyone to be encouraged to grow wings? Robbe-Grillet's actual fiction, in its very attempt to limit itself to words which are 'content to measure, to situate, to limit, to define', manages to subvert the plan comprehensively. Here's a characteristic passage from *Le voyeur*, which seems at first sight to be doing exactly what Robbe-Grillet is urging:

> On the draining-board, the three glasses were almost dry. The woman took them one after the other, wiped them with a swift flick of the dishcloth, and put them out of sight beneath the counter, in the place she had taken them from at the outset. They were placed once again in line, at the end of a long row of other copies of the same object, invisible, like them, to the customers.
>
> But the placing of the different ranges in lines not being convenient for serving it was in rectangles that they had been put away on the shelves. The three aperitif glasses had been put away beside three similar glasses, finishing a front row of six; a second, identical, rank was located behind them, then a third, a fourth, etc. . . . The sequence was finally lost to sight in the darkness towards the back of the cupboard. To the right and left of this range, as also above and below on the other shelves, other rectangular ranges were drawn up, varying by size and the shape of their parts, occasionally by colour.
>
> Alterations of detail, however, were to be noted here and there. One glass was missing from the rear rank of the sort which was used for the majority of wine-based aperitifs; two glasses, moreover, were not of the same manufacture as the others, from which they were distinguished by a pale rose tint. This non-homogeneous row comprised therefore (from west to east): three units conforming to type, two rose units, one empty space. The glasses in this range were not footed; their outline recalled in miniature that of a slightly convex barrel. It was from one of these – a colourless one – that the traveller had just drunk.[7]

I suspect that there is an element of teasing in this mind-numbing account. And in context the passage is not as boring as it appears. We are in fact observing this dull world through a particular human consciousness – of a salesman who (as we are never told but are allowed to infer) has just raped, tortured, and murdered a young girl. The very dullness of the world as he sees it around him is compelling. Precisely because Robbe-Grillet *doesn't* describe the crime, or the frenzy that led up to it, or the panic, remorse, or delusion which we might suppose to be seizing the murderer's mind afterwards – precisely because he limits himself to describing the world of mundane objects which continues to exist around the murderer, and in the murderer's consciousness – we read our own horror and tension into it.

What the description fails to achieve, however, is what Robbe-Grillet claims it is intended to do – establish the presence of its objects in their own right. Indeed, after a few pages of this kind of thing our eyes become glazed against the existence of the objects in themselves. We scan them purely for their human significance. The unmentioned glasses behind the bar in a Simenon novel have more presence, a more incandescent autonomy, than those glasses of Robbe-Grillet's. He cannot make the fact (or 'fact', since it is, of course, like everything else in the book, a fiction) that one glass is missing from the rear rank matter in itself, independent of any effect on human affairs. All he can establish by this obsessive concern for peripheral detail is a general impression of obsessive concern for peripheral detail.

There is a similar effect in another of his novels, *La jalousie*. Here we slowly discover an undisclosed aspect of the situation that is even more fundamental: the existence of an extra character, never mentioned or alluded to in any way, through whose consciousness we are following events – and not a detached observer, either, but one with a passionate personal interest in making sense of the apparently disconnected banalities that ostensibly constitute the narrative.

Curiously enough, Robbe-Grillet's theories also seem to recognise the centrality of the observer. In the same essay as the earlier passages he describes perception as the 'permanent invention' of the world:

> In dreams, in memories, as in the way we look at things, our imagination is the organising force of our life, of *our* world. Every man in his turn has to reinvent the things around him. These are the real, clear, hard and brilliant things of the real world. They refer to no other world. They are the signs of nothing other than themselves. And the only contact man can have with them is to imagine them.[8]

This seems to me an admirable attempt at the truth of the situation, and to be borne out in practice by his fiction. But how it sorts with the doctrine of the 'exteriority' and 'independence' of the objects of the physical world I have no idea.

What happens when we think and talk (as we do so much of the time) about something which is neither in words already, nor, as philosophers say, 'present to our senses' – when we are recalling earlier experience? How do we . . . *get hold of* things remembered?

I sit here in my room and I think about . . . well, some other room, say. The room that I lived in as a small child, perhaps. At once I feel I can see it with my mind's eye. I'm remembering it. But what does this mean? Am I simply taking images out of some mental store, as I might turn over the pages of an album of snapshots? Or running an old videotape? What exactly would those snapshots or clips be recording? Would it be a series of discrete encounters with the room? But my recollection isn't like this at all! What I'm experiencing is not a series of pre-set scenes. It's both less and more than that. Less, because the concrete details are not present – not, at any rate, until I opt to see some. More, because I'm not restricted to any particular selection of material. I feel that I can look in any direction in the room and know that I will find something – that I can open the drawers and cupboards, the doors and the windows, and that, wherever I turn, things will suggest themselves. If I want people to walk into the room – in they walk. If I want them simply to *be* in the room, without the trouble of entering – in the room they are. If I want things to happen, they happen. The room has a past and a future. It has hypothetical possibilities. It's as if I'm *inhabiting* the room, rather as I inhabit the room I'm in now.

So am I really remembering all this, including the things that might have been there but weren't, and might have happened but didn't, or am I simply making it up? Well, some things I'm pretty sure I'm remembering, and some I'm absolutely sure I'm inventing. But with a lot I really don't know. I think they're a kind of mixture, and where the remembering stops and the inventing starts I have no clear idea. What I seem to be doing is *characterising* the room to myself, in the same way that I might characterise a room in a story I'm writing.

Is it all that different, though, even when we are actually gazing at an object rather than thinking and talking about it? Here before us, duly present to our senses, simple and unpuzzling, is the humble brown penny that in the old days was so beloved of phenenomenalist philosophers. When we say we are seeing it, according to the phenomenalists, what we mean is that we are having some kind of elliptical brown experience (because we're not usually seeing it straight on as a circle, this is one of the things that worries them). We are entertaining some kind of elliptical brown somethings – phenomena, sense-data, qualia. Oh, sure. But what we're *seeing* is a *penny*! We're able to talk about it as a *penny*! It is in some way bodied forth to us. We know (and this is part of our seeing it) something about what we can do with it, what it would feel like if we touched it, what it would taste like if we put it in our mouths. An indefinite range of possibilities opens from it. It has a character. And our grasp of this character we can't 'justify' empirically as knowledge, because it's not knowledge in the strict sense – it's something we have constructed with the help of our imagination. Informed, researched imagination, but imagination none the less.

Justified or not, that is the experience I am having. Even if I know intellectually that what I'm experiencing as whole and continuous is derived from little broken bits and scraps, if I *experience* it as continuous and whole then that object *is* for me continuous and whole. I know that the moving picture in the cinema is made up of separate still pictures, that the image on a television screen is produced by a stream of discrete and consecutive digits in the signal. But this knowledge doesn't in any way affect the fact that, when I look at them, I see them as continuously moving pictures. An orange can be analysed into its chemical components – but this doesn't in any sense alter the fact that I taste and smell it as an orange. It's only when the projector breaks down, or the orange turns out to be poisonous and our perceptions wildly illusory that the mechanics of the thing are revealed.

So if you tell me that really, when I look at the penny (or what you *would* call the penny if you didn't know any better), all I'm doing is having elliptical brown sensations, then this is a *different* experience from the one I was actually having. A part of that experience (and an important part, which coloured everything else) was my understanding that I was looking at something, and that it was a penny. It's as if I'd thought I was talking to a friend, and now, when I look again, I see it's a stranger. So this person, it turns out, has the character not of my old friend, which I know quite a lot about, but of a stranger. This is an altogether thinner and less interesting characterisation. All the same, it *is* a characterisation. And similarly, when I look at the object that I once in my ignorance characterised as a penny, I now somehow manage to imagine myself as having only bits and pieces of sensation, which I then nevertheless seem to be able to assemble into a whole coherent enough for me to be able to say of it that it's an *experience*. A brown and elliptical experience.

Could I do that if I hadn't had other experiences? Other brown experiences? Other elliptical experiences? If I couldn't recognise the similarities between this experience and those? If I wasn't bold enough to commit myself to this identification (if only to myself, if only wordlessly)? If I didn't use my powers of imagination and invention? Once again I am supplying the background research, naming the character, and casting it. Once again I am doing all the work. It seems to me that if I have to make all this effort just to write my austere little narrative about the elliptical brown experience I might as well go the whole hog and indulge myself with the full romantic saga of the penny.

What the novelist does to characterise the locale of his story he does – has to do – even more thoroughly to characterise the protagonists themselves. He has to know not only how they look now, and what they have done to date. He has to be able to find himself knowing – and this is at the very heart of storytelling – how they will look tomorrow, and what they will do the day after. How they will look when they are told about something that has not yet happened. What they will do *if* this or that should occur.[9]

So too do all of us have to characterise (and again even more strongly than we do with things) the real people we know, so that we can think about them when they are out of our sight. You're off on your travels again. Is the remaining hold that I have on your existence restricted to your phone calls, and to my last glimpse of you as you vanished into Departures? Of course not – I project you forward into the Australian bush, or wherever I think you might be. (And my guess at your whereabouts is an act of imagination in itself!) I project you into fantasies of wishful thinking, where you enjoy imaginary triumphs and happinesses; and into the fantasies that proliferate from anxiety, where you face imaginary perils and disasters. I bring you back to be with me again. I cannot think of you otherwise than through the imagination, because of your present existence I have no present knowledge.

And now at last you really are with me again. I can see you, put my arms around you, hear what you're saying. So are you now simply the sum of what I see, touch, hear? Are you suddenly, as you walk into the room, *reduced* to that? Have you become uncharacterised? Is the side of your face that is turned away from me now a featureless blank? Don't I laugh or cry because what you're saying is so characteristic of you, or so uncharacteristic? Don't I go on thinking all the time about how things will turn out for you tomorrow – how you will cope, how you will succeed, how you will fail?

The story surely continues, even though new material towards its construction is coming into our possession, just as the novelist's story continues to develop in his head even as he does new research which changes its course. To have any real conception of the person standing in front of us, any possibility of saying or thinking anything about him that is not entirely obvious or entirely trivial, we have to reconstruct him as a kind of fiction inside ourselves.

In all these stories there is another character, too – the narrator himself. I am present to my senses more fully and continuously than anything or anyone else. And yet what an insubstantial figure I should cut for myself if I were just what I was immediately conscious of – the pressure of the chair on my bottom, the faint consciousness of breathing, of the word-processor humming, the elusive burden of the thought I am trying to think. I come complete with all my memories, expectations, fears, plans, strengths, weaknesses, just as you do. Since I know far more about myself than I know about you (or are more thoroughly and grandiosely deceived) – since one of my functions is to imagine you, and all the rest of the world around you – I occupy an inordinately large role in my own imagination, I bulk very large in the narrative. And to have any real conception of myself, to have any possibility of saying or thinking anything very much about myself, I have to do as I do for you, and out of all this material construct a workable character.

I can't help being struck by the jealousy with which playwrights and novelists attempt to defend their prerogative in this respect. They turn very nasty when their protagonists try to create characters for themselves, to offer the world personalities of their own construction. The professionals at once expose the hollowness of the pretence. They insist that the whole performance is a sham, a falsehood, designed to cover up unacknowledged social origins, undisclosed sexual orientations, complete spiritual desolation. But then I suppose all of us, even the non-professionals, warily observing other people from inside the carapace of character we have constructed for ourselves, feel defensive when we realise that we are being warily observed in return from inside some very similar carapace.

The medium for all this character creation, whether professional or amateur, is narrative. Daniel Dennett is very eloquent on this:[10]

> We . . . are almost constantly engaged in presenting ourselves to others, and to ourselves, and hence *representing* ourselves, in language and gesture, external and internal . . . Our human environment contains not just food and shelter, enemies to fight or flee, and conspecifics with whom to mate, but words, words, words. These words are potent elements of our environment that we readily incorporate, ingesting and extruding them, weaving them like spiderwebs into self-protective strings of *narrative* . . . Our fundamental tactic of self-protection, self-control, and self-definition is not spinning webs or building dams, but telling stories, and more particularly concocting and controlling the story we tell others – and ourselves – about who we are. And just as spiders don't have to think, consciously and deliberately, about how to spin their webs, and just as beavers, unlike professional human engineers, do not consciously and deliberately plan the structures they build, we (unlike *professional* human storytellers) do not consciously and deliberately figure out what narratives to tell and how to tell them. Our tales are spun, but for the most part we don't spin them; they spin us. Our human consciousness, and our narrative selfhood, is their product, not their source.
>
> These strings or streams of narrative issue forth *as if* from a single source – not just in the obvious physical sense of flowing from just one mouth, or one pencil or pen, but in a more subtle sense: their effect on any audience is to encourage them to (try to) posit a unified agent whose words they are, about whom they are: in short, to posit a *center of narrative gravity*.

You'd be surprised by my behaviour, I think, if I didn't have this sense of my own character and identity, if, in order to know who I was, I had to rely purely on my current perceptions of myself, my me-shaped, me-coloured sense-data. 'Aren't you Michael Frayn?' you ask uncertainly, meeting me in the street and not knowing me very well. I look down uncertainly in my turn at what I'm wearing. 'I think I am,' I say. 'These trousers are the sort that Michael wears. Though quite a lot of people have got trousers like these . . . My hands look familiar . . . Have you got a mirror? I'm very bad at names but I'm quite good at remembering faces.'

You expect me to have a sense of myself that goes beyond trousers and facial features – a sense against which I can check my identification even when I'm wearing a Father Christmas costume and mask. Locke says that identity depends upon the memory. And of course memory is part of this sense of oneself. But it surely goes far beyond that – far beyond anything that relates to the past. When you ask me who I am I don't have to pause before I answer while I check that I remember the things that someone with my name remembered yesterday, any more than I have to look down at my trousers. I am familiar with myself. *Very* familiar with myself, since I'm the central character in the story. I suppose that the ability to remember from one moment to the next is necessary to understanding a narrative, to having any sense of a character. All the same, the amnesia would have to be very deep before one lost all sense of onself. Even if I couldn't remember anything at all of what happened before the present moment – or somehow remembered only things that had happened to you, or to Napoleon – I should still think it was me to whom these bizarre things were happening.

'I' – 'me' – the word that names and thereby brings into being the irreducible core, the bare quick, at the heart of all our narratives.

Is It True About Lensky?

The truth-functions of fiction, the fiction-functions of truth

A lot of apparently factual statements, then (particularly ones about our thoughts, feelings, and reasons for doing things, and about ourselves and each other) have rather the same relationship to the world as fictitious ones. Which raises problems, because it's difficult to know quite how the logic and semantics of fiction work.

The subject has never much engaged the attention of philosophers. What they have worried about down the centuries is truth: whether and how propositions about the way the world actually is can ever be true in the way that the propositions of logic and mathematics can be true; what it *is* for a proposition to be true; whether statements have any meaning at all unless they express a proposition which we know, at any rate in theory, how to establish as true or false; and so on. All this leaves fictitious statements, which certainly appear to have some meaning and function, but which are by definition not offered as true, in a philosophical junkyard, functionally indistinguishable one from another. Even in the case of fictions created with serious artistic intent, the statements that express them seem to be merely pale ghosts of factual statements, and have meaning only in so far as they ape the real thing, rather as actors express real emotions by mimicking the expressions and gestures that real emotions give rise to.

Seen from a historical point of view this is curious. No one knows how language was first used in the millennia before there was a written record, and it may be that the Flintstones didn't grunt to themselves, or utter warning cries and monosyllabic commands to each other, as is usually supposed. They may have confined themselves to well-formed factual statements, each of which expressed a proposition with a clear truth-value. But, from the point where the written record begins, it wasn't like that at all. Language was used in all kinds of ways that were not even assertory. To list things, for example – measures of corn, Pharoahs and their benefactions – without making any ostensible statement about them at all (though the writers may have understood them to be making assertions that the corn was delivered, the Pharoahs born and died). And when we do come to assertions in sub-ject-predicate form, capable of carrying a truth-value, many if not most of them seem to be mythic – to be offered without any attempt at the verifiable one-to-one representation of particular states of affairs of which philosophers would have approved. Statements that may be intended factually are mixed among fictitious statements without any attempt to differentiate the two – perhaps without any understanding that the two *can* be differentiated. In Genesis the story of Adam and Eve, which plainly has no historical basis, moves seamlessly into the story of Cain and Abel, which conceivably might have done. And even when we reach plausibly historical figures like David, he is not only fighting battles which may well have been actually fought, and for which independent testimony may exist, but in the same tone of voice having conversations with God for which there could be no historical evidence at all.[1]

So fiction has a venerable ancestry, and if we could understand how it relates to the world it might cast a little more light on how factual language works.

Is fiction, for a start, subject to the laws of logic, which have always been regarded as the laws of thought (and which are the basis of math-ematics and the functioning of a digital computer)? To operate the procedures of truth-functional propositional logic as they are normally understood (by computer programmers, among others), we require a language of propositions (or one that can in theory be analysed into propositions), each of which can be unambiguously characterised as true or false. (It doesn't matter *which* they are, so long as they are one or the other, because what logic does is to study how they can be com-bined and operated upon to generate further propositions whose truth-value derives entirely from the truth-value of the input state-ments.[2])

I suppose you could try to argue that fictitious statements are mean-ingless – that they don't really express propositions all. They plainly *do*, though. They're plainly just as meaningful as factual statements – and indeed often indistinguishable from them if you don't happen to know their provenance. If fictitious statements *do* express proposi-tions, though, and these propositions are by definition not true, then it seems to leave no alternative in logic but to classify them as false. But if the proposition expressed by the sentence 'Onegin shot Lensky' is false, even though, in Pushkin's poem, Onegin *did* shoot Lensky, then it can't be distinguished from 'Lensky shot Onegin', even though, in the poem, he *didn't*. It can't, for that matter, be distingushed from 'Lensky shot Stalin', or even a false factual proposition – 'Attila the Hun shot Abraham Lincoln'.

It won't work, in any case, in terms of the logic whose demands we are trying to satisfy, because if P is false then not-P is true. So if 'Onegin shot Lensky' is false, then 'Onegin didn't shoot Lensky' is true. But it *isn't* true, in any sense at all, whatever view you take of fictitious propositions. It's not true within the context of the poem, which 'Onegin shot Lensky' *was*; and if 'Onegin shot Lensky' is false in a general context then so, *pari passu*, is 'Onegin didn't shoot Lensky'.[3]

All this is ridiculous enough – though perhaps not quite as ridiculous as what comes next, which is that if Pushkin had written 'Onegin didn't shoot Lensky', then by the same argument it would be true that he *did*. And where does it stop? Even if Pushkin *didn't* say 'Onegin didn't shoot Lensky', and you in your ignorance of the poem – or I in my desire to confuse the issue – did say it, it surely remains a fictitious statement, and if fictitious then false, so that its negation is true. Between us, therefore, we have managed to change history. We have inserted two more real people into it, and made one of them shoot the other. We could go on and write the whole of the world's fiction into the historical record.[4]

In practice, of course, there's no great mystery about what we do with statements like 'Onegin shot Lensky' and 'Jupiter is a god'; we assign some kind of unofficial truth-value to them on the basis of their consistency, or lack of it, with what Pushkin wrote and with what is said about Jupiter in Greek literature and mythology in general. This still doesn't make the statements true in terms of truth-theory. But at least it differentiates 'Onegin shot Lensky' from all the other statements about Onegin we looked at, and 'Jupiter is a god' from 'Jupiter is a clockwork mouse'. It enables us to fail exam candidates who get it wrong, and to advance others, who get it right, towards well-rewarded careers. And it makes it possible to combine fictitious statements logically in a way that preserves at any rate the same kind of unofficial truth-value.

All kinds of problems naturally arise about what exactly constitutes consistency. What are the (unofficial) truth-conditions for saying that Onegin was a covert freemason? Consistency with the written text again, certainly – but here it's not cut-and-dried, if Pushkin doesn't say. Judgement and historical knowledge are necessary. Genuine argument is possible. You might say that the same is true of 'Pushkin was a covert freemason'. But here, in theory, evidence could be found that would put the matter beyond reasonable debate. Could a statement like this about a fictitious character ever be shown to be in any sense, however unofficial, true or false?[5]

None of this, however, touches the real heart of the problem. There is another profound difference between the statements 'd'Anthès shot Pushkin' and 'Onegin shot Lensky'. The former is true (or false) whoever utters it, and on whatever occasion. (Even if a factual statement contains references such as 'I' and 'now', these can be separated out by analysis to isolate a proposition which is common to all utterances of it.) The logical status and truth-conditions of the latter depend entirely upon who utters it and when. If I say it, or if Pushkin says it himself in the years after he has finished the poem, there is the possibility of establishing some kind of truth-value, even if an unofficial one, through its consistency with what Pushkin says in the poem. But the first time that Pushkin utters the statement, when he is actually writing the poem, there is no existing statement for it to be consistent with. The form of words is the same, but the truth-conditions of the statement (if indeed there are any) are completely different.[6]

So what *is* going on when Pushkin first utters the statement? Could we perhaps translate it into a statement about the contents of its author's imagination, and say that it's true if it corresponds to some event that has taken shape there? Could we suppose that it is to be understood as beginning with the principal clause 'I am imagining that', omitted in normal usage like the formatting codes in the text produced by a word-processor? So that, if we selected the 'reveal codes' option for analytical purposes, we should discover a (theoretically) verifiable proposition?

Pushkin, though, isn't the only person who can imagine things about Onegin and Lensky. Even as he is truthfully reporting his imagining that Onegin shot Lensky I might be reporting, no less truthfully, my imagining that Lensky shot Onegin. There's no contradiction between the two statements; whereas there surely is between the two states of affairs that each of us is imagining. In any case, who is doing the imagining that Rapunzel let down her hair for her lover to climb up, or that God took a rib out of Adam's side to create Eve? What mental events are we checking these against?

Or could we translate the apparent assertions into some other grammatical form?[7] Could we see them, not as real assertions at all when they are made by their original author, but as what philosophers call performative utterances (like promises, and declarations that the new bridge is open), whose function is not to record a state of affairs but to bring it into being? Something in the nature of invitations, perhaps? 'Let us both imagine that . . .' Or even polite imperatives: 'Imagine that . . .' After all, we understand at once what's going on when we read in raised cursive print that Harold and Celia Shrubb request the pleasure of our company. We don't think they are reporting, truly or falsely, something they have observed going on in themselves, a requesting process, as detached from their own volition as peristalsis or the formation of plaque. Even when the authors of the invitation have disappeared behind a passive formulation – 'The pleasure of your company is requested . . .' – we understand that the words represent the *act* of invitation itself.

So that, if Pushkin says 'Onegin shot Lensky' and I say the same, then I am simply either repeating Pushkin's original invitation, or accepting it, or both. And if I say 'Lensky shot Onegin', I am not making a contradictory statement but either mistranscribing the invitation or issuing another and incompatible one. I am inviting you to dinner on an evening when someone has already invited you to dinner elsewhere. You can't accept both invitations. *Logically* can't? No, reasonably can't, behaviourally can't, because you can't be in two different places at the same time.[8]

This does seem to catch something of the transaction underlying a story. The storyteller is attempting to enter into some kind of compact with his hearers. 'Buy a ticket from the box-office – sit down – listen quietly – don't ask questions – and I'll tell you something that will make you forget your troubles for a while.' Why should the story interest you? The storyteller creates the reasons as he tells it; he constructs the demand as he goes along for what he is offering; he makes reasons for why you should be interested, like a car salesman creating for the customer the world of luxury and importance in which the car he is selling will enable the customer to move.

This, it seems to me, is a reasonably coherent and philosophically respectable account of fictitious propositions. The only thing wrong with it is that it's not true. It's as fictitious as Onegin's and Lensky's duel. There is not the slightest reason, apart from preserving the tidiness of logical taxonomy, to believe that any such mechanism of invitation or proposal was ever being covertly operated by Pushkin or anyone else. If you entertained the idea in your imagination for a moment it was only for the same reasons as you give house-room to any other fiction – because it had a kind of internal narrative dynamic.

We need to be a little bolder. A fictitious proposition, when first uttered by its original author, is a kind of performative act; but what it's enacting is not some ancillary function, such as an invitation to imagine a state of affairs – it *is* that state of affairs itself. Pushkin creates the situation which 'Onegin shot Lensky' states, within the frontiers of the story, by the very act of enunciating it.[9] He *declares* the shooting of Lensky within the confines of the story in the way that a government declares a state of emergency within its borders. A state of emergency exists in a nation if its government says it does; Lensky is shot if his author writes him shot. And when I in my turn say that Onegin shot Lensky, it's true in the sense that it fairly reiterates what Pushkin enacted, just as it's true when a historian says that a state of emergency existed in a country where a state of emergency had been declared.

You protest. Pushkin's statement didn't come into being out of thin air! He uttered it precisely because it expressed something! Something that he was imagining, some pre-existing state of affairs in his mind. The crucial act of imagination had already taken place. Well, possibly. But then again he might have had nothing in his mind at all until he wrote the words down. He might even have had some other state of affairs in his mind altogether – Lensky's shooting Onegin, for instance. And even if he had imagined the event that his words express before he wrote them down, and we decide that this is where the situation originated, then it's *this* act (whether in words or wordless, whether a sudden decision or a slow process of shaping) that brought the state of affairs into being – and it's *this* act that we are interested in.

You object again. Supposing Pushkin knows, even as he writes the words, that he intends to reveal two or three stanzas later that Onegin didn't shoot Lensky at all – it was a dream, a misunderstanding, a hidden marksman in the bushes. Isn't he uttering a falsehood (no doubt for the best of literary reasons) in the quite ordinary sense of saying something which isn't true? Isn't he in fact lying, in the quite ordinary sense of saying something that he knows not to be true, something which is intended to deceive? So his statement must be a mis-statement of some existing situation – a performative utterance surely can't be *false*!

No, but it might be misleading, either deliberately or by default. It might not be the act that it appears to be. The intention might be that it should appear to be directed towards a certain effect, but fail of it, like a dummy punch. If a government declares a ceasefire, and then orders its troops go on fighting, we conclude that its declaration was not false but fraudulent. It was intended to deceive by suggesting an enactment was taking place when it wasn't. It was not intended to create the state of affairs that it appeared to be creating.

It's curious how little interest philosophers have shown in fictitious statements. There *is* a literature on the subject, but everything that I've managed to find seems to me to miss the point. For a start there is often a confusion between the two fundamentally different kinds of statement that we noted earlier: ones made *within* a fiction, and ones relating to a fiction but made from outside it. The focus of the discussion, curiously, usually seems to be on the latter – on whether (for example) it is *true* that Sherlock Holmes lived closer to Paddington than to Waterloo, given that Holmes's address is fictitious, and that Conan Doyle didn't expressly say, or whether it's *false* that Holmes had three nostrils, given that men in general don't, though given also once again that in the case of Holmes the author didn't specify. All this has points of interest, but the first thing to get straight, surely, is what's happening *within* the context of the fiction – what's happening when Conan Doyle says that Holmes lived at 221B Baker Street.

David Lewis[10] and others[11] do discuss fictitious statements like these, and they treat them as 'make-believe'. And indeed there *is* a similarity between what the writer of fiction does and what children do when they play make-believe games like Mothers and Fathers. Just as the children take on the personae of parents, but without its entailing sleepless nights or anxiety about local educational standards, so the novelist takes on the personae of his characters and plays out the events of their lives, but with none of the practical consequences that those events would have in the real world. Kings and queens – but no actual mob at the gates, no assassin's bullet. A tour of nineteenth-century Petersburg – but no fuss over tickets and hotels.

All the same, the word 'make-believe' isn't right in either case. Creating belief is not really what either the novel or the children's game does, or is intended to do. The child doesn't *believe* he's an explorer; he simply, in his own mind, *is* one for the purposes of the game, just as on some days you're confident and attractive, and on other days anxious and plain. But then does an actual paid-up explorer, as he picks his way through the untrodden desert wastes, *believe* he's an explorer? Isn't he likely to be too busy exploring the territory to start exploring his own professional status? He just *is* an explorer! If the question arises in his mind at all I should think it's only on bad days, when everything goes wrong, and he isn't an explorer at all, merely a former local-government official who's pretending to be one. Nor is Shakespeare trying to make us (or himself) believe that Hamlet is seeing his father's ghost. He's simply telling us he is, and we neither believe it nor disbelieve it. He *is*, and that's all there is to it! The spectacle may make the hair rise on our necks, or it may make us laugh scornfully. Even if the former it doesn't incline us to subscribe to the Society for Psychical Research. Even if the latter it doesn't incline us to think we've seen through some kind of deceit.

Literary theorists have seized upon a complementary formula – Coleridge's phrase, 'the willing suspension of disbelief'[12] – to explain the audience's side of the transaction. Disbelief doesn't come into it, though, any more than belief does. For a start it suggests that one's initial attitude to anything one is told, certainly to anything avowedly fictitious, is disbelief. Is it? Disbelief, in any normal usage, is more than just an absence of belief. It's a refusal of belief, a rejection, and you can't reject anything until there's something to reject. You can't refuse to believe until you've been offered something to believe. But the approach to most fictions is littered with signs announcing their nature. 'Once upon a time', the storyteller begins. Or, 'There was this Irish fellow . . .' The author of the novel inserts an assurance that none of the characters or incidents has any relation to any real person or events. Why should I start off by disbelieving what follows, when I haven't been offered anything I'm supposed to believe? Why would I order ice cream and then complain that it isn't hot? And if I *do* order it and *don't* complain, does this mean that my enjoyment of it is to be explained as suspension of my objection to its being cold?

Well, maybe you have a particular talent for suspicion. You open my new novel and the thought at once comes to you that you're not going to believe a word of what follows. Now you've paid the money, however, you're anxious to enter into the spirit of the thing and show you've got a sense of fun, so you tolerantly suspend this natural disbelief of yours. You raise the portcullis and admit – what? Belief? You mean you *believe* what I'm telling you? All this farrago of sexual wishfulfilment and sadistic vengeance for imagined slights? Have you gone mad? Or are you just pretending? Are we kidding each other?

This is not how I behave when you tell me your wonderful story about a dog with a weakness for Guinness and gambling. I step readily and effortlessly, without any conscious mental shift, into the world I am being offered. Far from discouraging me, the specially humorous look on your face, the well-observed Irish accent, etc., raise my expectation that this is intended to be enjoyable, because it makes clear that its entertainment value (in one sense or another) is to be its justification. *Eugene Onegin* may arouse feelings in me – indignation, sadness, or simply an absorption in seeing how things work out – that are indistinguishable from the feelings aroused by the story of Pushkin's own fatal duel. Not for an instant, though, do I believe, even provisionally, that there was an actual Onegin who actually shot an actual Lensky. Not in the same sense as I accept that there was an actual d'Anthès who actually shot an actual Pushkin.

And in any case, my believing that d'Anthès shot Pushkin is not in itself a reason for finding the account of it interesting. In both the shooting of Lensky and the shooting of Pushkin the interest of the narrative, its ability to catch my imagination, is separate from the question of its relation to historical events.

There *are* occasions when one might be said to suspend one's disbelief, but these are quite different. A man tells a woman that he loves her. She knows perfectly well that he doesn't, but on that particular evening, with the warm breeze off the sea, and his eyes shining in the candlelight, she smiles and puts the knowledge away for an hour or two. You tell me an improbable anecdote in which someone I dislike behaves badly, and though I wish I could believe it, I can't really persuade myself. And yet at the same time . . . yes, I give it a little houseroom. I allow my wish to believe it some status alongside my rational assessment of its implausibility. I accord it some kind of private halfacceptance, and count it to his discredit, even though I know that I could never pass it on, and have a realistic understanding that I'm going to have to abandon it under the slightest pressure.

Now you tell me another story – an uncanny tale of strange presentiments and inexplicable events. Do I believe that it's even the *kind* of thing that could possibly have happened? Not for an instant. But I'm taken with it. It has a practical effect upon me – my scalp tingles. What I suspend is not disbelief but something more like any propensity I might have to divide the world into the believed and the disbelieved – even, in this case, into the believable and the unbelievable. What I accept, provisionally, is an alternative world. I recognise in this world something that remains half-hidden in the world I inhabit – not just the arrangement of particular circumstance that it's so hard to see beyond in ordinary life, but some of the underlying fears and longings that the tissue of actual circumstance obscures.

The storyteller, in other words, creates by his fiat, or web of fiats, an alternative world that I agree to inhabit for a while, and within this world statements are true or false just as factual statements are in the real world. It's like a social relationship: a story entertains you – *you* entertain the *story*. You invite it in, pour it a drink, and let it talk to you without interrupting. When I enter the world of the story I do it not with the narrowed eyes of a detective entering a suspect's house, but rather in the way I enter France on my holidays. I'm there to enjoy myself, not to show up the corruption and hypocrisy of French society. It's only a temporary visit – I don't give up my British citizenship and become a Frenchman. This is what makes it a noticeable experience – this is what makes it a holiday. For a Frenchman, being in France is as workaday and unremarkable as being in England is for an Englishman. But now, as an Englishman in France, I watch what's going on around me, and see the force of it. I catch something of the *savour* of life that eludes me in the life I actually live at home. Nor am I limited to simply observing. A story has an element of virtual reality about it. I live the life a little. I involve myself in it. I go into the *boulangerie*, I ask for a *ficelle* to go with my *pâté de campagne*, I pay for it in an unfamiliar currency; and there's something taking about it that doing the weekly shop in Sainsbury's entirely lacks.

When you play a game, too, you enter an alternative world, with different values and objectives. You have to accept that in this new world it is important to knock down three standing sticks, or to reach a piece of string before other people do. You know that these things have no importance in the context of ordinary life; but you also know what it is for things to be important, and by accepting these clear and simple objectives you create a context in which competition is given form and meaning, a world within whose altered surface you recognise some clearer and more definite representation of the confused struggles and contests of your own experience. By agreeing to the importance of knocking down the three sticks you create a whole consequent world of importance.

Even unintentional discrepancies in the story, as when Charlene's blazing blue eyes in Chapter One have somehow mutated into melting brown ones by Chapter Seven, we might glide over, if we're enjoying the story, without raising the question of truth or falsity. We understand that these little discontinuities occur. So what happens when sometimes the thread snaps, and, yes, we suddenly *don't believe a word of it*? We have faithfully followed the voluptuous but ruthless Charlene through not only her different eye colourations but also a series of improbable adventures involving time-warps and tumultuous sexual encounters with alien creatures in remote galaxies. Then, out of the blue, she renounces everything to save her little brother from terminal obesity, and we lose patience. It doesn't specifically contradict what the author has told us before, but it's at odds with the whole tenor of her character, and of the story. It doesn't fit, for that matter, with what we know from our own wide experience of the way that sexually alluring space-women behave in the real world. We are not so much withdrawing our assent from some particular statement as taking back our bat and ball and withdrawing from the game as a whole. We are falling out of love with France, or maybe just getting tired of being on holiday.

I said before that there was no logical difference, in the case of factual statements ('D'Anthès shot Pushkin', 'Charles I was beheaded'), between the first utterance of them and the reiteration of that first utterance. This was (as scientists say) a first approximation. There *is* a difference – and it is one not entirely unlike the difference that we examined between the first utterance of a fictitious statement and its reiteration. The first time that the eye-witness to the duel, or the execution, puts into words what he has just seen, his statement, like the storyteller's, has an element of the performative in it. This is the same point that was made in an earlier chapter,[13] in regard to statements about our thoughts and feelings and immediate experiences, and it's at the heart of understanding the role we play in the creation of a world. All these assertions, the first time they are made, are judgements, attempts at a determination as to what occurred, the unofficial equivalent of the official determination that a court might make (and which may, of course, like the determination of the court, be overturned at appeal). They are pronouncements, enactments of the facts in words. In all these cases there is some pre-existing . . . something . . . to be determined (which constitutes their difference from fiction); but what that something is remains undetermined until the words determine it.

The simple dichotomy between fact and fiction is misleading all along the line, just as it is between true and false, and between games and whatever the converse of a game is (a category so amorphous that is doesn't even have a single term to represent it). You can make a game into a matter of life and death – and you can make a matter of life and death into a game. There are infinitely many types of game, which we play in infinitely many different ways, and there are infinitely many types of seriousness. There are many different *levels* of game – games within games and games about games, as when someone gambles on the outcome of someone else's game, or modestly pretends to play cricket worse than he does, or uses a game of golf to stir trouble between rivals, or plays darts blindfold. Most of these games, when we come across them in our daily life, we understand immediately. When we listen to other people talking about themselves and their lives, similarly, we understand that we have to allow for elements of boasting and fantasy, of self-depreciation and self-deception. From our knowledge of the people and conventions concerned, we make the best assessment we can of the relation between the narrative we are hearing and the world it purports to describe.

And, conversely, when people seem to become a little confused about fictionality, and send letters of condolence for the deaths of characters in soap operas, are they really so confused? Aren't they like the children at a pantomime, who are persuaded to shout out to Jack that the robbers are behind him, or to save Tinkerbell's life by agreeing that they believe in fairies? Aren't they entering wholeheartedly into the story as one might enter France filled with eagerness to speak French and observe all the local customs? You might, after all, find your French holiday turning into something more than a holiday. You stay on for another week; buy a derelict house and do it up; spend more and more time in it; send your children to the local school; spend the winters as well as the summers; even take out French citizenship; die there. And yet in some sense it remains a performance, all this being French of yours.

Now I give you what seems to be a true account of some perfectly credible event – how I went with a Polish friend of mine to a certain cafe in Warsaw, etc., etc. You accept it without question, and you're intrigued. Then I reveal that there is no such cafe in Warsaw, no such friend, that I have never been to Warsaw. Now you have to revise your stock of knowledge and opinions about Warsaw, and no doubt about me. But does it make any difference to the force of the story? Does it cast doubt on the interest you felt in it at the time? Only if that interest had been a factual one, external to the force of the narrative. Only if you had listened in the way that someone might read *Eugene Onegin* purely because he was writing a history of duelling.

After confessing my deception to you I go to Warsaw. I find a cafe exactly like the one I described, and I visit it with a Pole I have met – who by the strangest coincidence happens to have the very name I gave the friend in the story. Does this retrospectively improve the quality of your enjoyment at all? Does it justify the interest you felt when you first heard it? Interesting though it would be if documents turned up showing that the court of some small German principality once did seem to have suffered from some kind of collective narcosis which preserved them, unconscious but unchanged, for a hundred years, or if actual cases were found of talking frogs, and cats walking round in boots, would this have any relevance at all to what happens when a mother sits a child in her lap at bedtime and says: 'Once upon a time there was a beautiful princess . . .'?

Most acknowledged fictions, in any case, contain non-fictitious statements among the fictitious ones, and we move effortlessly back and forth between them without the slightest change of gear. 'It is a truth universally acknowledged, that a single man in possession of a good fortune, must be in want of a wife.' Do we have to understand this opening paragraph in one way, as a statement about the real world, and then, when we reach paragraph three, where the fiction starts – '"My dear Mr Bennet," said his lady to him one day . . . ' – adopt a completely new mindset?

Do we even think of that first paragraph as factual or fictional, and, if factual, as true or false? Do we check it against our experience of the world before we risk a smile at it? If we do then at once a problem arises, because in our workaday modern world we're unlikely to have come across any single men in possession of a good fortune, and even if we have, to have heard any general truths expressed about them often enough to know whether such truths are universally acknowledged or not. Already we understand that the statement relates to a world which, even if real, is one that we do not ourselves inhabit, and to which we no longer have access.

We could of course embark on historical research. Does the thought enter our heads? I suppose it might enter the head of a postgraduate desperate for a research subject. Even the most desperate postgraduate, though, would understand that no amount of research among original documents could ever establish that such a truth was universally acknowledged – and that, in any case, proposing the observation as a universal truth is a comic exaggeration.

So it's *not* true? Yes! I'm sure it is, if we're trying to relate the novel to its period, in the sense that it expresses in a striking hyperbole attitudes which must have been very common at the time. In fact I accept it as true even if I know nothing about the period, because it seems to establish its own truth, in the context, by its sheer boldness and literary rightness, much as the conversation between Mr and Mrs Bennet does in paragraph three. That's why I smile.

I say this only because the question has now arisen. We all understand that in the confused borderlands where fiction and non-fiction meet it's sometimes hard to know exactly where the frontier runs, as in the Disputable Lands between great empires, where populations and customs become inextricably intertwined, and the laws perpetually change. Even with unambiguously factual assertions, unless the question is actually raised, I don't for the most part think of them as being true or false. Things are said, and information is thereby conveyed, impressions are thereby created. The truth of these remarks isn't usually at issue. When philosophers talk about it, truth-value seems to be a fundamental and inescapable quality of every statement, a perpetually live issue. When we go about our ordinary lives, though, it's only in particular circumstances that a truth-value is ever assigned – only once the matter has been called into question.

It's the same with monetary value, after all, or any other value. To an economist the potential exchange value of every object is no doubt its fundamental characteristic. But this potentiality remains purely notional until the question of its being bought or sold, hired or pawned, actually arises. If the rotten timber that has been lying in my yard waiting for someone to remove it stays there until the world's energy crisis gets bad enough, it may acquire a realisable value as fuel. You might say that in this case it has such a value already, even if for the moment that value is zero. What would be the *point* of saying this, though, except to prepare the conceptual ground for the future?

It's much the same with all the other qualities of the timber, for that matter. Does it have a weight, a colour, a shape? Of course it does, in the sense that values can be assigned for all these qualities if and when the question arises. Until then it has weight and so on in the same way that it has portability, flammability . . . friability, solubility . . . elegibility, lovability . . .

The opening sentence of the *Tractatus* ('The world is everything which is the case') is justly famous, because it perfectly expresses the unexpressed assumption of so much philosophy, both professional and home-baked – that the primary stuff of the world is its disposition in states of affairs. The picture that this brings to mind is of a stable rural world where nothing much happens, and where the propositions that give expression to the way things are, or fail to, go on down the ages, immovably true (or false), entailing and contradicting each other as immutably as the weathered gravestones in the churchyard incline towards each other or away. This is unfair, of course. It is just as much the case (or not) that events occur as that states of affairs exist (or don't). It is the case that the lightning strikes and the smile touches the corners of your lips and is gone. Somehow, though, when events are frozen into cases, their essential dynamic seems to drain away. The brief appearance of your smile at 7.54 one bright and hopeful summer's morning, fixed in the true proposition that enshrines it, approaches as solemnly as death through the first half of eternity, flashes into life for an instant, and then recedes as funereally through the second half.

Well, we've already noted the elusiveness of states of affairs, except in so far as they are given form in the statements that express them. We've also remarked upon the great variety of purposes for which statements are made – not just to express states of affairs but to persuade, console, amuse, amaze, hurt, flatter, and so on. Or just to sound nice. Why not? So the ubiquity and meaningfulness of fictitious statements, and the impossibility of dismissing them as poor relations of factual statements, surviving only on the strength of their family connections, is a further reminder of the primacy of human expression. First comes grub, said Brecht, then come morals. First comes the statement, likewise, then comes the state of affairs. First comes doing, then comes being.

You look sceptical. You argue that, at the very least, facts must take factual statements into dinner ahead of fictions and the shady assertions hanging on their arm, because you can't understand fictitious entities unless you have some acquaintance with their real counterparts. – But we *did*, we historically *did*! We understood Genesis long before we understood *The Origin of Species*. – What you mean, though, is that before we can understand about God making the greater and lesser lights we have to know something about human beings making pots and pottage. We have to know what real death and grief are before we can understand *Hamlet*. Oh sure, and it works the other way, too. We have to see *Hamlet*, or something like it, before we know quite what we're feeling about our own stepfather. We have to know that God made man before we can understand that man made God.

Let's be bolder still: *the story* is the paradigm. Factual statements are specialised derivatives of fictitious ones.

You smile that sceptical smile of yours. You have read somewhere that I write fiction for a living, so my announcing that fiction is the archetype of truth sounds suspiciously like an armaments manufacturer insisting that war is the way to peace. You feel that there's something not quite respectable about storytelling. You're far too sophisticated to think it's just a form of lying, and you know that it can be used with the most admirable artistic and moral purpose. All the same . . . it's not *disinterested*. Not in the way that the careful reports of historical research and scientific experiments are. There is an element of huckstering, of show-business. The point of telling a story is to seize the attention of the hearer, to divert, to entertain. The storyteller is forever digging you in the ribs. The biologist who describes the structure of cells in the digestive system of nematode worms doesn't care whether anyone's listening or not. He's describing it for its own sake, because this is how those cells are. Isn't he? Whereas you might say that a story is an inherently commercial enterprise, even if no money changes hands. It's function is to *sell* you something, and even if it's not some moral lesson, or some opinion or viewpoint, it has at the very least to sell you *itself*.

I admit the truth of this. What *is* a story, after all? It's an account of something that might catch someone's interest. Why might some accounts of things catch people's interest? – Because, surely, they have some relationship to the world with which their readers or hearers are confronted. Think for a start of what makes a newspaper story. A revelation relating to someone the reader knows, either personally or publicly, or to some event at which he was present or in which he was somehow involved. Something that confirms his prejudices. Something that contradicts his prejudices. Something that arouses his fear, scorn, amusement, sympathy. Something that touches on his anxiety about death and suffering, or his longing for love and happiness. Something that reveals the secret and the forbidden, or seems to. Something that offers a mystery, or purports to solve one. Something that explains the reader to himself, or else takes him out of himself in all his banal incomprehensibility.

There's an element of concealment in stories, and of revelation. It is of the essence in stories that the cowardly turn out to be brave, and the brave cowardly. And when we've got used to this apparently universal paradoxicality, then it becomes a surprise, and a workable narrative event, that the brave are sometimes brave and the cowardly cowardly. It is a surprise when reason, ingenuity, courage, and enterprise turn the tide of apparently ineluctable events. It is another surprise when, expecting that this is why you're being told the story, you find they don't.

But then even the biologist with his nematodes is angling for *someone's* attention (if only the employer or funding body who will count his list of publications.) The nematodes he is writing about are familiar personalities to his readers. He has revelations about them. His results will confirm or challenge prejudices; arouse surprise, jealousy, scorn; attract a research grant. The flatness of the style doesn't destroy the narrative. Nothing could be flatter than the local paper's report of a wedding or a funeral, which consists largely of a simple list of those present, but which is eagerly read by every one of them – and by every-one else who wasn't invited and who thinks he should have been. And the flatness of the prose serves a function. The author is establishing that he is a disinterested narrator. He is distancing himself from his material, like the malicious gossip who begins, 'It gives me no pleasure to tell you this, believe me, but . . .'

There was a fashion some years back among literary and social theorists for saying that the text, the narrative, was itself the reality, so that differing texts and narratives were all equally valid. There was no reason, apart from the desire to assert dominance and control, for preferring Western medicine to voodoo, Copernicus to Ptolemy, your sober account of life in West Norwood to my deliberate caricature of it. This is an attractively tolerant idea, which promises to save us all a lot of tedious argument and research, and offers a hope of avoiding the doctrinal differences that threaten to destroy us. It's true that there are various ways of coming at the same aspects of the world, and we should be cautious about dismissing unfamiliar ones out of hand. What makes any text or narrative of interest to its audience, though, is that it has some relevance to their actual experience of an actual world. This is, after all, *why* there are different texts and narratives in the first place – because the people who compose them have had different experiences, and learned different ways of dealing with them. The different versions find different audiences for the same reason – because those audiences have themselves had different experiences and learned different ways of dealing with them, and because it is against these experiences and techniques that the version will be tested. Without some external reality to give rise to them and test them against, even fictions, let alone histories, are incomprehensible. There's no escaping the traffic!

One of the distortions imposed by the story form, it once seemed to me, was that the natural tense of the novel, and other forms of written story, was the past. Part of its attraction, I felt, was that it suggested an enclosed, completed world, or at any rate a world which is on its way to a completion already safely established in the later pages of the volume that one holds in one's hands – a small non-nature reserve, from which the appallingly unresolved openness of the real world is reassuringly absent. No sooner had I observed this, curiously, than I began to write novels set in the present, to see if I could disprove my own thesis and suggest the open-endedness of our experience. It was made more natural by analogy with the plays that I had begun to write at the same time – because the natural tense of the play is the present. Not that the question usually arises, since in most plays there is no overt narration of events (except, intermittently and privately, in stage directions – and these are universally written in the present tense). But in the theatre one feels (or one should, surely) that these events are occurring *now*, that things might go one way or the other (or, as in Greek tragedy, that the future is, paradoxically and therefore interestingly, foredoomed). Another curious thing: as my novels have moved into the present so my plays have turned towards the past. Several of them have been framed in a narrative making clear that the action belonged to a world which had already closed, and whose end was settled before the curtain rose.

Jokes are told in either the past tense or the present. 'There was this feller on a building-site. Said to the foreman . . .' Or, equally: 'There's this feller. Says to the wife . . .' Even when they start in the past, though, jokes and other spoken anecdotes seem to have a yearning to relocate themselves in the present, which bursts out in the middle of things in the way that suppressed sexual feeling might. 'So I went up to her and had it out with her. And you know what she turns round and says? She says, "Oh," she says . . .'14

Another oddity: when people give an account of a story they have read – in a review, for example, or even simply talking about it to a friend – they almost always move it into the present tense, as if it were a film or a play. 'So God says to Saul . . .' 'Whereupon Alice starts to shrink . . .' The story has become the event itself, and this event is present to us even now.

If fiction is make-believe, as has been claimed, or if we entertain it by suspending our disbelief in it, this seems to imply that our understanding of factual statements involves belief. The position is supposed to be this: if I tell you with truthful intent that the weather is wonderful then I am saying what I believe to be the case (and for that matter believing what I say). My purpose in saying it is that you should believe it as well.

But what exactly *is* all this belief? What exactly am I doing when I believe something to be the case? And if I say what I believe to be the case, am I doing two things? First believing that the weather is wonderful, and then saying 'The weather is wonderful'? And if I believe what I say, am I first saying it and then believing it? Or am I somehow doing the believing and saying simultaneously?

I can't *detect* any believing going on, before, during, or after.

Probably I'm just glancing up at the sky, and out the words pop. And even if I'm speaking after some considerable study of cloud patterns and temperature levels, even if I feel I'm issuing a carefully deliberated statement upon which important decisions about picnics and invasions will be based, none of this great performance seems exactly to be *believing*. So I don't believe what I'm saying? Of course I believe what I'm saying! Since you've asked me.

And even if I could detect some believing going on inside my head before or as I speak, it's got to be doing more than happening to occur at around the same time, surely. I might also be tying my shoelace just before I open my mouth, or scratching my head as I speak, without it seeming to be of any logical relevance. There has to be some kind of linkage between the believing and the saying. So I'm saying it *because* I believe it? Well, yes, in the sense that I visit the Metropolitan Museum because I'm in New York – in the sense that I couldn't visit it if I *weren't* in New York. There are a lot of other things I could also be doing because I'm in New York that I'm not actually doing – and a lot of other things I am actually doing, such as cleaning my teeth and combing my hair, which don't seem to have any very necessary connection to my being in New York. Are there a lot of other things that I'm believing even though I'm not asserting them? Am I also believing, even though not currently asserting, that grass is green, that teatime is five o'clock, and that Oslo is the capital of Norway?

I could, of course, at any rate suggest the existence of some of these beliefs without asserting them, by some other form of behaviour. I put on sunblock before I go out; I usually have tea at five o'clock; I go to Norway to lobby members of the Norwegian parliament about fishery protection and I look for them in Oslo, not in Stavanger or Bergen. Daniel Dennett notes that even animals with no capacity for language can have beliefs (though not opinions);[15] a dog expresses its belief that a cat is up a tree by scrabbling at the trunk and barking, a cat its belief that a dog is at the bottom of the tree by arching its back and staying where it is, etc.

The question remains, though, whether these beliefs have some existence separate from their expression. Are they stored inside my head (or inside the animal's head) in the way that we feel memories are? There do seem to be a very large number of them to be accommodated. The number is even larger than you might suppose at first sight, because, when you think about it, each assertion and disposition seems to express and require not only the one belief but a host of others as well. I can't properly assert that the sky is blue unless I also believe that 'the sky' is the name of the canopy of space above our heads, and that 'blue' is the name of the colour that it is. Perhaps I also need to believe that my colour vision is not grossly abnormal, and that I have not been deceived by some optical illusion . . . Even the cat, if its refusal to come down from the tree is to express its belief that the dog is waiting at the bottom, also needs to believe that dogs are bad news for cats, and that they cannot climb trees.

You might think that my simple assertion that Oslo is the capital of Norway, which is merely the rehearsal of something already expressed in language, needs no battery of supplementary beliefs to connect it with the world. Doesn't it imply, though, that I also believe that Stavanger and Bergen are *not* the capital of Norway? And that I have similar beliefs relating to Ankara and Tokyo and Welwyn Garden City? You look sceptical. Are you telling me that I can assert, sincerely and indeed truthfully, 'Welwyn Garden City is not the capital of Norway,' without any of the concomitant belief that seemed so necessary in the case of my contrary assertion about Oslo?

You will argue, I imagine, that what my assertion in regard to Welwyn expresses, albeit in a different form, is simply my original belief – that the capital of Norway is *Oslo*. I'm prepared to concede this – but only if I am allowed another belief to make the implication justifiable – that Welwyn Garden City is not Oslo. Now you're looking a little impatient. Your impatience is misplaced, though. The same object can go by different names – Oslo has been Christiania and Kristiania in its time. I surely need to have, if only at the back of my mind, the belief that the name Welwyn Garden City has never figured among its aliases. Is it for that matter reasonable of me to give Welwyn a miss in my search for Norwegian legislators, unless I have other beliefs about the likelihood of finding a national assembly in a nation's capital, and the unlikelihood of its being housed seven hundred miles away in some other nation altogether?

Then again, if my assertion that the sky is blue expresses my belief that the sky is blue, does *this* assertion ('my assertion that the sky is blue expresses my belief that the sky is blue') express my belief that my assertion that the sky is blue expresses my belief that the sky is blue? Because if I need to believe that the sky is blue in order to assert that it is, then surely not only do I need to believe that I believe the sky is blue, but I need to believe that I believe that I believe . . .

And does the almighty yawn I see overtaking you as you plough through this ballooning schedule of my beliefs express *your* belief that this view of belief has become extremely tedious?

Don't you quite simply . . . *find* it to be tedious? Isn't there just the schedule, and your wandering attention, and then the yawn?

Don't I, for that matter, quite simply . . . *find* the weather to be wonderful? Isn't there just the wonderful weather, and me noticing it, or remembering it? And if I go on to remark upon it to you then it's because I have some purpose in mind, something relating to you – and something more substantial than an unexplained intention that you for reasons unknown should give house-room to the same belief. I want to suggest we may be able to eat outside after all, or to cheer you up, or to reassure you that I'm feeling cheerful. I can also imagine saying it for no practical reason at all, quite dreamily, not to you or to anyone but myself. But in this case isn't it because the wonderfulness of the weather has struck me? Aren't I exclaiming rather than asserting? I could equally well say, 'Oh, the wonderful weather!'

I go to Oslo because I want to meet Norwegian MPs. My journey doesn't *express* anything – neither beliefs nor desires nor the existence of flight tickets nor the money to pay for them – though all kinds of implications can be read into it if the need arises.

You might ask me if I really believe what I'm saying now, and I might sincerely put the question to myself. How do I answer it? By detecting, or failing to detect, a belief lurking somewhere inside me? No, I ask myself whether I wasn't oversimplifying or exaggerating the situation, whether I wasn't speaking thoughtlessly, or purely for effect, or with the intention of deceiving; whether I should be prepared, now that I have been put on my guard, to say the same again.

The belief that is supposed to characterise our understanding of factual statements, and to differentiate it from our understanding of fictitious ones, slips through your fingers like water when you attempt to get hold of it. But we do sometimes claim to believe something in ways that seem to imply something substantial about our attitude, something over and above the assertion itself. It's funny stuff, though, this belief: it can be quite contrary things. Sometimes it suggests doubt, or at any rate modifies the extent of my commitment to the assertion: 'I believe the train leaves at 11.14' (but I haven't checked – I may have misremembered – they may have changed the timetable . . .); 'I believe I'm right in saying . . .' (I'm certain I am, but I'm politely suggesting readiness to consider any idiotic alternative that you might want to offer).

Other usages of the term 'belief', however, appear to imply precisely the opposite: certainty or some kind of commitment. When you declare to me in ringing tones that you believe in privatisation of the Royal Family, or the health-giving effects of smoking, I understand at once that you're doing something much more muscular than merely offering an observation. You're nailing your colours to the mast; perhaps expressing solidarity with my own attitudes, or challenging me to dispute with you. You're putting on this performance precisely because there can be no certainty about the object of your belief – because there is more than one view possible, because the evidence is open to dispute – and because in the end your choice comes down to your personal interpretation of the world before you, or your personal preference for how it might be improved. And when you whisper nervously to me, as the strange phosphorence flickers through the darkened chambers of the old manor house, that you don't believe in ghosts (or that you do), you are arming yourself against (or surrendering to) fears that are probably beyond rational examination.

What you're doing, though, when you think about it, is not after all so totally unlike what you're doing when you say tentatively that you believe the train leaves at 11.14. A similar uncertainty clouds the situation; but now, when you believe in *this* kind of way, you are closing your eyes to it. If you believed in the same way in the departure of the train at 11.14 you would be by implication agreeing that the service was notoriously unreliable, but jutting your jaw defiantly and saying that you didn't care – at 11.14 you were proposing to step off the platform in the sure and certain hope of finding a train between you and the rails.

Well, there *might* be a train standing there when the moment comes. Financing the Royal Family by commercial sponsorship *might* improve the nation's constitutional arrangements. Scientific evidence *might* be found to show that phosphorescence and chill drafts are detectable in old houses where murders have occurred. Now you tell me that you also passionately believe in personal freedom and animal rights. I understand you to mean by this something rather different: that you are committing yourself not to a view which might turn out to be demonstrably the correct one, but to goals which you think are desirable in themselves, regardless of evidence or consequences. Once again, though, the implication is that different views are possible, and that you are resolving the objective uncertainty of the situation by a wilful act of personal certainty. Whether this certainty is carefully thought out to be consonant with your other tastes and beliefs, or whether it is a leap in the dark, it originates in you, and implies nothing (or nothing very definite) about the current state of the world outside yourself.

The catalogue of your beliefs is not exhausted yet, however. You tell me that you are a believer in the religious sense. Either the content of your beliefs, or the act of believing them, or both, is so important to you that you see it as identifying you in a kind of vocational way, as a baker is identified by his baking or a painter by his painting. Are *these* beliefs of yours like your belief in private finance initiatives or the dignity of man?

Well, what exactly do you believe? Let's say, for simplicity's sake, that the answer in your case is very straightforward and definite. You are a Christian, and a member of the Church of England. So what you believe is laid out for you in the Apostles' Creed, which you recite in church every Sunday along with the rest of the congregation. The list of beliefs in this is very specific. Some of them are historical assertions (that Christ was born of the Virgin Mary and suffered under Pontius Pilate) and are as theoretically verifiable as any other historical assertion. Some are predictions, and I suppose it's just possible to imagine that one day, as with the emergence or non-emergence of scientific evidence for ghosts, Christ either will or will not come to judge the quick and the dead, and that your body either will or will not be resurrected.

So is this why you announce each week that you believe these things to be the case, because there is potential evidence for them? Why don't you also tell us once a week that Mary Queen of Scots was born of Mary of Guise, and suffered under Elizabeth I, or that the melting of the ice-caps will one day wash away New York? Will you cease to recite the relevant passages if a historian turns up a birth certificate showing a different name for Christ's mother, or a Roman intelligence report revealing that Pontius Pilate had been secretly assassinated by the CIA and replaced by a lookalike? How long do you think the world should wait for the Last Judgement or the resurrection of the dead before inserting a few ifs and buts into the document?

And what about some of your other assertions each Sunday morning? You also tell us that you believe in the Holy Ghost and the communion of saints and the forgiveness of sins; and that Christ ascended into heaven and sitteth on the right hand of God the Father Almighty. Are you suggesting that evidence might turn up for these? Or are you saying that, as with personal freedom and animal rights, you approve of the idea of a Holy Ghost, and think that it would be nice if sins were forgiven?

Are you implying, as with the declarations of belief in ghosts and private finance and animal rights, that there is objective uncertainty about these things which you are resolving by your personal fiat?

You tell me that there is plenty of evidence for all these assertions. They derive from highly reputable authorities – they repeat or follow from what was said by Christ himself, or by his biographers, or from what was decided by conclaves of venerable churchmen in the past. In other words, the assertions you make in the Creed derive from a text, like the assertions made about Onegin and Lensky by critics and examination candidates. But what about when Christ, or St Matthew, or the early bishops[16] *first* asserted them? You tell me that they were instructed or inspired by God. But this is to go round in circles. It's as if Pushkin said he knew about Lensky's death because he was told by Onegin. You might do better to argue that statements like these don't need evidence – that Christ, or St Matthew, just *knew*. You might tell me that this goes for you, too. You don't need to trace back the assertions in the Creed to their original sources before you believe them. You just *do* believe them. You just know them to be so, because some inner prompting tells you, or some feeling that they make emotional and aesthetic sense.

Faith is what believers in this sense have, and it seems to be a free-standing condition, often admired and recommended, like grace or charm, which you just *possess*, without its implying any particular connection with external reality.

So assertions about the Holy Ghost are in origin like the assertions about Onegin and Lensky when they were first uttered by Pushkin in the poem. They are enactments which in some sense create the situations that they appear to report. This is no more to derogate them than it is to derogate Pushkin's verse. Is this to say, though, that latter-day believers have the same relationship to their beliefs as Pushkin's readers do to his narrative?

Well, they obviously don't value them for their psychological acuity, the elegance of their expression, or their entertainment value. They feel that the assertions possess, by their very nature, some kind of moral authority – an authority which imposes of itself the obligation to accept them. The assertions are felt to have a kind of moral force field around them, which it is wilful and wrong to resist. There's no great temptation here, in spite of the parallels with ordinary fiction, for philosophers to characterise the original assertions as make-believe – they are by definition do-believe. Nor for literary theorists to suggest that later adherents to the doctrines 'suspend their disbelief'; even if disbelief is what they began with, its removal is supposed to be not temporary suspension but permanent expulsion. Believers believe!

Or do they?

The unreligious sometimes make the mistake of thinking that the religious can't understand how oddly the fervour of their believing sits with the factitiousness of their beliefs. This is implausible. The ranks of the religious plainly include people no less intelligent or perceptive than anyone else. They can see perfectly well how unlike factual statements their assertions are. When they say they believe them, they don't mean that they believe them in the same way as they believe what they say when they tell you the weather is fine, or that there is a train to Birmingham in ten minutes' time – certainly not in the same way as they believe when they say they *believe* the weather will be fine later, or *believe* that there is a train to Birmingham.

Quite the contrary. It is the pure unevidenced and unevidenceable factitiousness of the beliefs that gives them their force, and makes the true believer so eager to believe; it is the very fact that the assertions have been conjured out of thin air, or taken on trust from others who got them from the same source, that makes them worth asserting. Believers are like people who are asked to provide an alibi for someone they love. If the alibi is false then to provide it is to offer even greater testimony to their love and loyalty than they would be demonstrating if they were required only to tell the truth. And to persuade oneself that one does in fact *believe* a false alibi is an even greater testimony. It is a testimony made before oneself as well as before others. Statements of faith are *tests* of faith. They are difficult feats which believers undertake precisely because they are difficult. Believers don't go around devoutly believing easy beliefs – they take those for granted. They often have to struggle with themselves, like soldiers training to overcome their natural scruples about inflicting suffering. They claim defiantly that their beliefs are not beliefs at all, but knowledge ('I know that my Redeemer liveth . . . ', 'In the sure and certain hope of resurrection . . .'[17]), in the way that I might insist in the course of an argument that I *know* what I'm saying is so, even though I can't explain why (then wake in the night and know just as certainly that I was wrong all the time). But believers, not knowers, are what they call themselves, and believers, not knowers, is what they are.

Some beliefs – in universal love or the perfectibility of man – are further specimens of that fabulous creature in the logical zoo, the uboama; they are breaches of the fundamental logic of language and thought. The worlds they project are not so much like the one imagined by Pushkin as the one created by Lewis Carroll in *Through the Looking Glass*, where the White Queen boasts of sometimes believing as many as six impossible things before breakfast. The challenge, not just of implausibility but of impossibility, might be an added encouragment to the faithful. '*Certum est quia impossibile est*,' as Tertullian, one of the early Fathers of the Church, famously declared – 'it is certain because it is impossible.'[18]

Systems of belief often have an impressive stability. The less rational they are the less subject they are to change, and the more stubbornly they endure down the centuries. Saussure notes a similar persistence from one generation to the next in the systems of signs that make up the different human languages.[19] This, he argues, is because they are *arbitrary* – there is no reason why a sign takes any one form rather than any other, except that it always has. There is nothing about the morphology of the word 'dog' that reflects the reality of a dog, or our corrigible knowledge of an ever-changing family of breeds; nothing about the sound or shape of the word 'atom', for that matter, that is affected by all the profound shifts that have occurred in our understanding of what it refers to. But if there is no reason for a word's being what it is then there is equally no reason for changing it. The more arbitrary a belief, likewise, the more removed from observable reality, then the less it can ever be affected by argument or experience, or by changes in our understanding. The stability of religious belief, in which so many believers take pride, is a testimonial to its irrationality, to its insulation from the world.

In any case, systems of belief tend to carry a built-in defence to protect their adherents from rational debate. Freudians, Marxists, and

Christians all know that resistance springs from sources already predicted by their respective theories: denial, class interest, pride. You see the knowing, forgiving smile cross their faces even as you raise your objection. They don't have to listen to it – its nature is unimportant, since the motive behind it has already been explained by the system. Imagine a scientific theory that included a built-in technique for dismissing any alternative theory (quantum mechanics, say, extended to include a doctrine classifying the concepts as themselves quantum phenomena, and therefore unexaminable without reducing them to classical mechanics).

Not that all believers get even this far in the rational defence of their systems. Fearful that faith may be as vulnerable to the normal procedures of human debate as cities are to nuclear weapons, they feel that their best hope, as against the missiles, is to terrorise their opponents into never getting their weapons off the ground. Certain thoughts are never even to be put into words, certain subjects are never even to be mentioned in a profane spirit, on the pain of shattering retaliation.

Declarations of faith, like declarations of anything else, are uttered with some purpose in mind. This is not usually to catalogue articles of belief out of academic interest, as you might list different species of beetle. The intention is to secure an effect, upon others or upon oneself. At their lowest and most depressing they have something in common with the chants and taunts that supporters utter at football matches, where the intention is not to express any particular states of affairs, or hopes for their alteration, but to induce feelings of solidarity and right-mindedness among the supporters themselves, and to incense their opponents.

At their highest, though, I suppose you might understand declarations of faith as being like a performance of one of the great heroic parts in the theatre, which offers the possibility of expressing human truth by a dramatic fiction. Some of the same difficulties and opportunities arise. The believer, like the actor, may impress by the sheer force of difficulty overcome. On the other hand he may force his performance, and become unconvincing. The performance may vary from the crude and obvious, with much weeping and tearing of the hair, to the subtly understated and inward, with nothing to show on the surface but the occasional shadow of emotion passing across the features, a certain way of going about things, a style.

Whether done well or ill, the believer is exploiting the *possibilities* offered by this thing that he has taken upon himself. Roles and beliefs have lives of their own, implications which lead us on to destinations we never dreamt of when we signed the contract for the show. (Hence, perhaps, the tendency of actors' performances to get coarser during a long run, and for believers to go off at tangents from one another, into ever more extreme sects and heresies.)

Religious belief is a sensitive area. There is something embarrassing about it which makes it very difficult to examine or discuss. It is embarrassing to the unreligious, who feel that they are called upon to look away, and that failure to do so is as intrusive and insensitive as peering at people deprived of their normal self-control by pain or grief.

But it is embarrassing to the religious as well. They tend to maintain a decent discretion about their beliefs, as people do (or did) about their sexual feelings, which makes it possible for us all to get along socially. If they are obliged to talk about them they find it difficult to look you in the eye – even to look each other in the eye. Or if they do it's with a special fixity and determination which is intended to conceal their unease. No wonder, when the religious catch people looking at them with a mildly sardonic raised eyebrow, that they sometimes want to kill them.

There's a special tone of voice, too, which priests adopt, much mocked, and intended (I think) as a kind of mask to hide behind. I'm always struck, too, by the heroic intensity with which street-corner evangelists bellow their message to the indifferent passers-by. Both the intensity and the indifference seem to be inherent parts of the procedure. If someone were to stop and listen (which no one ever does) the ranting preacher wouldn't moderate his volume, or address his hearer personally. The performance is directed not outwards, at a possible audience, but inwards. It is a demonstration of fear and doubt overcome, a testimony of belief in belief. The words being uttered at such loudspeaker pitch are often entirely uninflected and incomprehensible. They are being uttered on the basis of having been uttered before, and they are uttered again for utterance's sake.

Why the mutual embarrassment? Isn't it because we know, in our heart of hearts, that the religious don't *really* believe in any normal sense what they say they believe – that an announcement of belief in this sense is really a kind of announcement of non-belief? We also know that they know this perfectly well themselves. And that they know that we know that they know. So we are all, religious and unreligious alike, caught in a false position, and are embarrassed to know how to proceed, like a hoaxer and his victim, when the hoaxer knows that his hoax has failed to deceive, but the victim is too polite to admit the fact.

We all, as Erving Goffman says, tacitly agree to help each other preserve the performances through which we present ourselves to the world,[20] and it requires remarkable aggression to abandon this policy. The less reasonable the performance, the more collaboration the performer requires from the rest of us to keep it going. The greater, too, the efforts that he has to make himself. It's like a farce, where the more preposterous the falsehood someone has committed himself to, the more desperate the efforts he has to make to sustain it. In real life, where we all tend to be afflicted with niceties of feeling that are missing in theatrical characters, this goes for the other characters, too. They also have to make huge and desperate efforts to enable the culprit not to lose face.

The ontological argument for the existence of God (God is defined as having all possible qualities, or as being perfect; and one of these qualities, or one aspect of his perfection, must be existence) has an engaging boldness. There's something about its breathtaking cheek that reminds me of the plea offered by the man convicted of murdering his parents, in the traditional Jewish definition of *chutzpah*, who throws himself upon the mercy of the court because he's an orphan. And it does catch something of the self-creatingness and self-containment of religious belief, its ability to suspend itself in space by its own bootstraps. It might indeed be usefully adapted to demonstrate the existence even of more modest characters in fiction. I have only to describe the characters in my next novel as having all possible qualities, or as being perfect – and, lo! there they are in the real world. Do we even need such elevated characteristics? Nowhere, so far as I can recall, does Jane Austen make the same claim of Mr Darcy, for example. But she does describe him as having a fine, tall person, handsome features, noble mien, and reportedly ten thousand a year, which must surely imply his existence just as clearly, because it's difficult to see how anyone can have a fine, tall person and all the rest of it unless he exists. If God is summoned into being by words then so is Mr Darcy. And indeed they both are so summoned. This is the truth – the truth of fiction.

I have the feeling that all this was simpler once – that the Children of Israel, for example, were as *undeceived* about the stories in the Pentateuch as the Children of Britain are by the age of four or so about the story of the tooth fairy. What went through their minds when they read (for instance) the account of the wonderful Dutch auction that Abraham conducts with God before the destruction of Sodom and Gomorrah?[21] What did they think about its *provenance*?

And Abraham drew near, and said, Wilt thou also destroy the righteous with the wicked? Peradventure there be fifty righteous within the city: wilt thou also destroy and not spare the place for the fifty righteous that are therein . . . ?

And the Lord said, If I find in Sodom fifty righteous within the city, then I will spare all the place for their sakes.

And Abraham answered and said, Behold now, I have taken upon me to speak unto the Lord, which am but dust and ashes: Peradventure there shall lack five of the fifty righteous; wilt thou destroy all the city for the lack of five?

And he said, If I find there forty and five, I will not destroy it. And he spake unto him yet again, and said, Peradventure there shall be forty found there. And he said, I will not do it for forty's sake . . . .

Peradventure . . . thirty . . .

I will not do it if I find thirty there.

. . . Twenty . . . ? . . .

I will not destroy it for twenty's sake.

. . . Ten . . . . . ?

This magnificent scene of heroic haggling in a noble cause reads to us as classic high comedy – perhaps the first comic scene ever written, and the precursor of all the great novels and plays where some little man gets the better of a tyrannical superior with no weapons but his native wit, shrewdness, and courage. I imagine that the Israelites took it more seriously than we do. (They were the ones, after all, who were going to be subjected to an early prototype for Dresden or Hiroshima at the end of all this.) What did they think, though, not about the historicity of the scene in general, but about one small specific aspect of it – the accuracy of the dialogue? Did they think of it as a word-for-word verbatim report? If so, how did they suppose that this had been achieved? The meeting seems to have been a private one, with no secretaries or aides on either side. Presumably Abraham himself kept some kind of record that was handed down and used by Moses when he wrote the report. (It seems unlikely to be based on a briefing from God's side, since he emerges from the story so badly.) Did readers suppose that he took a shorthand note? Perhaps they assumed that he wrote it all down immediately afterwards, as journalists often do in interviews so sensitive that they have not dared to produce a notebook earlier. This is more plausible, but if so, however accurate Abraham's memory, however deep his devotion to the highest standards of journalistic accuracy, it most certainly cannot have been a precise word-for-word transcription of what had been said.

Did this worry people? Did they discuss around the camp-fires how difficult it was to make a completely accurate assessment of the spirit in which God had conducted his side of the negotiations if they were not certain *exactly* what words he had used? Supposing God had said not 'I will not destroy it,' but 'I will give serious consideration to not destroying it.' Supposing the negotiations had not apparently broken down after Abraham had whittled God's minimum requirement down to ten, but had continued into some final late-night session, when both sides were too exhausted to be very clear about what they were doing, and reached some fudged formula that might somehow have been interpreted as meaning four. An important detail for the citizens of Sodom and Gomorrah, you might think, because there *were* four righteous people there – Lot and his family.

Then again, accuracy apart, a lot depends upon the tone of voice in which God was speaking, and there is no indication of this at all – a serious omission. The sophisticated modern political commentator, I think, would want to know whether God had spoken with growing irritation at Abraham's salami tactics, or with growing amusement at his cheek; and whether he was genuinely touched by Abraham's advocacy, or whether he was merely thinking about the public-relations fallout. Did he hesitate at all as the concessions increased? It would be difficult to make a serious estimate of the long-term implications of this conversation for the Children of Israel without knowing these things.

And yet I doubt if any of these questions were raised at the time. I imagine that the distinction between accuracy and inaccuracy, between completeness and incompleteness, was not at issue. The dialogue was what it was. It was complete in itself, as the courtroom dialogue in *The Merchant of Venice* is without reference to Venetian judicial records of the period. The general sense of both texts express-es something of the relations perceived to exist, or hoped to exist, between human behaviour and human destiny, between the warring virtues of justice and mercy. No questions arose about what was story and what was history, because no division was perceived to exist between the two. Nor were there, for that matter, even in compara-tively recent historical times, about the precise verbal accuracy of reports of proceedings in Parliament – not until Mr Pitman came upon the scene, and made technologically possible the distinction between the verbatim and the non-verbatim.

Even when Talmudic scholars began to pick over the ancient texts word by word, their concern was not with the relationship between the words and the world they purported to describe, but between words and words. They studied the text as a thing complete in itself. The words related to each other. In so far as they related to the world at all it was to the world that might be regulated by reference to the words – the world of the present and the future. The words, the writ-ten precepts and prohibitions, were the given; the raw material of subsequent human behaviour had to be fitted to them. Does a string stretched upon poles around a suburb of London count as a house-wall of the sort that exempts the people inside it from observation of some of the laws that control behaviour on the Sabbath? Is the Orthodox population of present-day Israel large enough to count as an extrapolation of the ten who would have saved Sodom and Gomorrah?

There is also a way of believing things that we ourselves, in our everyday life, understand as unrelated to the world of ordinary expe-rience. Children, even before they've reached the age when they for-mally renounce belief in Father Christmas, wouldn't actually expect to see an old man with a white beard and red coat putting presents at the bottom of their bed in the small hours of Christmas morning. If they did wake up and see such a person they would scream with terror. They might think it was a burglar, or a terrorist planting a bomb, or the waking dead, but the explanation that it was *Father Christmas* would come very far down the list of possibilities. And yet . . . they don't disbelieve. They go along with it. They carry it in a separate compartment of their mind from the practical matters of the world.

I had the most alarming demonstration of this ability to both believe and not believe when I was a child. It also concerned Father

Christmas, though in this case I didn't see him – I *was* him. I appeared, with the most benevolent intentions, round the door of my small sister's bedroom one evening in December wearing a Father Christmas mask and asking her in a Father Christmas voice what presents she wanted. Did I really hope to persuade her that I actually *was* Father Christmas, and expect that she would be overcome by delight? I suppose that all I really hoped was that she would go along with the pretence in the spirit in which it was offered, and what I feared was a scornful refusal to. Her actual reaction came as a complete surprise. She screamed. She screamed and screamed and screamed, hysterical with terror. I snatched off the mask and showed her that it was only me; she went on screaming, and continued until my parents came running, and then screamed some more. After I had been suitably scolded for my cruel behaviour, and things had settled down a little, she said she'd known perfectly well all the time that it was merely me wearing a mask. This is what had frightened her, she said – that it was me, but that I had turned into something else, with a bulging pink cardboard face half-wrapped in cotton-wool. The knowledge that she was watching a fiction had not for a moment taken away its unforeseen power.

It's not only children who think like this.[22] We all sometimes loyally cling to the idea we have of certain people – as being benevolent and brave, for example, even when they have proved themselves malevolent and cowardly. The idea is not a *practical* one – it doesn't raise in us any great expectation that they will help us out or stand up for us. This is something that often happens in love, and fills the refuges for battered wives.

Stories that we believe, stories that we don't believe, stories that we half-believe, stories that we believe and don't believe, stories that we say we believe because we don't . . . As we tell them, and listen to them, and understand them, we bring into their various forms of existence all the receding ontological planes of the world we inhabit.

IV: Words

Ricefiring

Sense and syntax

Stories, beliefs, explanations, laws, the causes that we trace and the motives that we unearth – they all have something in common: the medium in which they are generally expressed, which for many millennia now has been language. This is how most of our traffic with the world is given outward form – certainly all of it generalised and explained, even to ourselves. Most of the time we take it almost as much for granted as the air we breathe; and then, when we stop to think about it, we see what a puzzle it is. How do permutations of a handful of different sounds, or of a few marks on paper, make it possible to enter into a relationship with the things and people around us, even perhaps to understand something about ourselves? How does language say what it says?

There shouldn't be any great mysteries here. We all learn to use language early in life, in one form or another, unless we've been raised by wolves, and we go on using it until our breath ceases or our brain fails. We speak it, we read it, we write it; we shout it, we murmur it; we sing it, we pray it, we worry it; we study it, we interpret it; and we think it and think it and think it. We have language on our lips or in our heads from when we get up in the morning until we go to bed at night, and then we dream it. It's picked over from the standpoint of phonetics and linguistics and semantics; of etymology and orthography; by philologists and philosophers, translators and teachers, literary critics and leader-writers.

It's so ubiquitous that it's difficult to see. It surrounds us like a great forest. Its venerable trunks and branches carry our eye upwards from the humble mud underfoot and shut out the light of the sky above. That mass of timber, though, is the mere residue of its living function. Just as the function that characterises a tree, that makes a tree a tree, is photosynthesis – the bringing of the intractable mud below into relationship with the fugitive light of the sky above – so the function of perception and language is psychosynthesis, the bringing of the intractable materials of the world that confronts us into relationship with the fugitive pursuits that enable us to retain our footing in that world.

How do words . . . *catch hold of* things? This is what's so elusive; as elusive, perhaps, as all the other things that elude us. But then how would the answer be expressed? There's only one way it could be: through language. Once again we have to pull ourselves up by our bootstraps.

Language is what makes us human. Genetically we are very close to various other extant and vanished species, as the ethnologist Jared Diamond[1] explains:

> We share ninety-eight per cent of our genes with chimpanzees . . . The Africans making Neanderthal-like tools just before our sudden rise to humanity had covered almost all the remaining genetic distance between us and chimps, to judge from their skeletons. Perhaps they shared 99.9% of their genes with us. Their brains were as large as ours, and Neanderthals' brains were even slightly larger. The missing ingredient may have been a change in only 0.1% of our genes. What tiny change in genes could have such enormous consequences?

And he concludes, like other scientists who have speculated about this question: 'I can think of only one plausible answer: the anatomical basis for spoken complex language.'

Now that we have it, language gives form to the world around us – and gives us our experience of ourselves. Philosophers have in the past talked as if it were some kind of freestanding contingent natural phenomenon, like a sheet of water that happens to reflect the scenery around it, more or less clearly or confusedly, something that could exist or not exist without its in any way affecting either the scenery or us.[2] As Daniel Dennett says, though,[3] 'language infects and inflects our thought at every level. The words in our vocabularies are catalysts that can precipitate fixations of content as one part of the brain tries to communicate with another. The structures of grammar enforce a discipline on our habits of thought, shaping the ways in which we probe our own "data bases" . . .' He quotes Justin Leiber:[4] 'One could say that natural language was our first great original artifact and since, as we increasingly realize, languages are machines, so natural language, with our brains to run it, was our primal invention of the universal computer.' But Dennett then goes on to put it even more strongly by voicing 'the sneaking suspicion that language isn't something we invented but something we became, not something we constructed but something in which we created, and recreated, ourselves.'

Views of this kind have not gone undisputed. Russell dismissed linguistic philosophy, which first began to study the world through the ways that we do in fact talk about it, as being concerned merely with 'the different ways in which silly people can say silly things'. Ernest Gellner was positively vituperative about it. Even Chomsky, one of the founding fathers of modern linguistics, says he is sure that 'everyone who introspects will know at once that much of his thinking doesn't involve language'.[5] The remark is quoted approvingly by Bryan Magee,[6] who insists: 'I can only affirm . . . that life and the world as I experience them are completely at odds with what language-oriented philosophers and literary critics customarily say or seem to assume, in that both are largely inexpressible in language.' It's true that people who work with words (philosophers among them) sometimes become a little hypnotised by them, and Magee is surely right in thinking that a lot of our experience can be expressed only in music, or only in graphic form. It's also plain that there is more in heaven and earth than can ever be expressed in all the various modes available to us put together.

All the same, in so far as our experience is communicable at all, it has to be expressed in one symbolic form or another. Among these various forms, language is unique in one particular way which can't help but recommend it to philosophers and scientists alike: it is the only one that is capable of expressing propositions. Only in language can we assert that something is or is not the case; only in language can we say things that are true or false.[7] Not even the most profound and expressive music can assert that something is thus and so; not even the most naturalistic picture, or the most impressionistic.[8]

Our present culture, it's frequently said, is replacing words with images. You think of films where the story is told with a minimum of dialogue, of commercials that consist of brief visual allusions to disparate moments of action, or of magazines where the minimal text is merely the largely unread decorative framework to pictures of famous faces and attractive bodies. These juxtapositions of images generate affective contexts of fear, anticipation, desire, etc. They evoke feelings by association. They are like the lion's roar and the peacock's tail. But they assert nothing, or nothing specific. They are no more *replacements* for language than hamburgers or electric beard-trimmers are.

Not even the most graphic news photograph asserts what is the case. It shows (unforgettably) the soldier holding the gun at the terrified prisoner's temple. The image records a certain arrangement of things in space at a particular moment of time, and does it more specifically than any verbal description ever could. But it doesn't actually claim, as the words do in the caption, that this is a soldier, and a gun, and a terrified prisoner. It doesn't even assert that someone is holding something at someone's temple. It leaves us to place our own construction upon what we see – as we should if we were seeing the original scene – from the knowledge that we have of weapons, or human expressions and human relationships.

A length of film can show an event actually unfold. We see the President grasp his chest, then slump forwards, and this is crucial evidence for the investigators who are trying to understand what's happened; but it doesn't *assert* that the President grasps his chest or slumps forward; this is left to us to do when we see the film. The only way in which the film, like the photograph, has a semantics that the original event lacked is that it selects one small area of the scene for us to inspect. To have any idea what the film is *saying*, we need to know *that* it refers to persons and events that have already been identified and described. A car collector who knew nothing about modern American history might see the same sequence and regard it as suggesting, if anything, something about the spaciousness of a Lincoln convertible. The silent antics of its occupants might say nothing to him. The photograph, and the film, may be accurate, fair, graphic, and unretouched, but neither is *true*. They may be misleading, posed by actors, airbrushed, intended to deceive, but they can't be *false*.

You might object that what your bank manager or vicar certifies, on the back of your passport photograph, is precisely that this is 'a true likeness' of you. Yes, but the truth really resides in the identification that this endorsement is making of you with the image, not in the image itself. It is the endorsement that declares the status of the image to be a likeness of someone, something that the photograph itself cannot say. A 'true likeness' is an image of which the assertion is truly made, or could be truly made, that it is a likeness.

– But there must be something about the picture, some objective quality, that makes your bank manager think this!
– Of course, but what that something is he could attempt to specify only in words.

Is this pin a true likeness of that pin? Certainly, an absolutely true likeness – now that you've raised the question in so many words. You've bought a whole boxful of true likenesses! Or you have now I tell you so; until then you'd bought a box of pins. Were they even a box of pins until I or the manufacturers told you they were? Well, it wasn't *true* that they were pins. Or only in the hypothetical sense that *if* the assertion had been made it would have been true.

– Oh, come on! Were they pins or weren't they?
– Of course they were – now that we look back, the question having been put.
– But there must be something about them – there must already have been something about them – something inherent in the pins themselves – which makes it possible for you and the manufacturers to say truly that they're pins!
– All kinds of things, certainly. The fact that each one is (I quote) 'a sharp-pointed piece of wire with a rounded head for fastening clothes, etc.'; the fact that each one was designed and manufactured so as both to function as a pin and to advertise itself as a pin; the fact that the manufacturers labelled them as pins . . . All things that have to be expressed in words. You can't go round the back of the words!
– But I could hold out a pin to you without saying anything, and you could look at it and see, without saying anything, that it was a pin.
– I shouldn't be 'seeing that it was a pin' unless the question of what it was arose. I might of course understand the question from the interrogative look on your face, and from knowing that the identity of the object you had found was in some way significant. Otherwise I should be thinking something more like: Why is this idiot holding out a pin to me without saying anything?

If we could invent a method of projecting our thoughts into other people's minds, though, couldn't we dispense with language?

I switch on the thought-projector, then think – not of anything involving words, of course – but of a strawberry ice cream, say. Into

your head, without my saying anything, comes the thought of a strawberry ice cream. Remarkable, certainly. But not, so far, of much significance. You still need to know, if this is supposed to be a
communication between us, that the thought has been put there by
me, and hasn't just popped into your head out of nowhere. You also
need to know what I'm *telling* you about it. Am I telling you that I'd
like an ice cream, or that I wouldn't? Am I offering *you* an ice cream?
Am I warning you against indulgence in one? Am I reminding you of
the taste of strawberry ice cream in general? With nostalgia? With dis-
gust? If I haven't managed to communicate the *function* of the transmission then the idea of the ice cream is as vacuous as the famous '42'
arrived at by the computer Deep Thought, after seven-and-a-half million years' contemplation, as the answer to the great Question of Life,
the Universe and Everything,[9] when it turned out that no one had ever
exactly formulated the question. A thought, in other words, cannot
really be transmitted unless it forms part of some assertion, question,
offer, warning, or allusion. Could any of these things be expressed
without some form of language, even if it was transmitted by the same
strange telepathic means?

Well, I suppose it's just about imaginable that a wordless craving for
an ice cream in me could somehow evoke a wordless craving for an ice
cream in you. But my craving and your craving are not the same thing
at all! I could perhaps supplement the wordless thought with a wordless gesture or expression. I could raise my eyebrows interrogatively,
when the thought of the ice cream appeared in your head, at which I
suppose you might possibly understand that I'm offering you one, just
as you do if I gesture towards an actual ice cream and raise my eyebrows. I can see that this could be handy, if there was no ice cream
available to gesture towards, and we happened to be in a Trappist
monastery. But isn't this combination of eyebrows-and-gesture, or
eyebrows-and-thought, the equivalent of *saying* 'Would you like an ice
cream?'? Just as it would be if you asked the question in sign language? And still would be even if in sign it happens to be possible to
express it with a single handstroke?

In any case, this thought that I've planted in your head isn't the
thought of a strawberry ice cream unless you know that it is – unless
it has some kind of label on it identifying it as a strawberry ice cream,
even if this label is in some kind of submerged pre-linguistic form
which would translate as 'strawberry ice cream' only if you had to

express it, to yourself or others.

The question arises as to what the thought of the strawberry ice cream, unattached to a statement or a question or other functional context, is like. Would it be the mental *image* of a strawberry ice cream? This sounds solid and self-explanatory, because we're so familiar with images in the external world, and we're so familiar with looking at them and recognising them. And it's true that we can do something that's often described as picturing things in our head. But these supposed mental images are not remotely like physical images. However vivid they seem, they are entirely indeterminate. I'm seeing a strawberry ice cream in my mind's eye at the moment, and yes, for a moment it does seem quite precise. The illusion vanishes, though, as soon as you ask me a question about it. What sort of container is it in or on? At once I rush to supply one. But it wasn't there, this absolutely obvious and unavoidable feature of an ice cream, until you asked, and I invented it. The particular features of the image are determined only when we select them. Even its central identifying characteristic, this vivid pink, slightly deliquescent coolness that I can almost feel as well as see – even this comes only when the question arises, and slips away as soon as the question turns to the dish, or the taste.

Thinking of an ice cream is not like painting a picture of one, or creating another physical object in the world which is not in itself an ice cream but which somehow represents one. The parallel is much simpler. It is with *seeing* an ice cream. Here, once again, there is a single detail that we foveate at any one moment, while everything else lurks indeterminate at the edge of our vision, only to be brought into the spotlight once the question has been asked, or our attention redirected.

And how do I know that this mental picture is not the image of a raspberry ice cream rather than a strawberry one? Or of a Hunka Munka ice cream, made from plaster? – I *don't* know until you ask, and then I know because I decide what it is and what it's not – because I declare its nature, I put my own label on it. Doesn't this strawberry-ness, this non-raspberryness and non-Hunka-Munka-ness, if it is to have any standing in the world at all, have to be expressed in words, or at any rate in a kind of chrysalis just on the point of metamorphosing into words?

The forms of language reveal something about the forms of the world it deals with – if not quite in the way that pictures reveal them, then in the ways that the blades and surfaces of our tools reflect the nature of the material they have been designed to work. They also reveal something about ourselves, rather as the handles and eyepieces of those tools reveal something about the limbs and organs with which we operate them. Modern linguistics has become interested in this second aspect of language, and specifically in what it reveals about the physical functioning of the brain. What concerns Chomsky and his followers is the syntactical structure of language, which they see as determined not by the world confronting us, or by the nature of our confrontation with it, but by the fixed possibilities of our neurology.

They adduce as evidence the fact that the complex procedures of language are acquired astonishingly fast at a certain stage of a child's development (when it's around two), and are impossible to learn at all after a certain point (around the age of ten), which suggests that the window of opportunity occurs as a result of specific physical changes taking place in the child's brain. Even when people who learnt to speak and understand a language during this fertile period are obliged to develop a new language *de novo* as a common ground for talking to people who were brought up to speak different languages, it's too late to construct a new common syntax, so that pidgins have relatively little grammatical structure; whilst the children of these pioneers, if they begin to acquire their parents' pidgin at the age of two as their native tongue, creolise it, and by unspoken consent develop some of the same kind of grammatical structures as natural languages have the world over.[10] This is the central argument: that all natural grammars have an underlying similarity so profound that it can be accounted for only by postulating a common neurological source.[11]

The nature of this deep grammar can evidently not be expressed in so many words; it can only be inferred, like the Platonic ideal or the *Ding an sich* behind the phenomena we experience, from the nature of the various emergent natural grammars. Chomsky and his followers are not saying that all natural grammars are the same, but that there are various possible alternatives for them offered by the universal grammar, which they either take up or do not; though Derek Bickerton[12] goes further, and argues from the experience of creolisation that the human mind is programmed to take up one particular set of grammatical possibilities unless this is overridden by local variants.

All revelations of occult causes underlying the outward appearances of the world have an immediate charm, whether they explain politics or neurosis, and they all need to be looked at with some suspicion. To what extent are they capable of being falsified? (Popper's test.) What are their implications? Can the phenomena be explained in some other way, without resort to the occult?[13]

Language may be the profoundest human activity, and the one that most clearly defines our nature. But it is not the only skill that we develop, or fail to, at a certain age, and it is not the only activity that takes a similar form wherever it appears, even when it appears spontaneously, without any apparent cultural continuity. Do we need to postulate occult innate rules for the non-linguistic crazes that sweep through communities of children at certain ages – conkers, marbles, trainers, transfers, and so on? Or do we think they are more likely to start with one or two particular children in some particular place trying out some particular new thing, which is then copied by the children around them, extended, systematised, and radiated outwards by the natural geometrical progression of human acquaintance? Some of these crazes reappear from generation to generation – breaking out again and again from nowhere like underground fire. Does *this* suggest innate rules? Or does it suggest another constant – the availability of the same materials (conkers, marbles), the unchanging possibilities of which keep catching the imagination of new groups of children afresh (with the seed of some new outbursts perhaps being sown by one particular parent who remembers the game from his own childhood)?

Most of our skills, like language, are harder (or impossible) to learn after certain crucial ages, and (unless they depend upon an idiosyncrasy such as double-jointedness or total recall) display an even greater uniformity than language across all races and societies. Some of these skills, such as playing the piano and balancing a billiard cue on the tip of your nose, are acquired only by certain individuals. Others, such as crawling and walking, we mostly take for granted.

Think about walking, though. It's a remarkably complex activity, involving the coordination of a wide range of muscles and responses. Yet all human beings walk in much the same way, by putting one foot in front of the other; there is a common syntax of walking. Within this syntax many different types of task can be performed, many different moods or character traits can be expressed. You can walk quickly or slowly, purposefully or aimlessly, eagerly or reluctantly. You can walk with your bottom stuck out, or march like a guardsman. But all of this depends upon the basic grammar of putting one foot in front of the other, of transferring the weight from foot to foot, of continuously varying the geometry of the ankle in the process so as to keep the sole of the foot aligned with the ground, and so on.

Where does this grammar come from? It surely developed over the millennia in parallel with the development of our legs and our need to move around to find our food. Does it reappear afresh in much the same form in each one of us as we learn to walk, as the ever-renewed expression of the same basic relationship between the ground beneath us, our purposes, and the muscular equipment we are endowed with? Or is it the outward expression of a deeper set of instructions, already waiting in our brains at birth? If we need a universal underlying grammar to explain the syntax of language, then we need another one to explain this no less puzzling homogeneity of behaviour with our legs, from the commodity traders of New York to the head-hunters of New Guinea.

A deep grammar of walking . . . Certainly, you say – why not? So, I'm walking down the street, unconsciously translating this inaccessible inner code into the familiar outward pattern of footsteps, when I see you ahead of me, and break into a run to catch up with you. Am I revealing the existence of another universal grammar inside my brain? Or is this a natural extension of the walking grammar? So delighted am I to see you that I skip a few steps in sheer high spirits, which causes me to stub my toe and hop a few steps in agony. Am I still operating within the syntax of walking? Are skipping and hopping also hard-wired into me? Have they, like walking and talking, been selected because of the advantage they offered my forebears in surviving and propagating their genes?

Well, possibly. Hopping and skipping might have been attractive to the opposite sex; or children might have become more cheerful, and therefore more likely to survive, if they saw their parents hopping and skipping. It turns out that you're going ice-skating, though, and I decide to join you. You, of course, glide effortlessly about the rink in the manner of ice-skaters in all countries that have access to ice,[14] and there is no great mystery about this – it's because you must have tapped into the deep grammar of ice-skating at the right age. I missed the chance, however, in my underprivileged childhood, and instead I go stumbling ludicrously about the rink. All the same, there is something painfully familiar and predictable about my antics . . . Yes, I believe that I'm stumbling around in exactly the manner that people all over the world who can't skate stumble about an ice rink! I am revealing to any watching structuralists the existence of a deep grammar of non-skating!

Are the deep grammars of skating and non-skating hard-wired into our brain?[15] It's difficult to believe that the ability to glide about sheets of ice on steel runners has proved of any great selective advantage; more difficult still to believe that the ability to wobble about on it and fall over has. But then the deep grammar of walking was presumably not hard-wired into us until we had legs to walk on. Not, at any rate, the grammar of walking on two legs. Before that we must have had a four-legged grammar – a remarkably different grammar, if a horse is anything to go by. Is this four-legged grammar still dormant in our brains? Of course, you tell me – how else could babies learn to crawl? So when one of my daughters began to crawl, and did it by sitting sideways and sliding her legs along behind her, was this because she had been born with another, aberrant or alternative, grammar of crawling in her brain? Or was it simply because we happened to have polished wooden floors, which offered her the possibility of creating a different grammar of crawling from children brought up on fitted carpets or beaten earth? Do we have to think that fitted-carpet-beaten-earth crawling reflects some more profound inner reality than polished-floor crawling? And when someone's leg is amputated, does it reveal yet another deep grammar, of walking with one leg and two crutches, or with an electronically controlled prosthesis, which was lurking somewhere in the brain all the time, just waiting for the eventuality?

Language shades into related activities as imperceptibly as walking does into running, skipping, and skating. I use words – but at the same time I flesh out my talking by making gestures, by the look on my face. I point. I draw figures with my finger in the air, and with a stick in the dust. I leave implications to your imagination. I may say formally 'The pen of my aunt is on the table of my uncle,' but, if you're already acquainted with my family's furnishings and personal possessions I may equally well convey the same essential information by nodding towards the table. Is the syntax for that somewhat dubious set to the mouth as you read this hard-wired into you? For that expressive sigh?

Language, as a symbolic form capable of expressing a proposition, includes not just words but any combination of snorts, grunts, whistles, frowns, mimings, and pokes in the ribs that will serve the purposes in hand. A pictorial representation is incapable of expressing an assertion in itself, but it might become part of an assertion where your words or some other behaviour on your part make it clear that an assertory sense is intended – if, for example, you put the snapshot of me and your wife before me with an accusing look in your eye; or where the context does – if, for example, you hold up a pin, with no look at all, after circumstances have made it clear that the finding of an inconspicuous sharp object will solve the mystery of how the botulin got into the baronet.

In the appropriate context, silently pointing at someone asserts that he is the man who committed the elaborate fraud which it takes a skilled barrister three weeks to explain to the jury. Shining a spotlight on someone else asserts that he is the best male supporting player under twenty-five in a musical comedy produced by a provincial theatre. What could Δ possibly assert? – What *couldn't* it? On a lift it asserts that the lift is going up; in a mathematical formula that the following variable is to be read as a difference or increment; on a map that a trigonometrical survey point is to be found at that location; on a care label that the garment can be dry-cleaned . . . And if you shine the spotlight on the right actor then you are expressing the truth; if you put the triangle on a garment that can't be dry-cleaned then you are uttering a falsehood.

Walking (whether or not it is controlled by a code of rules) has emerged as the solution to a complex equation of many terms, involving manifold trade-offs between length of leg, food input, the values for the load-bearing indices of bone and the elasticity of muscle, between speed and endurance. The equation goes on being rewritten all the time, like a weather map, as new local elements are factored in and out: ice underfoot; snow; snow plus skis; precipitous ground; longer and shorter legs; Chinese foot-binding; stilts; blisters . . .

The tool is shaped to fit the hand that holds it; the hand is calloused by the tool it holds. Man makes the clothes; clothes make the man. Syntax determines our thought and relations with the world; our thought and relations with the world determine the syntax. Do we need to possess, or develop, a set of rules to play the piano? Each of us has a selection from a somewhat indefinite range of possible manual and cerebral capabilities (developed by selection, certainly) which some of us then painfully refine and adapt, according to a tradition gradually developed over the generations, in order to play an instrument which was in its turn gradually evolved to exploit first those same basic capabilities, and then the refinements and adaptations of them that the generations have found to be possible. Piano-playing is a three-way contract between the player, the piano-maker, and the composer; and this main contract is buttressed by a series of sub-contracts – between the player and his predecessors at the keyboard, the maker and *his* predecessors, the composer and his. The various sub-contractors deliver the techniques that they have developed over the centuries for training fingers, connecting fingers to hammers and strings, and selecting the notes that fingers and strings can between them produce; the player then delivers his ten fingers, modified by inputs from his sub-contractors and his own efforts; the manufacturer his machine likewise; and the composer his selection of notes.

Then comes the First World War, and one of the parties to the contract, the pianist Paul Wittgenstein, gets his right arm blown off. Now he can deliver only five fingers instead of the specified ten. So for him the contract has to be rewritten. And since no one can think of a way to adapt either the technique or the keyboard so that ten-finger music can be played with half the number of fingers, it's the musical input that has to change, and Prokofiev, Ravel, Strauss, and Hindemith agree to supply concertos written for the left hand only.

Are the grammars of different languages as fundamentally alike as writers on linguistics always claim? On the face of it there are considerable syntactical differences between even the tiny handful of Indo-European languages that I'm familiar with.

In all West European languages, for example, it is possible – and important – to differentiate by means of a definite or indefinite article between an unspecified member of a named class and one particular member of it. In Russian, however, no such distinction is apparently felt necessary, and no such distinction is possible. *Муж убил жену* (husband killed wife) expresses the staggeringly different assertions (to a West European way of thinking) that some husband or other killed some wife or other, that the particular husband we have been discussing killed someone's wife, that some unspecified husband killed the very wife we have been talking about, that some unspecified husband killed his own wife, that this very husband killed this very wife, and that this very husband killed his own wife.

Then again, every West European language has an elaborate machinery of either inflections or auxiliary verbs or both to express conditionals and subjunctives, and so to distinguish between what did happen and what would happen and what would have happened; between what is the case, what isn't the case but could be, and what isn't the case and couldn't be; or (in German) what is said to be the case; or (in French) what might be made to be the case by commanding, wishing, praying; etc. In Russian none of these shades of possibility emerges at all. One single monosyllabic particle distinguishes the entire universe of subjunctives and conditionals from the actual universe. *Муж убил бы жену* tells us that a/the husband would have killed a/the wife, or that he would still kill her; and if that little бы is moved forward one place in the sentence it even suggests a wish that he should kill her, or an obligation upon him to do so. Or to have done so. On the other hand, Russian contains elaborate possibilities, unavailable in West European languages, for distinguishing various modes of moving from one place to another so as to indicate whether it is by one's own efforts or by the efforts of some animal or machine,[16] and so as to separate out in each mode the difference between making the journey once, being in the process of making it, and being in the habit of making it.

So presumably the universal grammar must contain the conceptual mechanisms for all these things – for definite and indefinite articles, for conditionals and subjunctives, for what Russian philologists call verbs of determinate and indeterminate motion.[17] Some of these submerged archetypes have been realised in particular natural grammars, and some have not. It would help to establish this provenance if one could think of some plausible reasons for the really quite profound divergences that have emerged, in the way that we can for why white fur should have had an adaptive advantage in the Arctic, or why a long tongue should have developed in an anteater but not in a sheep. Could the failure of the article to survive into Russian be accounted for by the notorious lack of respect for the sovereignty of the individual which seems to have marked Russian political life down the ages? Has the traditional Russian resistance to reform atrophied the wealth of grammatical mechanism that enables West Europeans to think about the contingency of the past and present, and the alternatives available for the future? On the other hand, does the poverty of our own inheritance from the invisible stock of grammar to do with travelling perhaps reflect the insignificance of journeys from London to Leicester, and even Antwerp to Augsburg, compared with being in the process of marching in chains, or of riding in an unsprung cart, or with making a habit of either, from Petersburg to Khabarovsk?

Well, possibly. It's true that one of my Russian instructors was a former major in the Soviet army who as a result of some administrative cock-up had spent a substantial part of his life commuting between Moscow and Vladivostok on the Trans-Siberian Railway, a journey which then, I believe, took fourteen days; and I did begin to see why it would be natural for him to want to distinguish between the miseries of being in the midst of it, and the larger miseries of being in the midst of a lifetime of being in the midst of it.

Are you convinced by this? If so, mightn't you be equally convinced by the converse argument – that the Russians failed to develop a respect for the sovereignty of the individual because they didn't have a definite article? As long as the police arrest *person*, who's going to notice whether it's what pedantic Westerners would call *the* person, i.e. the particular person who is suspected of an offence, or merely what they'd regard as *a* person? Or that they never discovered much of an interest in reform because they didn't have proper conditionals or subjunctives? And, by the same token, that West Europeans failed to travel very far because they didn't have a full deck of verbs of determinate and indeterminate motion? In which case, why didn't they develop them when they began to establish empires in the New World and Australia? Or, alternatively, how did they ever manage to found overseas empires without the verbs?

Russian, English, etc., are what Bickerton calls 'cultural languages' – languages developed according to local traditions in different societies – so even a Chomskyan expects some variations. The syntax of creole languages, however, he argues,[18] is developed direct from what he calls the 'bioprogram', the neurological operating system in all our brains upon which local cultural languages are constructed, and upon which we can fall back if the local linguistic tradition is destroyed, as it was in the case of the polyglot societies created by slavery. Some creole syntax does seem appropriately 'natural' (at least to English-speakers) – for example, the apparent ubiquity of the subject–verb–object word order. Other usages seem as profoundly idiosyncratic as any of the usages we have noted in 'cultural' languagues. Bickerton records,[19] for example, that in Hawaian Creole English there are two 'sentential complements', *fo* and *go*, both of which are used with an infinitive form in the same way as 'to' is in standard English. But *go* is used with realised events (*ai gata go haia wan kapinta go fiks da fom* = I had to hire a carpenter to fix the form) and *fo* with unrealised events (*pipl no laik tek om fo go wok* = People don't want to employ him). You can see a distant cousinage here with the familiar indicative/conditional marking of verbs, but Bickerton records no trace in any other language of the distinction being made here with the infinitive. So are we to suppose that the rule for this *fo–go* distinction is lurking somewhere inside all of us, but in otherwise universal abeyance? Do we need another rule to explain why *this* apparently basic rule has been so pointedly dropped?

If the really quite profound dissimilarities between the various emergent natural grammars are not enough to falsify the theory of a common universal grammar, then the question arises as to what could *ever* falsify it. There must be *some* possible evidence that could count against it if it is to pass Popper's test. Could one invent a grammar whose origins would *not* be explained by the rule?

I strain my imagination to think of something that might serve for the purpose. All right, let's look at certain aspects of Middle High Neverlandish. In this language there is a *putative case* which indicates that the noun in question only supposedly names what it appears to name. There are *oblivative pronouns* referring exclusively to things and people mentioned earlier but now forgotten about. Verbs have an *aspirational mood*, as in 'The cat o-sat on the mat', which indicates that the statement is made in a devotional frame of mind, and also an *improper mood*, as in 'The cat a-sat on the mat', which indicates that some indecent meaning is to be read into it. Then there are the so-called *insinuative conjunctions*, used for linking names with purely scurrilous intent . . .

Useless, though. All these features can be found in the existing great universal grammar, and if they have failed to surface in any language except a fictitious one then this must be for a number of purely contingent reasons, such as their having no discoverable usefulness in human affairs. Or so I believe. And if you believe otherwise, and you can't produce any more evidence for your belief than I can for mine, then we may be reduced to settling the matter in the same way as Onegin and Lensky.

*

Another question suggests itself: how truly fundamental are the supposedly fundamental features that actual languages are said to share?

Much is made of the 'natural' word order, subject–verb–object, which is said to have appeared many times quite spontaneously in the formation of completely different creoles in different parts of the world. If this really is fundamental, though, why don't all the natural languages of one's acquaintance collapse when the order is changed, in the way that a house collapses if you take out structural walls and beams? In various Romance languages the accusative pronoun normally precedes the verb as a matter of course, and no Frenchman or Italian raises an eyebrow. In more highly inflected languages, like Russian and Latin, the case endings make grammatical function clear even if you scatter the words around like confetti (*Arma virumque cano, Troiae qui primus ab oris / Italiam fato profugus Laviniaque venit / Litora . . .* – Arms man-and [I] sing, [of] Troy who first from shores / [to] Italy [by] fate driven Lavinian-and [he] came / Beaches . . .) Even in staid, pragmatic, relatively uninflected English those structural joists are thrown on the skip without a second thought if the house would look nicer without them. *In the beginning was the word . . . This I believe . . . Fridays I can never manage . . . Touch me who dares . . . 'Says who?' says he . . .* Standard English word order is in any case changing in front of our eyes. Diamond, noting that many languages, including creoles, preserve the subject–verb–object order in questions, which are distinguished merely by a different tone of voice, says 'the English language does not treat questions in this way', but instead inverts the order either of the sentence or of a phrase using the auxiliary verb such as 'do'. Really? The example he gives of a non-English creole question is 'You want juice?' But 'You want juice?' is now standard colloquial British English (having entered, like so many other usages, from American English), and in my observation more commonly used than 'Do you want juice?' Tone of voice rather than inversion is the indicator for a huge range of English questions – and was even in the past. ('You off?', 'The bus leaves when?', 'I'm supposed to believe that?', 'The pound is 1.9 to the dollar?')

Then again, linguisticians are very impressed by the convention that allows qualifying clauses to be embedded inside one another in an infinitely extendable hierarchy. ('Once upon a time there arrived in the little town of N, which was in those days thought to be, by all who lived within its walls, which had been erected by a former ruler who believed that the cause of freedom, which he had espoused in the hope, a hope never to be fulfilled, such being the common fate of hopes entertained by mortal man, who . . .' And so on, deeper and deeper into the forest of qualifications, until at last we reach the centre and begin to emerge again, layer by layer, through the completions of the clauses that were broached on the way in, and we get back to: ' . . . is doomed to see all his hopes burst like soap bubbles, that one day all men would be brothers, was more important than the wishes of his subjects, to be the centre of the world, a TRAVELLER, who dimly recalled arriving in the town 397 nested clauses earlier, but who had by this time completely forgotten what he was supposed to be doing there.') All human languages, we are told, share this convenience.

Well, it's also theoretically possible to go on qualifying the same noun with infinitely many adjectives, or modifying the same verb with infinitely many adverbs, or using the same subject to govern infinitely many verbs, just as it's theoretically possible to balance infinitely many tins of baked beans one on top of another. Does this reveal further aspects of the deep grammar of language, or of baked beans?

In practice shopkeepers don't pile more than a dozen or so tins on top of one another, because they would fall over too easily. In practice we don't pile up more than three or four adjectives, or adverbs, or verbs, because it would make it difficult to follow the thread of the sentence, nor more than one or two embedded clauses, for the same reason. So the infinite extendability of such a hierarchy is purely notional, and quite beside the point in explaining the utility of the convention.

Does the embedding of clauses express anything that can't be expressed without this piece of occult machinery? Any good copy editor would break down the complex of embeddings in the example above, if an author had been fool enough to attempt it, into its constituent parts. What does the original assert about the world that is not asserted by the sum of these parts unconnected by any articulation? 'Once upon a time a traveller arrived in the little town of N. The town was in those days thought to be the centre of the world by all who lived within its walls. The walls, incidentally, had been erected . . . etc.' What possible colour or shade of meaning has been lost in the translation, except perhaps a misleading suggestion of interconnectedness, or an only too accurate one of confusion and incomprehensibility?

It's possible to embed brackets in mathematical expressions in a very similar manner, again with no theoretical limit to the number of levels, again with a very low practical limit. Does this represent something profound about the structure of our minds? Then again, a qualifying clause can be expressed as a footnote. Is the grammar of footnotes innate? To reproduce the effect of multiple embedding we could extend the convention of footnotes to include footnotes upon footnotes:

'Once upon a time a traveller arrived in the little town of N.[1]

1. The town was in those days thought to be the centre of the world by all who lived within its walls.[2]

2. The walls had been erected by a former ruler[3] . . .

Is the grammar for this new way of using footnotes already present in the depths of our minds, waiting to be put into practice? And what does our patent inability to follow more than a layer or two of embedded clauses tell us about the working of our brains? Is this limitation also the expression of a rule in the deep grammar?

Embedding clauses has the air of being more grammatically structural than listing them, just as inflection seems more profound than the combination of word order, prepositional constructions, auxiliary verbs, and guesswork upon which English in its decay tends to depend. But 'Man's inhumanity to man' is no more structurally complex an expression of the idea than '*L'inhumanité de l'homme envers l'homme*', even though it has one of our rare inflections to express the genitive, any more than it is less structurally complex than the Latin '*Homini hominis inhumanitas*' because the latter also has the benefit of an inflected form for the dative. The embedding, and the inflection, are mere conveniences, short cuts, just as the embedding of brackets in mathematical expressions is a neat way of showing the order in which the various parts of a mathematical operation have to be performed. There is no more structural significance in these modifications than there is in the hooks and loops used in shorthand to render common prefixes and suffixes.

When we are agitated, or the situation is urgent, the grammatical form may be simplified to the point of vanishing altogether. 'Man! Bag! Gun! Run!' I scream. And you run. Did you have to pause for a moment first while you translated my unstructured words into: '(A man [who is carrying a bag {which contains a gun}] is approaching), which makes it advisable to run'? Is Mr Jingle translating out of well-articulated forms before he says, 'Of course – she don't like it – but must be done – avert suspicion – afraid of her brother – says there's no help for it – only a few days more – when old folks blinded – crown your happiness'?

Pidgins have little or no grammar. The lack of grammar makes them cumbersome, but they work. If they didn't work they wouldn't be spoken. Jared Diamond gives a good account of how he and an Indonesian fellow scientist, with whom he had no common language, began developing a private pidgin when they were dropped in a remote part of Indonesia together to study birds:

> Within a week we had evolved a crude pidgin, based solely on Indonesian nouns, to communicate about camp chores: for instance *rice fire* meant 'to cook rice', while *bird binoculars* meant 'to watch birds' . . . [We] were just in the process of reinventing prepositions when the helicopter picked us up and terminated our experiment in pidgin evolution. We had begun to assemble word strings that functioned as locative prepositional phrases but were still composed solely of nouns with concrete referents – strings such as 'spoon top plate' and 'spoon bottom plate', to mean that the spoon was on or under the plate. Many virtual prepositions in Neo-Melanesian, Indonesian, and other creoles are similarly constructed.[20]

Diamond says that 'rice fire' *meant* 'to cook rice' and 'bird binoculars' *meant* 'to watch birds', which is a convenient way of explaining to English-speaking readers the use of the phrases. And maybe, in those first few weeks of their collaboration, he did have to translate the new language in and out of English to himself. But in time the English would have dropped out of the picture, and it would have been plain that what the phrases *meant* was not the English, but what the English also meant. So that what he and his colleague did during the day was to birdbinocular; what they did when the day's work was done was to ricefire. After they had finished ricefiring they put rice top plates. Is 'top' still a noun in this context, simply because it can in other contexts indicate an object? Or has it become a preposition? Or is it a noun functioning as a preposition? Does it matter, provided both parties understand that it relates the rice to the plate in such a way that gravity keeps it there?

What makes this new language work is that its two speakers share not a grammar but a world. In this world rice and birds are important.

So are cooking the rice and watching the birds. So, therefore, is *getting* the rice cooked and the birds watched. Ways have to be improvised to achieve this, whether with words or with matchsticks and rubber bands; and because they have to be they are.

You might say that there is still a grammar lurking beneath the surface. We still need to understand that the fire is heating the rice and not the rice extinguishing the fire, that the binoculars are trained on the birds and not the birds perching on the binoculars. But the means by which we understand this have nothing to do with grammar. We can make a pretty good guess that 'ricefire' signifies heating the rice and not extinguishing the fire because we have both used many fires before to heat much rice, and have never used rice to extinguish the fire. Our interpretation is reinforced because we are both hungry and want to eat, and because neither of us has any reason to want the fire out.

Now you object that before the concepts of ricefiring and bird-binoculuring have any practical utility we still need to understand various other things that can be indicated only by some kind of grammar, such as which of the two men is doing the ricefiring and the bird-binoculuring. We need to know whether his colleague is telling Diamond *that* he is birdbinoculuring or *has* birdbinoculured or *will* birdbinocular; whether he is asking Diamond *if* he is birdbinoculuring or has birdbinoculured or will birdbinocular; if he might, could, or should; might have, could have, or should have; or whether he is requesting Diamond to birdbinocular. True, but if all this is not made clear by the convenient gadgets of a developed grammar, such as inflections, particles, and auxiliary verbs, then some other way has to be found to do it – and *will* be found. Context and plain common sense, mostly, once again, supplemented no doubt by gestures and facial expressions.

So perhaps Diamond's colleague says 'ricefire,' and at the same time hands over the chef's hat and apron, which probably suggests to Diamond that the activity in question is in some way being transferred to him this evening. The action indicates, even if only ambiguously, a request or command, or perhaps, if the colleague raises his eyebrows at the same time, some kind of enquiry: how would he feel about doing the ricefiring tonight, or what are his opinions on ricefiring in general?

Do these gestures and expressions constitute a grammatical construction? Or is the grammar external to the gestures and expressions

– a non-linguistic, non-gestural entity that the gesture ambiguously expresses, and which a developed natural language might express more precisely? An entity that it might express more precisely still if used by an educated speaker who fully understood the shades of meaning implied by the correct usage of 'will', 'shall', 'should', 'would'?

Some of these grammatical distinctions are vanishing from the English language in front of our eyes – the nuances of 'will' and 'shall', for example, and the inflected accusative 'whom'. Are the distinctions still *there* somewhere, disused but resumable, like mothballed aircraft-carriers? There in our heads? There in the world? In logical space?

Could I for that matter reasonably write a Letter to the Editor complaining that important grammatical distinctions are being obliterated these days because of the sloppy way young people raise their eyebrows?

Isn't some rudimentary syntax necessary to express an assertion, at any rate in verbal form? A proposition seems to require at the very least the presence of a verb. The dynamic quality of the verb, its doingness, does seem to make a claim that something is going forward in the world. In fact you can imagine an assertion being made, and a proposition expressed, by a verb and nothing else. Stage directions are often written in this way: (*Rises*), (*Thinks*), (*Exits*).[21] Here, of course, a subject is understood from the context. But an assertion is certainly made by an impersonal colloquial construction such as 'Hurts', or 'Raining again', where the understood subject is merely a grammatical fiction.

The verb itself might be reduced to a very low energy state. Think of the modest copula in 'Socrates is a man'. No activity is asserted here, nothing is going forward. The 'is' merely attaches subject to complement, an individual to a class. (Imagine finding it as a stage direction: '*Nonchalantly sips his cocktail. Is.*') But it is surely what carries the assertory weight of the proposition. It is the minimal verb, the verb reduced to its grammatical essence. It is a scrap of nothing that we *understand* to be a verb; a grammatical function without semantic flesh. Without this irreducible minimum there would surely be no proposition. 'Socrates man' would surely assert nothing.

It *might*, though. It would, perfectly well – if we understood it to. It does in Russian, which mostly makes do without a copula: Сократ - человек. Do we have to understand the dash as being some kind of unconjugatable verb? And if even the dash is left out, as it most frequently is, do we first have to understand a dash before we understand the dash as a verb? Well, if the participle 'Raining' expresses a proposition, because we understand the 'it is', could the same proposition be expressed by the bare noun, 'Rain'? In the appropriate context (you lift your head and listen for a moment before you speak, for instance) I certainly understand you to be diagnosing exactly the same state of weather as I should if you said 'Raining'. Do I quietly have to add an '–ing' to the end of the noun inside my head before I grasp this?

Does 'Man! Bag! Gun!' really assert that a man who *is carrying* a bag *is approaching*, and that the bag *contains* a gun? It asserts exactly the same as the long form, if we understand from it that a man with a gun in his bag is coming! And it's true if he is, false if he isn't. Do we have to understand it in fully syntactical language? No, if we run then we plainly understand it. If we run and there *is* a man coming with a gun, then we have been warned just as successfully and specifically as if the sentence had been fully articulated. If we run and there *isn't*, then we have been deceived or misinformed.

There are two very good reasons why the grammatical structures of different languages have so much in common. The first is that the creatures who use those different languages have so much in common quite apart from the language centres of their brains. The second is that the worlds confronting those creatures also have so much in common.

If subject–verb–object frequently seems to be the 'natural' word order in uninflected languages, then it's surely because it expresses the natural sequence, the logical (and often chronological) sequence, of most actions. There is an originator of the action, who or which is usually in existence before its commencement; then the action occurs; then the action exerts an effect upon its object. This 'natural' order is often varied for rhetorical reasons, to focus the interest of a sentence upon a particular order. Linguisticians believe that this can only be explained by positing special 'movement rules'.[22] Do we then need to suppose that we also have rules to explain how the 'natural' order of any other set of human actions might be varied to dramatise them? How, for instance, a dramatist shows in Act One the gun that will be used in Act Four? How the author of a detective story conceals the identity of the murderer until the end of the last chapter?

Pinker says that we should be surprised at the universality of the division between subject and object. I don't see why. It reflects a division that exists in our actual relations with the world at large, not just in our minds, and which long predates language and grammar: the profound and objective difference between the 'I' who wields the club and the 'him' it descends upon, between the 'he' who decides to retaliate, and the 'me' who feels the resulting blow on my skull.

To explain why there is a genitive case in so many languages, for that matter, we don't need to postulate some special piece of neurological hardware that manufactures genitives. It's explained perfectly well by the fact that the relationship of possession is important to so much of our activity, and it's natural for us to develop a shorthand to express it. There are locatives because the position of things is so often relevant to the way we deal with them, and cases to express motion towards because the point of most of our movement is to go towards some objective; instrumentals because we need to explain by what means we perform actions. There are adjectival clauses because the people and objects that we deal with are identified by, or need to be associated with, particular actions and states of affairs. Verbs because so much of our life consists in doing things and having things done to us; past, present, and future tenses because our actions lie in the past, present, and future; first, second, and third persons because all my communication and yours is between a first person and a second, and may concern a third.

Maybe the Chomskyans are right after all, though, and one day neurological exploration will actually identify a single common human grammar, evolved in the way that everything else about us was evolved, from a series of random genetic modifications.

I shall stand abashed. But will it really change the logic of the situation? However and whenever the syntax of an activity was acquired, whether it arose from mutations that occurred a hundred thousand years ago or whether we are constructing it purposively even now, much as we work out a way of using our new electric eyebrow-trimmer, the reason for its taking up residence in our procedures is the same: the fact that it works. It was or is adopted if it succeeds in relating the functions that we need to perform and the possibilities for performing them that are offered by the material at our disposal (including our own physical and mental equipment). If we settle on holding the eyebrow taut with our left hand as we trim it with the right, this is because it's the most convenient and effective way of doing the job. If our predisposition to form genitives and locatives is innate, then this is because creatures with brains which happen to have evolved such a tendency have survived well and propagated in a world where genitives and locatives have a useful application – and if they have a useful application then it is still for the simple reason that they reflect objective aspects of our world and of our dealings in it.

There is a traffic, in other words, between you and me, between me and the world around me, between you and me and the world around us both, between me and myself, and between you and yourself; and the syntax of our activities, language among them, is a reflection of the activities that constitute the traffic.

2

The Rule of Rules

Which comes first, the soup or the recipe?

Rules. These are what constitute the fundamental fabric of language, in the view of Chomsky and other writers on linguistics. Rules upon rules upon rules. A gigantic storeyed structure of rules, lodged inside our brains, like a Piranesi dungeon.

Rules on some of the levels govern the physical formation of words, on others the way in which the words are assembled syntactically to convey meaning. And not only is the grammar of each of the different languages that people speak governed by its own particular set of rules, but each of those grammars is derived by a further set of rules from the universal 'deep' grammar, shared by all human beings – which is in its turn governed by yet another stratum of rules.[1]

There is plainly an analogy between the rules of language and the scientific laws that are held to be constitutive of the physical universe; but also a fundamental difference. The laws of science, if they really are constitutive in this way, must be independent of mankind, and must have in some sense pre-existed both us and the fabric that they have woven; the problem, as we have seen, is in understanding what substance they have outside the human formulation of them, which came very late in the history of the universe. The rules of grammar, on the other hand, relate to a human activity, and presumably evolved (in one way or another) only in tandem with the human mind.[2]

All this seems very logical and computer-friendly. But the concept of rules, like the concept of laws, is more complex than it appears, and more elusive. There are various sorts of formulations, arrived at in different ways and with different functions, that go under the same name. If there are rules governing the formation of words and their syntactical assembly, for example, then they are not like the rules governing the use of equipment in a municipal playground. The latter are a guide to our conscious behaviour which we have the option to follow or to ignore, and which have to be imposed on us by sanctions, cajolery, conscience, etc. The former are procedures over which we have no conscious control at all. They are like the instructions that computer programmers write (except that they have not been written!), which the machine has no choice but to follow.

Leave aside for the moment the rules in and around the universal 'deep' grammar. What sort of rules underlie the use of our familiar natural languages? Take an elementary concrete example. There is a rule in English that to form the negative you use a single negating term, such as 'not', 'nowhere', 'no one', etc.; but never more than one ('I never saw no one', 'I don't want nothing'), which is held to negate the negation, like a double negative in arithmetic, and to make the sentence positive. There is a rule in French, however, that to form the negative you require not only a negating term, such as *'pas'*, *'point'*, *'jamais'*, etc., but *also* a negative particle, *'ne'*, in front of the verb. What is the derivation of these rules, as copied from generation to generation of English and French textbooks? In the first place, presumably, observation. The authors of the earliest English and French grammars simply wrote down the way that they and their friends made positive sentences into negative ones. The achievement of the grammarians was to see and to codify the similarities between one negative sentence and another in each of the languages. The rules they wrote down are simply descriptive – and like other descriptive laws they fail to cover a mass of anomalies and exceptions.

If there were regularities of this sort for the grammarians to articulate, how did they develop? Was it some kind of structural similarity in the brains of Mr and Mrs Flintstone that was there even before they attempted to communicate with each other? Well, was it some kind of similarity in the brains of Jared Diamond and his Indonesian colleague that produced the regularities in their private pidgin – the consistency in the meaning of 'ricefiring' from one day to the next, and the understanding that nouns could have a prepositional function? Only in the general sense that they also had the same kind of digestive systems to feed and the same kind of limbs to do it with. The specific cause of the regularities, though, was the range of common purposes for which the language was being used, and the need for consistency to make the phrases function. Isn't this how the Flintstones developed the grammar of negatives? – Because sometimes they had to find a way of marking the absence of firewood in the cave (say) and doing something about it. Once Mr and Mrs had hit upon a way of indicating it to each other, Mrs used the same method for passing the information on to her children, who had to go out and find some; and the children explained to their friends in the next cave along by repeating the technique why they had to go out looking for firewood instead of coming for a sleepover.

So that was one bit of negating done. Now Mrs Flintstone wants to stir her husband into a little economic activity by telling him that, although there's now firewood in the house, there's no food to cook on it. How does she manage it? Well, perhaps by adapting the same kind of formulation that they established for communicating the lack of firewood. And so on over the years, adapting the same means to discuss the absence not only of household necessities but of their deceased relatives, hope, the magic in their marriage, the meaning of life . . . So the same technique spreads from one person to another, from one particular application to another, with random errors in transmission occurring from time to time, just as they do in the transmission of genetic information.[3]

Whether it happened exactly like this with Mr and Mrs Flintstone I don't know – I wasn't there. But Jared Diamond and his colleague *were* present as the grammar of their language began to take shape. And I've seen modern additions to English grammar evolving around me during my lifetime, in much this sort of way, spreading from person to person like a craze, and used in more and more different situations. For instance, the use of a particular conjunction, 'like', to incorporate an illustrative performance, usually of some set phrase of dialogue, into a spoken sentence. I can't remember when I first heard odd individual uses of this – perhaps twenty years ago – but now it seems impossible for anyone to recount any narrative without the construction. 'And she comes up to me and it's like, "You're not going to believe this!" And I'm like, "Oh, *no*!" And she's like, "Did I say something wrong?" And everyone in the room's like, "*Oh* oh! *Trouble*!"'

When grammarians write down as rules the regularities that they have observed, what function do these rules serve? Well, it might be a straightforwardly descriptive one, like the rules themselves. They are useful to a philologist, for example, who is trying to compare one language with another and trace its evolution. As description, though, the rules are often wrong – they don't describe the way the language is actually used. Even though the rule in the grammar books says that the English negative is single, we all know that huge numbers of English-speakers use a double form, like the French ('You don't know nothing', 'He ain't going nowhere').[4] Why does this usually go unrecorded in the rules? Because the people who write them are proposing them not only descriptively but *prescriptively*, to suggest how the language *should* be spoken. They know that most of the foreign students who will use the grammar are not motivated simply by an academic interest in English semantics. They want to learn to speak the language – and not, on the whole, with the intention of working in the kind of trades where the double negative is commonly used, but in business and the professions, where the single negative is regarded as standard.[5]

The function of the rules is not all that different when they are used in England, where they are taught not to small children first learning the language, but to older ones (and to upwardly mobile adults) who speak the language already, in order to guide them towards the particular forms of the language that are regarded as standard. Once consciously learned it is expected that the rules will change their nature – that they will become internalised, and settle into being unconscious procedures. They are not *necessary* as teaching aids. Foreign students can (and often do) perfectly well dispense with them, and learn English in the way that the English learn it, by the 'direct' method of hearing it spoken and imitating the usages unsystematically; the rules are simply a set of practical instructions for use as a short cut. Nor is it absolutely necessary to observe them, consciously or unconsciously, in the way that it is to observe the rules for long division, which doesn't come out right if you don't. The rules of English grammar can be, and often are, ignored or flouted, and you can still make yourself understood reasonably well even with a very defective command of it. Indeed, their use with English-speakers implies that they are often breached even in England, or else there would be no point in enunciating them. What need would there be of a law against murder if no one ever committed it?

So much for the rules of grammar in natural languages. What about the rules of the deep grammar, and of the transformational grammar that regulates its expression in particular natural grammars? Are these also descriptive? Are they derived from observing how some language is actually used? How can they be? There is no language of which these grammars codify the rules. No one could transcribe them from seeing them in action because no one ever has. They are inferred from the usages of natural languages.

What is the origin of these pre-existing grammars? If they are genetically transmissible, as they must be if they are hard-wired into our neurology, then their source could not have been the regularities created by their users (or the users of the natural grammars derived from them). They must have evolved in the same way as the rest of our neurology – as the effects of random mutations selected because they happened to be useful for survival and propagation. If this is how grammars are acquired, though, why did the process stop once the original deep grammar had been acquired? The random neurological changes must have continued from generation to generation. Why didn't they go on being adaptively selected to form new grammars? At the biological level the same process has led, not to a single ineffable Ur-form, but to a branching family of species, as random changes were selected under adaptive pressures that differed from time to time and place to place; it has directly created the diversity that we see today. If there had been a branching family of deep grammars then our present natural grammars would presumably have arisen from the different branches, and we should be unable to acquire any natural grammar other than one arising from the particular species of deep grammar that we were born with, whereas the natural grammar that we do in fact acquire is the one that we are brought up with.

So the situation must presumably be this: the grammars of our natural languages are derived from an Ur-grammar, impossible to observe or describe directly, that was evolved by natural selection as a one-off form, after which this particular line of evolution ceased. The rules of this Ur-grammar, unlike the rules of natural grammars, have no descriptive function, because there is nothing for them to describe. They are purely prescriptive. Not, plainly, in the sense that natural grammars are prescriptive: of instructing non-deep-grammar users how to learn it, or native deep-grammar users how to use it to the best social advantage. So prescriptive in what sense?

John Searle, who believes, like Chomsky, that language is a rule-governed activity, distinguishes two sorts of prescriptive rules, constitutive and regulative,[6] which he tries to elucidate by analogy with the rules involved in various non-linguistic activities. As an example of regulative rules he offers the rules of etiquette, which codify existing practices, and as an example of constitutive rules he proposes the rules of football or chess, which he says 'do not merely regulate playing football or chess, but as it were . . . create the very possibility of playing such games'.

So, regulative rules for a start. This is the sub-set into which the rules of English grammar presumably fall, when they are used prescriptively. But this sub-set really should be divided again, into rules that are descriptive in origin (as I claim the rules of English grammar to be) and purely factitious ones, such as the dietary laws of Leviticus and the latest government directive on tax allowances. These are not based upon observing the eating habits of the Israelites, or the taxpaying practices of taxpayers – they have been invented precisely in order to modify these habits and practices. Rules of this sort don't seem to be what Searle has in mind, because the example that he picks, the rules of etiquette, are surely descriptive in origin.

They are also interestingly elusive. Although the phrase seems so familiar, what *are* these rules? Where are they formulated? Well, there are books and magazine articles that tell you how you should behave socially, and you might say that these are a kind of social Leviticus or tax guide. But they aren't the *source* of correct behaviour. They're a description of existing practice which (like grammatical rules) can be used prescriptively for anyone unfamiliar with it. *You* haven't read them, and nor have I, any more than we have read the rules of our native language, yet we both do observe some sort of common social code. Even if rules can be written down that attempt to *describe* this code, is the code itself a set of rules?

Well, what's the rule that regulates the saying of 'please' and 'thank you'? We certainly try to get our children to learn the practice by telling them in so many words. But does this constitute a *rule* exactly? One of the things that any rule governing the usage of 'please' and 'thank you' would have to specify is the occasions on which they should be said. You might argue that, yes, the nature of the occasions, at any rate, *could* be formulated and is: you are to say 'please' if you are asking for something, and 'thank you' when you have been given something. That seems clear enough. Until little Wilhelmina puts her money into a slot machine and is given a bar of chocolate. 'Thank you,' says Wilhelmina politely to the machine. So kind Papa amplifies the rule to make plain that it refers only to transactions with the child's fellow human beings. 'Don't go round thanking machines,' he says. 'People are staring.'

'Do you mean,' says Wilhelmina, '"*please* don't go round thanking machines"?'

Papa gives her a funny look.

'Thank you for that funny look,' says the obedient child.

You brush all this impatiently aside. Exceptions to the rule like these are so obvious that any child understands them without explanation. But how, on what basis, does it understand them, without further amplifications of the rule? And where would these amplifications come from? Are they at the back of your mind all the time, just waiting for the right moment to be articulated?

And where would the amplifications stop? Does the child have to say 'please' if it wants to go to the lavatory? Probably, if it's going to disturb the class at school. No, if it's in its own home.

'Please may I have a second helping?' – certainly. But, 'Please, now you've put the second helping on my plate, may I take a spoonful of it?' Unnecessary. Do I need to add 'Thank you for asking, though'? If I do, does the child have to thank me for thanking it? Does Wilhelmina have to thank Granny for the kiss she greets her with? The smile she gives her when Wilhelmina tells her her new joke?

Well, perhaps all this could be codified in an enormous set of rules, and the rules could be learnt by children like the multiplication tables. I can't imagine that many parents attempt to do this, though. The usual procedure is surely to instruct or remind the child to say 'please' and 'thank you' on each particular occasion where it seems to us to be suitable, until it begins to get the sense of what those occasions are. So what *is* the sense? How do *we*, in our wisdom as adults, know when to prompt the child? And when, for that matter, to say 'please' and 'thank you' ourselves in circumstances far removed from any that we were taught by our parents to recognise when we were children?

Take what seems on the face of it to be a very clear unwritten rule, at any rate in middle-class British society, that you write a thank-you letter after you have been entertained to a meal. Open and shut. You can write it down in four words: you eat – you write. The next time you're invited out, though, it's not to dinner – it's to drinks. Should you write to thank? You consult the rule. No, no need to write, because no eats. Hold on, though. There *were* canapés – most delicious ones. You reach for your pen after all . . . Not so fast! There seems to be another rule floating in the air somewhere that modifies the first, and says that you don't need to write to thank for drinks, even if food was offered as well – if this was consumed standing up. You put the top back on your pen. Then another modification comes to mind: if consuming the food, even standing up, required the use of an individual plate, together with a knife and fork, then this might well constitute eating serious enough to require a letter . . .

Or possibly not. It depends. Depends on what? Well, not on any rule that *I* can begin to formulate. On a complex assessment of the seriousness of the occasion, the nature of the hosts and your relationship to them . . . Even the simple extensions of the rule that I *have* managed to formulate, distinguishing between fingers and silverware, I have never consciously attempted to formulate even for myself before. How did I do it? By introspection? By dredging some obscure entities up from the depths of myself, as one might an old classroom mnemonic for the kings and queens of England? Or by remembering what I have myself done and seen others do? And how did I manage to go on considering the problem, when all hope of distilling further rules circumstantial enough to meet particular cases seemed to have evaporated? By imagining what I might do in a particular imagined situation.

So how did I know what to imagine? How did I manage to behave in any particular way in the first place? Wasn't it simply by observing the behaviour of others (if only through the medium of generalisations expressed in language) and attempting to imitate it? By observing it, and imitating it, in ever finer detail and in ever more various situations? And at the same time adapting it to my own use and understanding – improving upon it here and toning it down there? If there are rules in all this then I am writing them as I go along, as judges write case law, by the series of individual decisions that I make in adapting the basic statute to particular circumstances. Why should I need to postulate some innate but ineffable set of instructions, translated out of a universal code of politeness into local middle-class British usage by the application of some further set of rules, distinguishing among other things between smoked salmon served on pumpernickel and smoked salmon served on a bed of wild rice?

I suppose I might have some general motivating principle hardwired into me by evolution – the desire to please, the urge to fit in with the mores of the tribe . . . But then how do we account for *your* behaviour? *I* cravenly try to follow every nuance of the rule, it's true. *You*, however, with your noble disregard of pettifogging social niceties, wouldn't dream of writing a thank-you letter even for a six-course banquet consumed while sitting on a genuine Chippendale chair with the help of six different sets of Georgian silver. In fact you go way beyond mere passive indifference to the rule – you send your hosts a letter making clear your absolute lack of any sense of social indebtedness. Do we need to postulate another rule to explain how you managed to subvert the first one? How you managed to establish what after all might be adopted by other free spirits as a new rule?

Even the most straightforwardly regulative rules are of various sorts, and shade into other sorts of constraint. The principles that give form to the vehicular traffic on the roads, for example, are very heterogeneous.

There are first of all the juridical rules imposed by the authorities for good practical reasons, such as the laws relating to standards of equipment and training, of sobriety and reasonable eyesight;

then (slightly different): the purely arbitrary rules that arise from chance or forgotten and no longer relevant historical precedents, but also juridically imposed because they settle questions that have to be settled, such as which side of the road we drive on, and whether the 'Stop' light is above or below the 'Go';

next, entirely different in origin but still affecting what happens on the roads: the laws of physics, which we formulate from what we observe, but which we cannot change at will, and which affect such things as acceleration and stopping distances;

slightly different again, the rules relating to the behaviour of the human participants: reaction times, ability to perceive colour and peripheral movement, and so on;

another class of constraint altogether: the unofficial conventions that drivers develop themselves in practice, purposefully but probably unconsciously, such as the uncodified rules of headlight-flashing to allow precedence and acknowledge the granting of it that are widely (but not universally!) understood and observed;

not entirely dissimilar: the personal styles and tastes of individual drivers – their choice of vehicle, their general inclinations to observe or break the law, their particular choice of laws to observe or break, and situations in which to do it;

a related class: the unconscious considerations affecting individual drivers in deciding when and whether to overtake, when and whether to accept being overtaken, etc.;

certain principles of traffic management derived from the preceding factors, or perhaps from some combination of them which is more complex (as in chaotic situations) than the sum of its parts, and dictating such things as the amount and placing of advance warning for contingencies, the direction of the vehicle flow into narrower channels, and the combination and separation of the flow.

Perhaps you could locate another rule, too – the rule of drift, the unceasing slow shift of things for no assignable reason, the unending flow of small random mutations. The rules governing the appearance of cars have shifted, for instance, in much the same way as the vowels of spoken English have. (Compare the clipped vowels of speakers recorded in the 1930s with the sprawling dipthongs of modern English – even though recording itself must have had a stabilising effect.) The conventions of driving have shifted in some ways towards the more courteous, in other ways towards the more aggressive. Some of these changes can be traced, no doubt, to particular sources – the influence of American culture, the shift away from and back towards political consensus, the changes in the technology of road-building. I suspect that even in a system isolated from all outside influences, the drift would continue.

Regulative rules, in other words, derive from some kind of existing practice, and are used either to standardise or to modify it. What we need to examine is the other sub-set of prescriptive rules proposed by Searle – constitutive ones, that 'as it were . . . create the very possibility' of the activity they seem to govern.

The examples that he offers are football and chess. In the case of football he presumably doesn't mean that the rules were *historically* constitutive of the game, either in their present form or any earlier one. I don't know how football began, but I imagine it was as a series of disparate events – various confused struggles in which one lot of people tried to get something somewhere, and another lot tried to stop them and perhaps get it somewhere else instead. Do you need a rule, in any sense at all, to constitute what was happening on each of these occasions? Do we need to imagine that anyone said to a new player:

'The rule is that you score by getting a sheep's head out of the slaughterhouse and into the doorway of the Goat and Compasses'? If they felt the need to explain anything at all they would surely have put it in terms of objectives. 'What we're trying to do is . . . What *they're* trying to do is . . .' The rules, when they came later, were regulative. They were introduced piecemeal to make the struggle more coherent, safer, faster, fairer. The means of getting the sheep's head into the pub was restricted to kicking. The object of the kicking was restricted to the sheep's head rather than the heads of the opponents . . . The sheep's head was replaced by an inflated pig's bladder . . . A particular number of players was settled upon . . . And so on, down to highly technical rules about offside. But the offside rule is not *inherent* in the nature of football, nor is the size of the ball or the number of players. They are all arbitrary restrictions imposed upon an existing activity in order to make it more enjoyable. What constituted the activity in the first place was simply an arbitrary and formless expression of conflicting purposes.

All the rules of football can be abandoned, and regularly are. You can play it with six players instead of twenty-two, in a backyard a tenth the size of the regulation pitch; with a single goal marked by a couple of coats. You can play it with a tin can instead of a ball. You can dribble a tin can round a street on your own, with no goal and no opponents at all, and still have a bit of fun. Do you need to postulate some last pale ghost of a constitutive rule to explain what's going on here? 'The kicking of the tin can must result in a sense of fun, where fun is defined as an enjoyable engagement in some activity for its own sake . . .' What would you *do* with the rule? Well, I suppose it would enable you to program a computer to kick a tin can round the street, once you'd managed to program it to enjoy itself.

Now, despondent about the difficultes of finding a constitutive rule to create the possibility of what you're doing, you kick the tin can around without getting any fun out of it at all – and it's *still* a meaningful activity, in a dreary sort of way . . .

And you give up on football, and turn to Searle's other example, chess. Now, here's a much more plausible candidate! Not, of course, that chess is constituted by obviously late regulative additions like the castling and *en passant* rules. There's no game without the *basic* rules, though – the ones that specify the moves of the different pieces (even though these have changed over the centuries), of capture, checking, and mate . . . On the other hand, there are no rules without the game. They didn't exist before there was a game for them to exist *in*, as the laws of physics supposedly did. They evolved, partly no doubt by conscious design, partly no doubt by happenstance, in step with the game as a whole, because they were useful in shaping the conditions for an enjoyably complex and demanding contest. The emerging context gave them meaning; they in their turn shaped the context.

Let me propose a rather more hopeful candidate of my own for a game constituted by a constitutive rule: not an established game at all, but one of those ad hoc games, where children lay down in advance the conditions which define the nature of some as yet unperformed feat. ('You've got to hop from here to that tree and back blindfold without touching the wall.') Or one of those challenges where people try to invent a new category of world record (the largest number of people who can fit themselves into . . . not a telephone box, that's been done . . . but the upstairs back bedroom of 43 Spadina Road SW23). Even here, though, is the rule the spring that drives the machinery? Isn't the motor of the game a set of imaginative possibilities? The difficulty of hopping any great distance blindfold on the one hand, and on the other the similarity between that distance and the distance of the tree, together with the dangerous closeness of the wall? You could, instead of issuing a challenge, just set off without a word and hop the course yourself – perhaps without any clear idea of how far you are going or what you are allowed to touch on the way. And I might imitate you – still with no rule either stated or implied. The rule is a secondary device to give the activity more shape and make it more communicable.

Forget games. Think of a purely formal system like logic. Here, surely, constitutive rules are its very fabric. Are they, though? Does A entail B *because* of the laws of logic? You might refer to them to decide a particular case – once you've established what the laws are. But the laws are there because they codify and extend something more fundamental – the understanding we have about how different verbal and symbolic forms relate to each other. They formalise and generalise the decisions we have already taken about particular procedures.

Do we have to believe that a draughtsman who can draw a good freehand curve has somewhere inside his brain the equation that would generate that curve on a graph? Would it help him train another draughtsman if he worked out what this equation must be and wrote it down for him? And how would you choose to learn the tango? By reading the rules for what constitutes the tango in competitions organised by the National Ballroom Dancing Council, or by watching tango dancers tango?

There are shadowy borderlands in the world of rules here, too. Just as some rules can be descriptive or prescriptive, or both, depending upon the context in which they are used, so others can confuse the constitutive and the regulative. Take the rules that are laid down, in the form of recipes, for cooking particular dishes. What sort are they? Well, you might use a recipe in a highly constitutive way – you might blindly follow it wherever it led, not knowing – perhaps not even caring – what the end product was supposed to taste like, so long as the directions were faithfully adhered to; and if you are running a rather old-fashioned academy of housekeeping you might, I suppose, *define* the dish as one that was prepared in accordance with the recipe. On the other hand you might use the same recipe in a more regulative way – to help steer proceedings so as to produce the dish you know you want.

When Escoffier or the currently fashionable television chef first invented the dish, though, he presumably cooked it first and wrote the recipe afterwards. What rule of invention was *he* following? What rule are *you* following when you make one of your famous soups, and throw in a bit of this and a bit of that, depending on what's in season and what's in the store cupboard and what catches your fancy, tasting it as you go along, then throwing in a bit more of this and a touch more of that? I suppose you could try formalising even this into a set of constitutive rules: 'IF you have no leeks THEN use shallots; IF you have leeks and shallots but you haven't used shallots before THEN use them anyway, to see how it works out; IF all else fails THEN shove in some curry powder . . .' All very satisfactory and computer-friendly – except that the next time you make the soup you seem to be following a different rule entirely: 'IF you have both leeks and shallots THEN try the shallots again anyway; IF both the leeks and the shallots look a bit past it THEN open a can of soup instead . . .'

All this, though, is just the beginning of our problems in deriving activities from rules. Constitutive and regulative rules, says Searle, are not the only sort we need to explain a game. There must also be more general ones, such as a rule requiring that the players are playing in order to win.[7]

Now we are on a very slippery slope. If we need a rule requiring this then there surely has to be another that defines what constitutes really trying to win. Does it mean making a reasonable effort? A superhuman one? Being prepared to give up an hour to practise every evening? Two hours? Your job? Your family? Your life . . . ?

Eventually, though, after much argument, we get the rule defining 'playing in order to win' established. And we relax from our efforts by playing a game of chess with our small daughter, or kicking a football about with our small son. Naturally we let them win. So it *wasn't* a game of chess we were playing? It *wasn't* football? Well, it wasn't chess as it's played in a chess tournament, certainly, or football as seen on television. But they were games of *some* sort. So now we need a rule to allow this exception to the rule in the case of games with small children . . . and a rule to define how small a child has to be to qualify . . . and a rule to except child prodigies . . .

This is one of the problems with rules: the Talmudic effect – the perpetual need, once one has entrusted oneself to them, for further rules to elucidate their application. 'No dogs' – what could be more straightforward, more open and shut? But now here comes a blind guest with his guide-dog . . . and of course we never meant to exclude *guide-dogs*. Now the Queen honours us with a visit – and even as we obsequiously bob and smile we see that she is accompanied by a retinue of pet corgis . . . Of course, Your Majesty, bring them in! Never mind the sign on the wall! There's another rule – I've just composed it as Your Majesty walked through the door – that the No Dogs rule doesn't apply to the pets of reigning monarchs . . . Next comes an eccentric millionaire with his departed schnauzer pickled in formaldehyde . . . *Dead* dogs? No problem at all! Particularly if owned by millionaires . . . Now a child, tragically orphaned in a Canadian air disaster, with the wolf cub which has been his sole companion in the frozen wastes . . . A guest who obediently leaves his dog in the kennels we have provided outside, where it howls piteously all night and keeps everyone awake . . . Locked in the car half a mile away, where it will probably die of heatstroke and get us all arrested for cruelty to animals . . .

So we are driven to formulate catch-all rules like the famous section in Queen's Regulations punishing anything that anyone in any position of military authority considers 'conduct prejudicial to discipline and good order', and escape clauses about things being permitted or forbidden 'at the discretion of the management', which are not rules at all, but an abdication from rules in favour of the arbitrary and the contingent.

Back to the rules of language. Here, too, presumably, if Searle is right, on top of all the layers of constitutive and regulative rules, we need general rules about speaking with the intention of being understood, etc., together with all the subsidiary rules that govern their application. And it gets worse. If we need a rule to explain how we understand 'The cat sat on the mat', then we surely need another rule to explain how we know which cat and which mat are being referred to, and another one to relate the cat's posture to sitting, and another one to explain the 'on' relationship . . . and another one to explain how we know that these words are intended to be understood as English, and not as homonyms in some unknown language where they may mean something entirely different . . . to define 'English' . . . to define 'define' . . . and so on, in a regress which looks set to be infinite.

As a practical example of the transformational processes connecting surface and deep structures, Chomsky takes an apparently simple sentence at random, 'A wise man is honest,' and offers a complex map to show how this is derived from two separate propositions in the deep structure, 'neither of which is asserted, but which interrelate in such a way as to express the meaning of the sentence': *A man is wise* and *A man is honest*.[8] Everything about this analysis seems to me baffling – though Chomsky is certainly right about one thing: that these two propositions are not *asserted* by the first sentence. The adage might be offered, and might be true, even if no man is either wise or honest. But I can see no reason to think that the adage is in any way derived from the two propositions. Without a context it's hard to see what they mean – what they would be saying if they *were* asserted – but presumably something about some particular though unspecified individual, or pair of individuals. The adage, though, is saying nothing about any one particular individual (or pair of individuals); it is recommending honesty to mankind at large.

You could, though, express much the same idea with an alternative version of the adage which *does* contain at any rate versions of both the simpler propositions, and there would be nothing mysterious about the way in which they were 'interrelated': '*If* a man is wise *then* he will try to be honest.' But you could also put roughly the same idea in a great many different ways, involving quite different constructions. 'Look into the heart of an honest man and there you will find wisdom.' 'Honest? – Wise of you!' 'In honesty lies wisdom.' 'Honest = wise.' 'Wise man! At any rate you are if you're honest!' Do all these have to be derived from the same two (unasserted) propositions? Do we have to have a separate map – a separate set of rules – to make each derivation? Now we change the words as well as the syntax. 'Honesty is the best policy.' 'Dishonesty is unlikely to serve your own best interests.' More sets of rules?

Now you generalise the idea a bit and make a joke of it: 'It ain't no good being bad!' Is there a separate set of rules relating double negatives in the surface structure to single ones in the deep? Do I have to have a private rule to enable me to convert the double negative into a single one before I can understand it as a negative? If you habitually use double negatives, do you have to have a converse rule to enable you to convert my single negative to a double one? Isn't our power of analogy at work once again (often without our even noticing it), accommodating the incorrect formulation to the correct one intended; or our power of guessing the intended meaning from its context?

Nothing in Chomsky's view accounts for the way in which all the rules of language are endlessly extended, adapted, and stood on their head. If you subvert the rules of logic or mathematics, you end up with answers that are either incorrect or meaningless. If you bend some accepted convention of language – 'Don't nevertheless me!' 'Me saw he' 'You – me; lock – key' – the results can still be meaningful, still be as true or false as their conventional cousins. Oddly formed sentences, indeed, are a challenge to us to find meaning in them. As an example of a meaningless sentence Chomsky famously offers 'Colourless green ideas sleep furiously.' No sooner do we read this than we think of possible readings. Uninspired ecological proposals, we understand after a couple of seconds' thought, lie dormant in spite of the anger that gave rise to them. Or perhaps we're talking about burgeoning concepts of a racially non-discriminatory nature . . . or non-political, immature philosophies . . .

Even if rules could explain how we understood language, they wouldn't go very far towards explaining how we used it for any of the 'speech acts' in which Searle, like other philosophers, is particularly interested – promising, naming, apologising, pleading, etc. None of these makes any sense at all without the concept of intention – without roots in the feelings and purposive activities of the people who utter them. You can program a computer to inform another computer that there will be overcrowding on the approach to Heathrow in seven hours' time, and if the second computer then advises the controller at JFK to hold a departing flight you might be able to persuade yourself that it had 'understood' what it had been told. But by what conceivable hierarchy of rules could a computer be made to *boast* about its achievement in doing this, or *regret* its inability to allow also for overcrowding at the gate?

Rules explain some things, certainly – but a lot less than is sometimes supposed. Treating language as a system of rules simply shifts the point of interest back one stage, from how we understand a sentence to how we understand the rules that have generated it. The problem is in understanding how we understand things, and the problem remains.[9]

Our yearning to find that language is generated by rules, or the universe by laws, springs from our everlasting need to read significance into things by detecting pictures and patterns. It is comforting to see the patterns as primary, to believe that there is an order in the universe, and in our own activities, that transcends us and our passing needs and pleasures. It's as if, looking at the night sky, we saw the images of hunters and swans not as our projections upon the randomness of the universe, but as the explanations of it. Why are Mintaka, Al Nilan, and Al Nitak arranged in a straight line across the middle of Orion? – Because a hunter needs a sword, stupid, and a belt to hang it on!

3

Mailing a Cat

How words have meanings

In understanding how language gets hold of the world, in any case, the syntax is a secondary matter. First come the words that the syntax relates to each other. 'Man! Bag! Gun! Run!' may in the appropriate circumstances inform us that a man is approaching who is carrying a bag that contains a gun, which makes it advisable to run, but 'A [–] is [–]ing who is [–]ing a [–] that [–]s a [–], which makes it advisable to [–]' informs us of nothing at all in any imaginable circumstances, even though its clauses are properly embedded, and all the distinctions between animate and inanimate, indicative and infinitive, are correctly formulated according to the rules of English grammar.

So how do words mean what they mean? How do the gestures, grimaces, and so on, that supplement and sometimes replace the words? Is it some kind of contractual arrangement secured by yet another structure of rules?

One of the things that makes it difficult to examine this elusive relationship is that so much of our everyday world has been lodged in language so long and so securely. Cats have been 'cats', and pins 'pins', for so long that we move unexaminably fast from cats and pins to 'cats' and 'pins', and back again from 'cats' and 'pins' to cats and pins. Indeed most of the time we don't make the move between name and object at all – we merely repeat and re-order what has already been said without any fresh contact with the objects themselves. We pass on the gossip we have heard. We rehearse the lessons we have learnt. We tell the old stories. We recite the standard prayers and salutations and politenesses. We utter the opinions we have read – or negate them, or modify them, or translate them into more congenial terms. Even when we have hit upon them ourselves in the first place, we go on uttering them long after the last trace of the original thought has vanished from our minds. We spend half our life moving around inside language, and never looking out, like addicts of computer games inside their tiny man-made worlds.

Nevertheless, our use of language implies something about our relationship with the world of objects; it implies identification. In order to assert that the cat sat on the mat, even before any question of truth or falsity of the assertion arises, something has to be identified as a cat and something else as a mat, and the relationship between the two of them as sitting. This goes to the very heart of things. Without some kind of identification it's difficult to see how we could begin to *experience* anything very coherent, let alone talk about it.

To identify something we have to categorise it, and exactly how we do this will be examined more closely in the next chapter. What I want to look at here is one particular question: whether identification really is a freestanding precondition of language, or whether identification and naming are bound up with each other. It seems at first sight obvious that before we can put names to the cat and the mat, and thereby take them into the fabric of language, we must have already seen what they are – a cat and a mat. After all, the cat identifies the mat, if not the mat the cat, and does it without words, without so much as a miaow. It *must* identify the mat, just as it does the mouse, the kitten, and the saucer of milk, otherwise it would kill the milk, drink the mat, groom the mouse, and sit on the kitten.

True, true. But what precipitates the identification? Our visual field is full of material that remains unidentified, even when we are presented with things that we have seen every day of our lives, until we turn our attention upon particular details.[1] When we focus upon a particular aspect of the scene we enter into some kind of relationship with it – a purposive relationship, or a potentially purposive one – that takes what we are looking at into the traffic of our lives. We distinguish it because we could eat it, because it threatens us; because it reassures, frightens, pleases, disgusts, bores, arouses, amuses us; because we are for it, because we are against it; because we want to tidy it, destroy it, encourage it. Is it the distinguishing of the object that arouses the hunger, love, hatred? Or do our hunger, love, and hatred go out towards the world and establish trade relations? Both, surely. We and the world engage.

It's the same with the cat. The cat sits on the mat and drinks the milk because it identifies them as mat and milk, certainly. But also it identifies the mat *by* sitting on it, the milk *by* drinking it. The need for a warm place to rest and food to sustain life precipitates the identification, the identification makes possible the satisfaction of the need. In the human traffic there is also another need – to talk to others (and ourselves) about cat, mat, and milk. So this is one of the possible precipitants of identification in our case. And our ability to fix this identification in a symbol is what enables us to extend the range of our identification beyond what the cat can achieve. The cat's lack of words restricts the poor creature to identifying only objects that can be distinguished by its behaviour towards them, by sitting upon and drinking and killing and licking. It is this limitation that hinders its advance into the higher ranks of academe or the civil service, by making it unable to separate intertextuality from sodium nitrate, or devolution from dementia.

Bryan Magee, attacking linguistic philosophy in his memoirs,[2] talks about how distinct the familiar tastes of potatoes and custard are, and how incommunicable by language. It's true – they are as distinct as cats and mats, or as the tastes of mats and mice presumably are to cats. In our case, though, we can still use language as the instrument of distinction, to precipitate it and give it form, just as we do with cats and mats. When we take a forkful of something unrecognisable out of the stew in front of us, and taste it thoughtfully, trying to decide what it is, don't we say to ourselves, 'Ah, yes – potato'? When we remember the taste of potatoes, as I'm doing now, isn't it evoked by 'potatoes', and if the taste comes to mind first, unbidden, as tastes sometimes do, doesn't it evoke 'potatoes' in its turn?

It often happens, of course, that we taste something and know it's familiar, but we *can't* quite place it. We can't quite put a word to it – we know there is a word written on the label but we can't quite focus it. But until we have we don't feel that this haunting sense of familiarity counts as distinguishing it. Distinguishing it is precisely what we haven't done. Something has happened, certainly. The plane has left London. But it hasn't yet reached New York, and until it has, the purpose of the flight has not been achieved.

Then again, there is a first time for everything. I take my first mouthful ever of potatoes . . . I identify the taste as one unknown to me, distinct from everything I've eaten before. And in so doing I've begun to identify it. I've opened a fresh file, labelled 'potatoes', since that's what you tell me they are, about which I know at any rate that it doesn't contain any of the other tastes with which I'm familiar. Or perhaps the label is still blank; I don't know the name of what I'm eating. But that blank space is simply waiting for a name to be written in.

I have performed the first part of the mental act. I'm not going to get much further in my thinking about it before *something* has been written on the label, even if it's only 'unidentified specimen F/734/GQ'. If I have potatoes again the next day, and still no one has told me what they are, then I fill in the label with some perhaps half-thought formulation about the similarity with yesterday's lunch, or with some first inkling of familiarity. Without *some* usable label, even if it's only one I can use in communication with myself, I'm going to be like a customer in a hardware shop trying to buy one of those little . . . you know, sort of . . . well, they're about this size . . . and they go on the top of the . . . what's that other little thing called . . . ?

I suppose we might continue to label a taste not by a word, but by the wordless recollection of previous occasions on which we've had the same taste. How would we identify these occasions to ourselves, though? Wouldn't it be by words? Perhaps by some quite precise formulation: 'When we had lunch that day last summer in the riverside restaurant in Bamberg.' Or by some less precise version of the same: 'Summer . . . river . . . Germany . . .' And if there is just the uncaptioned feeling of something that's not quite taking form in one's mind as a summer's day on the river, or Germany, doesn't having that haunting but unplaceable sense of something just out of reach count as *failing* to distinguish the taste?

At any rate I know it's not custard! Yes – now you've raised the question, or I have; I can't honestly say that the thought was in my mind before. Nor can I easily imagine, even if it had been, what the thought would have been like if it hadn't been in words, exactly as the question and the answer to it are.

Could one imagine identifying all tastes for ever just purely as tastes? Well, the wolf-boy of the Aveyron must have done, since he didn't know any words. So did the wolves that raised him, and the cat that drinks the milk but not the whitewash. Dogs plainly have a whole dictionary of smells at their disposal. Can we imagine our way into their non-linguistic world? This is what Patrick Süsskind, in his novel *Das Parfum*, tried to do. He invented a character who spoke French like any other Frenchman, but who could remember and identify every smell he had ever come across, and who could do it without words. Naturalism is not the mode of Süsskind's novel, though. And how does he give us whatever access we have to the mind of his character? Through smells? Of course not – through words. Magee, like Süsskind, has chosen words as the medium for his book. Why didn't he try expressing his arguments about the limitations of language by cooking us lunch? Why didn't he paint them or whistle them?

In any case he is underestimating his own powers. I think that as a matter of fact he *has* managed to communicate the taste of potatoes and custard. I have the taste of the potatoes in my mind right now . . . and now the taste of the custard. How did he communicate them to me? – By naming them. By writing the words, 'the tastes of potatoes and custard'.

The tastes of potatoes and custard are not the only things that move wordlessly through Magee's mind, he says:

> I am thinking all the time. I am observing situations and perceiving movements and changes, noting contrasts and connections, forming expectations, making choices, taking decisions, feeling uncertainties, suffering regrets, and a thousand other things – but for the most part without the use of language. And for me this is all a matter of directly apprehended experience, known with a certainty as immediate as I can know anything.

Introspection is as difficult here as ever, but it's true, so far as I can tell if I look into myself, that I as well as Magee do have things going on that are not articulated in words. Some of these things – observing and perceiving, making choices and decisions – we have already tried to examine. I also have the kind of *feelings* that Magee mentions – uncertainties, regrets, and so on. But does having feelings really constitute *thinking*, as he claims, in any normal sense of the word?

We think *about* them – when, for instance, we know that we are having the feelings; but doesn't this involve identifying them, characterising them, in some kind of way? And doesn't that usually involve words?

I try to turn my mind to (I won't say think about) the least determinate and most apparently non-verbal of these mental events and states: moods of sadness or lightheartedness which I can't quite locate; free-floating anxieties and uneases, unidentifiable irritations and impatiences, unlocated feelings of hope and anticipation; and even less substantial wisps of something, like faint veils of cirrus in the summer sky, which have just about enough form for me to recognise them as longings, regrets, yearnings, etc. Or have now that I do so recognise them, now that I have to put them into words to tell you about them. Was it a regret I was feeling before I put a name to it? Or was it a yearning? It was a *something*, I know that . . . *Was* it, though? Was it even a *something*?

These nameless not-quite-anythings are like those first premonitions of illness that you sometimes get – vague unlocatable shifts in the internal weather that can't really be dignified as symptoms. Some slight . . . suggestion of unease around the middle of the abdomen . . . or at the back of the palate . . . Until finally a shape emerges from the mist. 'Of course – I'm feeling sick . . . I've got a cold coming on . . .'

Now the pain is agonising. Now you're feeling really sick, and your cold is so bad that everything inside your head is like minestrone. Intense aspects of our experience, certainly, for all of us. But would we normally call having a pain, or feeling sick, *thinking*? And isn't that what's at issue here? A pain, or nausea (or even a purely mental state like remorse or longing) can sometimes be so intense that we can't think about anything else. Is this because we are already fully occupied thinking the pain or the remorse? Or are we thinking *about* it? Or not really thinking at all, because what we are feeling is so intense that it precludes thought?

To communicate with others about pain and other strong feelings is notoriously difficult – even to give some impression of the location and nature of a pain to a doctor, who has spent his life diagnosing similar feelings in others from just such vague metaphors of sharpness and dullness as you are using now. But then it's also difficult to communicate with yourself about them. What can you do with them mentally? Well, you can wish they would stop, you can wonder how long they will go on. You can ponder possible courses of action, as you might about external matters – how to find a more comfortable position in bed, whether or not to take medicine. But doesn't doing even this imply that we are in some way naming the pain to ourselves? Ostensively, perhaps, without so much as thinking the word 'pain', identifying it to ourselves by the very fact of our attention to it.

Or when you ask yourself how much more of this you can bear, meaning the pain you are feeling, isn't the 'this' naming something, just as much as when you wonder the same thing about something external, and know (without the need to articulate it to yourself) that the 'this' refers to the play you are watching, or to your husband's behaviour? 'This' and 'that' are parts of language, even when they are thought without articulating them, as they probably are here. So are the names they constitute when they have been fleshed out with a reference, even when that reference is a non-verbal one such as pointing or sketching. Even when, as it probably is here, the reference is made simply by our being privately aware of what it is, without articulating it to ourselves.

With the much slighter and less definite colorations that pass so fleetingly through our minds, there's often too little substance to dignify them with even as much as an unarticulated 'this'. But until we do, can we think about them in any sense at all? And can we even then think very much about them until we have identified the final form that those first faint stirrings are prefiguring, or before, in the case of feelings with an external object, such as love and envy, we can name that object to ourselves?

I suppose sometimes one has an area of brightness or darkness in one's mind that one knows one *could* perfectly easily name a source for, and that one doesn't only because one doesn't need to. But the name is there, waiting. And often announcing itself. You feel irritated – and the words 'That bloody man . . .!' form themselves. You don't need to append his name and address to make it clear to yourself who you mean. Or your mind is suffused with a warm golden agony of anticipation, and it condenses into 'Another hour and twenty minutes . . .' You know perfectly well the event that this identifies, because you have a kind of feeling of her smile already emerging from among the crowds coming off the train. As soon as you try to put your mind to it, it condenses out into words once again: 'Another hour and nineteen minutes . . .'

Or you have a feeling of certainty, say. Certainty of what? Of it. Of them. Of her. (These pronouns name objects to you, if not to me.) You have the feeling of a decision having been taken. A decision about what? About Saturday. About what on Saturday? You know what on Saturday. You don't have to spell it out to yourself once again, any more than you have to when you announce to your fellow conspirators the long-awaited date of the assassination. Perhaps you don't even need to mention the date, either to them or even to yourself, if the date is the one expected. *You* know what that feeling of having made up your mind refers to; so do they when they see the look on your face. But then that feeling, that look, are the pronouns referring to what has been established before, to what has already been named.

Or maybe you never *have* referred to the projected assassination in so many words. You showed the conspirators the photograph of their victim; you had a certain look on your face. But in the context – given that you have the reputation in extremist circles that you have, and they have the reputation *they* have – this silent gesture and this silent look are themselves not just the pronouns but the complete name of the event.

I'm not sure, in any case, quite how common these glimpses and intuitions, these internal nods and winks, actually are. It seems to me that most of the thoughts I have that can be articulated at all are incarnated in language already in the moment of their birth. In some way the words come before my mind, and *that is the thought*. The word emerging from whatever the word emerges from is like the decision emerging from wherever decisions emerge from.[3] It might be anything up to the moment it emerges. Just as, in the end, the action itself is the decision, so the word itself is the language act.

All this relates to our *experience* of thinking, since this is what was at issue. It plainly can't be a complete account of what's going on inside our heads when we think; a vast amount must be happening somewhere behind the scenes to enable us to think the words we think, or to have those familiar tastes and vague premonitions, just as a lot must be going on backstage inside my word-processor to allow it to form these words on the screen. I don't know anything about the circuitry of the machine or the instructions in the program. Even if I did, I couldn't be thinking simultaneously about the mechanics of the word-forming process and the sense of the words themselves. The electronics engineers and programmers who set the thing up know about the mechanics, but then they know nothing about the words that form as my fingers move.

It's true that sometimes, usually when something goes wrong, the smiling icons and simple dialogues of the word-processing program are replaced by arcane references to files and procedures that are normally hidden from me, and I get some partial insight into the way the machine operates. In rather the same way, we sometimes see a little more of the mental processes that underlie language. As Magee says, we sometimes consciously hunt for a word, rejecting one alternative after another because it fails somehow to express something that lurks in waiting. (He even does a bit of linguistic philosophy at this point, by examining some of the common expressions we have for the experience.4) It's not only words that you hunt for in this way. You try out and reject whole phrases, even whole sentences, because they fail to give form to . . . something that you 'have in your mind'. To *what* that you have in your mind? But this is where our insight into the process ceases, because it's precisely our inability to get hold of this something that prevents our expressing it. Until we've settled on a word, or a sentence, that somehow seems right, the thought has not been born. The only examinable part of the thinking process is its completion, the embodiment in language of something which up to that moment had remained unexaminable. We don't *know* what we think in any normal sense of 'know' (here we go again – more linguistic analysis!) until we've found the words to express it. We don't even know what we're doing to find the words – or why the words we do find seem to express what they express.

All this is to treat language as an internal operation, whose function is a passive one, to express for our own satisfaction, for the clarification of our own thoughts, what's before our eyes, or in our minds and hearts. The primary function of language, though, and the one that determines its nature, is to communicate with others. And here the word-processor offers another simple analogy.

When you press the key for the letter *m*, the machine doesn't send to the screen or printer, or store in the saved document, a complete specification for writing an *m* – so many pixels to be lit or dots to be printed in this line, so many in that. This would be a slow and cumbersome process, and it would involve transmitting and storing a great deal of unnecessary information, because the letter *m* is standard – the *m* you require to write 'minuscule' is exactly like the *m* you require to write 'massive'. So the formula for screening and printing *m* can be stored as part of a font, the electronic equivalent of the metal slug in Gutenberg's tray, and *released* by a brief command formula sent when the key is pressed. The same shortcut is used by the prisoners in the old story who can raise a laugh by saying 'Five!' or 'A hundred and forty-seven!', because these are the numbers of the jokes they have all told each other so many times already.

This is how you know what I mean when I tell you that a cat sat on a mat. The meaning of each element in the sentence is already in some sense stored inside you, and is released by the word that refers to it. You have some acquaintance with cats and with mats and with sitting, or at any rate with the idea of them. I press the keys which direct you to the appropriate files in your directory. I can send you flowers in the same way, by Interflora. I could post you a cat, too, quite humanely and at ordinary letter rate, by sending you the money to buy one. We're trafficking once again. This particular traffic only works, of course, if you're living somewhere where you can buy cats, and if neither of us is concerned about which particular cat you end up with. But then, while '*a* cat sat on *a* mat' only works if there are cats and mats in your world, it doesn't matter if I'm thinking of black cats and you're thinking of ginger ones.

Well, then, supposing it *does* mattter which particular cat and which particular mat we're talking about. '*The* cat', I'm telling you now, 'sat on *the* mat.' The function of *the* is defined by the context. You may understand me to be referring to a cat and a mat about which you know nothing except that I have mentioned them already. You may understand me to be referring to an imaginary cat, or a metaphorical cat, or (as here) a notional cat whose only substance is as an exemplar of something referred to in language.

On the other hand, you may understand me to be referring to a particular cat of your acquaintance. About this animal, and the mat it habitually sits on, and the idiosyncratic way in which it does it, you may know a great deal – much more than me, perhaps – so that my bald statement is rich in the imagery it brings to your mind and the associations and emotions it evokes, all of which may give it a significance for you hearing it that is completely different from the significance it has for me uttering it. In the same way, when I type '*the cat sat on the mat*' in twelve-point black Times Roman and email it to you, it will quite probably come out on your machine, thanks to the idiosyncrasies of our respective software, in ten-point blue Arial, which may give it an air of quiet insouciance that it didn't have for me. The overtones of the typeface and the personal associations are secondary, though. For my email to mean anything when it arrives you must have English fonts of some sort on your machine. For my statement about the cat to refer meaningfully to a particular cat we must be living not just in similar worlds but in the same world (so far as this cat and its mat goes); or at any rate I must have made an imaginative incursion into your world, or you into mine.

You're very familiar, one way and another, with the idea of cats sitting on mats, and you're unlikely to be drawing out the meaning of my sentence word by word – as unlikely as you are to be drawing out the meaning of each word letter by letter. The meaning of my sentence comes to you as entirely and effortlessly as your complete letterhead does when you press the simple combination of keys that you've programmed to summon it. Now I move on to less familiar ground, though, and tell you that the mat sat on the cat.

Well, you can think of possible meanings for this, though they don't jump out at you quite so readily – you have to go back to the individual items. The sentence is like an unfamiliar but meaningful-looking single word – *positration*, let's say – which with a little thought you can guess several possible meanings for. Now I tell you that the cat sat on an armoured division, or on a warm westerly airstream. Various fantastical possibilities present themselves, of flying cats, of giant tank-busting cats, of cats crushing a column of army ants. Then we come to a cat that sat on the square root of five, or on the best will in the world, and at this, probably, no meaning at all comes readily to mind. But each of the elements seems as sharp and clear as the letters on the screen always are.

Nddlefyg, they say now, as distinctly as ever, but they don't speak to you. It doesn't look as if it exists as a word, but you never know – it may be a neologism (like WYSIWYG) which you're going to hear over and over again in the future, and whose meaning will become gradually clearer from the different contexts in which you hear it.

So now we go back to the situation of the speaker, when he uses words to express what he sees before him; or what he remembers of what he has seen; or what he feels in his stomach or his soul; or what he imagines or supposes; or what he denies to be what he sees, remembers, feels, imagines, supposes. Does the object evoke the word, as the word does the object? Or do we, at the sight of the cat, whack 'cat' upon it because once again the word evokes – evokes the cats of our experience? With something as recognisable and familiar as a cat the process goes so fast that it's difficult to have any idea what's happening. But how do we know, to go back to the searches we conduct in our memory for the right word or for someone's name, when we've found it? I am an expert on this, or should be, since I spend a large part of my waking life in just such searches. Even so, I can't tell you what happens. I don't find the suspect's dossier in records, with his photograph on it and a name underneath it. I don't compare the name I've found with some notional ghost-name in my memory. A name comes to me, if it does, and it somehow feels right. It fits. There is a sense of familiarity about it. (And this fit, this familiarity, may turn out to be deceptive.)

Or take something which, unlike cats and people's names, is right at the edge of our ability to capture in language – one of the vague physical or spiritual uneases that we were feeling earlier. Don't you try various categories over to yourself, with or without being conscious of their verbal labels? The category of feelings like *this* (what you would call twinges) . . . or like *that* (aches) . . . And is this elusively negative colouring to your mood a regret . . . ? No, somehow the analogies with the feelings already categorised as regrets don't seem very clear . . . A yearning? Yes, maybe. *Yearning* has the right sort of feel to it.

Or think of turning on the radio and hearing a piece of music that you can't quite place. The phrase *the latter half of the nineteenth century* suggests itself, and seems to bring the right analogies in its train . . . And yet *classical* feels right as well . . . Now it begins to seem as if *Bruckner* sits well on it . . .

But *why* it sits well you cannot say or think. The only way you could even begin to try is with words. You could explain quite a lot with words in this case, it's true, about Bruckner's characteristic tropes and procedures. But would even a whole volume on the subject really explain why Bruckner sounds like Bruckner and Brahms like Brahms? Why, after hearing a few bars, without any conscious analysis, the word 'Bruckner' or 'Brahms' comes into your head?

*

By the time we reached the street he was himself again, talking about buying a used Chrysler, one with a divider between chauffeur and passengers.

This is the first complete sentence of a randomly opened page of a book taken at random from the shelf. Something awkwardly specific and idiosyncratic and everyday. You can't completely understand it without knowing its context. And yet, phrase by phrase and word by word, even without a context, you have no difficulty with it. You feel that its meaning is plain.

So how, quite specifically, phrase by phrase and word by word, does it mean what it means?

Let's start with what seems the simplest word to make sense of: 'street'. We all know what a street is because we've *seen* streets – we've seen ten thousand streets. So how, precisely, do we relate this street to all the ten thousand we've seen? Do we recall, as we read the word, our memories of all ten thousand? Of a representative selection of them? Of a street in Mayfair, say, to cover the upper end of the market; a back street in a working-class district of Huddersfield, to make sure that the full social spectrum is covered; a street in Paris and a street in Calcutta to ensure that we're not being too parochial? This is ridiculous. A single street, then? But *which* single street? One drawn at random from your great store? A famous one: Downing Street, perhaps? A familar one: the street you live in? An emotionally important one: the street you grew up in?

The sort of street that flashed through my mind, now that the question has been asked, was a street in an American city (suggested, no doubt, by the reference to buying a Chrysler). Which street in which American city? I don't think it was a specific street. It was a generic American city street. Just a passing glimpse of it – just a suspicion, a feeling of it – gone before I could see any of the details. Did I perhaps notice a fireplug, say, or some other characteristic that might have identified it as an American street? – I don't think I noticed anything that might have identified it as a *street* even. So how did I know it was an American street? How did I know it was a street at all? – I think because I *decided* it was. I decided it was because I needed a glimpse of an American street to supply an answer to the question. But when I first read the sentence, before the question had been asked, I don't believe any picture of it, however fleeting, however non-specific, came to mind at all.

So what *did* come to mind? – Nothing, so far as I can tell. A sense of familiarity, of being at home with the concept. Not even anything as specific as that. A lack of unfamiliarity.

Well, how did the writer identify the street as a street in the first place? He seems to be describing a particular incident in a particular street. He may have recalled this one particular street, or he may have had no recollection of it all, and have meant something as generic and unspecific as what flashed through my mind. So how did he know, as he framed the sentence, that it was a street he was talking about?

Even if it was a particular street he was thinking of, it seems to me unlikely that he had to go to the corner of it in his mind and read the sign designating it as a street. Or that he checked his recollection against the definition ('a road lined with houses, broader than a lane . . .')

in either the dictionary on his shelf or one inside his head. Or that he had to look around the street he was picturing to himself and see the kind of thing you expect to see in a street – passing cars, urban-looking buildings. He just *knew* it was a street. How did he know it was a street? Because, just like us, his readers, he had seen a street – he had seen ten thousand streets. But, as with me, I don't believe he necessarily pictured any specific street or selection of streets from all that great store.

Well, let's suppose that it wasn't the street that they reached, but the *souk*. Now my end of the business, at any rate, becomes much clearer. '*Souk*', unlike 'street', is not a word I come across every day. I have to stop for a fraction of a second to understand it. And, yes, a picture of some sort does flash through my mind – and not just a visual one. Beams of filtered sunlight falling from openings in a vaulted stone roof on to displays of unfamiliar food; the sounds of an unfamiliar language all around me; the feel of heat; the smell of cumin; a sense of bustle and confusion . . . I know precisely where I am, because I haven't seen ten thousand *souks* – my life has been so parochial that I've seen only one. It is the *souk* in Jerusalem that I have in mind. I don't suppose that the *souk* they have reached, out of all the *souks* in the Arab world, is that particular one, but the *kind* of place they have emerged into I understand by reference to it. I assume that, since they are denoted by the same word, they have salient features in common.

How do I know what those features are? The *souk* in Jerusalem has two highly noticeable features of its own: the proximity of Christian holy places and the presence of armed Israeli police. Do I suppose that the *souk* into which the writer has emerged is characterised by memorials to the origins of Christianity, or to armed political conflict? I don't think so. Why not? – Because I know that the word is used to refer to the market in other cities where these features are absent. If I had visited the *souk* in Damascus and Dubai as well, some other specific features of my picture might be elided – the vaulted stone roof, perhaps, and the smell of cumin. I should have a better sense of what characterises a *souk*. I should have begun to acquire a general sense of the analogy between one *souk* and another.

Analogy, yes. This (with or without the images or the remembered smells) is what enables us to name what we name, to understand the words we understand, now that we have slowed the process down a little. What *precipitates* our categorisation of the world around us is our purposive relationship with it, our need to draw it into the traffic; what makes the categorisation possible is our ability to see the analogy between one thing and another. We hold up one familiar part of the world (or three, or ten thousand of them) against the scene before our eyes, and recognise the similarities. We hold up the name of that one familiar part (or of those three, or those ten thousand), and project its characteristics on to the scene we are trying to envisage.

It's the same when he turn from the street to the rather less familiar 'used Chrysler'. I suppose I've seen used Chryslers, but even if I haven't it doesn't matter, because I can easily extend the analogy to include used Dodges and Buicks . . . used Minis and 2CVs . . . *new* Minis and 2CVs. All I need, to make the analogy work, is to understand that a Chrysler is a car, and to have some acquaintance with cars, or, if I am an Amish, with what cars are said to be like. And so on, with the *reaching* of the street, the talking and the buying. We hold on to the thread of likenesses that we have caught between all our previous reachings, talkings, and buyings – and all our readings and hearings about them, and we follow it forward into the new territory.

The case with 'we' is slightly different. Here the writer certainly *is* referring to specific people and not generic ones, one of them being himself (or herself, or some particular fictitious character). The pronoun is intended to be proper in the same way that a proper noun is, and could be replaced by the names of the writer and the person or persons in whose company he is reaching the street. But, if we don't know who these people are, the proper reference is missing, and so is a lot of the circumstantial analogy. I have no idea, for a start, exactly how many people to envisage. I don't know whether one or some or all of them are women, or children, or one-legged. But I have plenty of precedents for the situation in which someone refers to himself together with another, or others, as 'we'. I have heard other people do it; I have done it myself. I have a structural analogy.

All this relates to our reading of the sentence as I have quoted it here, in isolation from its context. We do sometimes come across odd sentences like this – as translation exercises, perhaps, or as snatches from torn-up letters and the conversations of people passing us in the street.

Usually, though, words come embedded in a context. Our understanding of the sentence I quoted is different if we read it as it was intended to be read, as in normal circumstances we *should* read it, in the context of the book in which it occurs, which is Arthur Miller's autobiography, *Timebends*. Now 'we' becomes not just the writer of the sentence and one or more others; it designates, now as properly as the proper nouns themselves, Arthur Miller and his father. The street remains unspecified, but it is at any rate a street in New York, with all the associations that come with it. The Chrysler that his father is talking about buying remains notional, but the picture that comes to mind is of a model built in the late forties. His father's talking about it is not an emotionally neutral activity – it is a further example of his grandiose and unrealisable longings. A great structure of analogy is already in place. We know a lot about New York and the kind of cars that were being built in the 1940s. We know about people who have projects of this nature. We know, by this far into the book, a lot about Arthur Miller, and his relations with his father. We have been coaxed step by step, analogy by analogy, into a whole world of meaning.

And for me, at this point, the picture changes again. I added the last two paragraphs, identifying the source of the passage as *Timebends*, long after the earlier ones were written. Now that I come to check it in *Timebends*, much later still, I can't find the sentence. Perhaps it was from someone else's autobiography, or from a novel.

At once Arthur Miller's familiar face vanishes. So do his feelings about his father, and the streets of New York. (I have a hunch that it could be *Augie March*, and that the city is Chicago.) The world of meaning in which I thought I found myself collapses around me. The sentence is stripped to its bald minimum once again, and becomes just a snatch of narrative overheard, an exercise for translation into German.

Various problems arise in understanding how we use language – for example, how we sort out the meaning of homonyms and other ambiguous usages, and how we use different expressions to refer to the same object. Stephen Pinker, in *The Language Instinct*,[5] approaches these problems by taking the structuralist idea of an innate 'deep' grammar one stage further. He proposes a complete inner *language*, which he calls 'mentalese', which is internal and covert like the deep grammar, and from which all the usages of natural language derive:

People do not think in English or Chinese or Apache; they think in a language of thought. This language of thought probably looks a bit like all these languages; presumably it has symbols for concepts, and arrangements of symbols that correspond to who did what to whom . . . But compared with any given language, mentalese must be richer in some ways and simpler in others. It must be richer, for example, in that several concept symbols must correspond to a given English word [where, as in the case of *stool* or *stud*, it has more than one possible meaning] . . . There must be extra paraphernalia that differentiate logically distinct kinds of concepts . . . and that link different symbols that refer to the same thing . . . On the other hand, mentalese must be simpler than spoken languages; conversation-specific words and constructions (like *a* and *the*) are absent, and information about pronouncing words, or even ordering them, is unnecessary. Now, it could be that English speakers think in some kind of simplified and annotated quasi-English, with the design I have just described, that Apache speakers think in a simplified and annotated quasi-Apache. But to get these languages of thought to subserve reasoning properly, they would have to look much more like each other than either one does to its spoken counterpart, and it is likely that they are the same: a universal mentalese.

Knowing a language, then, is knowing how to translate mentalese into strings of words and vice versa. People without a language would still have mentalese, and babies and many nonhuman animals presumably have simpler dialects.

A universal language! And one that we all already possess, without any need for international conferences to formulate and promote it! This surely deserves a little examination.

What is this language like? The answer to this question requires none of the tedious study associated with mastering natural languages, no understanding of linguistics or phonetics, because there *is* no answer. No one has, or ever could have, the faintest idea what mentalese is like, since its only manifestation is through the natural languages into which it translates. All we can say about it is that it must be free of the ambiguities of natural languages.

So how do I know how to translate between English and mentalese? How do I ever learn that a certain word in mentalese, about whose form and shape and signification I can know nothing, has to be replaced by the word 'rhododendron' when I am communicating with others, or for that matter with myself? I imagine that Pinker would say that the translation is done automatically for me. You remark to me on the brilliant blue of the rhododendrons. My brain swiftly consults some internal dictionary inaccessible to me, and translates the word 'rhododendron' into its mentalese equivalent. The word 'rhododendron' evidently meant nothing to me when you said it. But now I understand. My brain can work with it. In some unconscious chamber a reply is drafted in mentalese. Immediately the machinery translates it into English for me to utter: 'I find the blue of these rhododendrons slightly acid.' Now the translating machine in *your* brain is whirring. So, possibly, is the one in mine, as a check on what I've just said, because I've no understanding of it until it's been translated back into mentalese . . .

Although we can know nothing about this hidden language, Pinker does claim to recognise one important characteristic: it is unambiguous. This is why we need it – to extract ourselves from the ambiguities of natural languages and make it possible to understand what we are talking and thinking about. He offers a selection of ambiguous headlines such as 'Drunk Gets Nine Months in Violin Case', and 'Queen Mary Having Bottom Scraped', and argues that if there are two different thoughts corresponding to one word, or one expression, thoughts can't be words.

What these expressions are expressing, though, or attempting to express, is not thoughts but situations, states of affairs in the world. The thought that's going through the sub-editor's mind as he writes the headline is irrelevant – he might be thinking of lunch. What we're interested in is the violin case itself, the thing in which the defendant may be either convicted or incarcerated. Why shouldn't the same words refer to two different situations? Even an apparently unambiguous version of the headline, 'Drunk in Violin Trial Gets Nine Months', has a vast number of possible interpretations. There are many drunks in the world, and many violins, and any of them could be involved with any of them. The headline is unambiguous only for readers who have some previous acquaintance with this particular case – and this knowledge comes from a situation in the world around them. It will become unambiguous for other readers only when they read the rest of the story (also written in ambiguous English).

The expression 'violin case' admittedly introduces an extra ambiguity. How do we settle it? By our knowledge of the world once again. Perhaps of this particular event, or perhaps from our general experience of life, which suggests that court cases are where people are likely to be sentenced to nine months, and instrument cases are not. Translating the English headline into one of two unambiguous mentalese expressions solves nothing. If we can't see the two meanings from the English words, how would we know which words in mentalese we should choose to translate them? And how would we know which of the two mentalese words then to choose as likely to express the intended meaning, and which as expressing the comic variant?

One of the benefits of Pinker's hypothesis, he claims, is that it explains how creatures without a natural language (untrained deaf-mutes, small children, animals) manage to think. Well, it's true, as we have seen, that those of us who *do* use language to communicate with others also often use it in our own private thinking. It's also true, though, as Magee and others have argued, that a lot of our mental activity (to do with the taste of potatoes, etc.) doesn't seem to involve language. Now we watch a creature without a natural language performing some sort of intelligent activity – a small child working out which shape goes into which slot, say, or a robin learning to expect when and where to find the food we put out for it. Which sort of thinking do we ascribe to them? Are we for a moment tempted to suppose that they, too, are thinking linguistically, that they must be secretly saying things to themselves, in some private language of their own – 'No, the cylinder won't go into the square slot, so let's try the cross-shaped one . . .' 'The lady of the house usually puts the bread-crumbs out after breakfast, at about 8.30 am . . .'? Fond relatives, of course, look at little Wilhelmina struggling with the post-box toy and say: 'Oh, she's thinking: "I don't know why these people expect me to post silly shapes into silly slots."' But they say this precisely because they know that little Wilhelmina isn't thinking any such thing. This is why it's a humorous projection to ascribe the thought to her.

Pinker's claims for mentalese go even further. 'If babies did not have a mentalese to translate to and from English,' he says, 'it is not clear how learning English could take place, or even what learning English would mean.'[6] But if little Wilhelmina already speaks mentalese (to herself) even before she speaks English, how did she learn it? The way she appears to learn English (with or without the help of mentalese) is by listening to people speaking English to her and around her, and by attempting to imitate it. She plainly can't have learnt mentalese in this way, since she's never heard anyone communicate aloud in it. So presumably she learnt it in the same way that she is going to learn English, by having *another* internal language, meta-mentalese, to translate it into and out of. And to learn meta-mentalese she had first to learn meta-meta-mentalese . . .

Or is mentalese innate, like the deep grammar? About something as abstract as grammar I might be persuaded to believe almost anything. But does Wilhelmina really have some kind of symbols stored in her brain that represent the Lord Privy Seal or a *souk* or the up quark? Does a child born in Uzbekistan, even though Uzbekistan doesn't have a Lord Privy Seal? Did a child born in the seventh century, when even the wisest man had no conscious notion of the entities now known in English as quarks? And if she is born already equipped with this wonderful complete language, which her parents, siblings, fond aunts, and fellow members of her nursery group also speak, why does she bother to learn English? Why doesn't she chatter learnedly away to them all in mentalese?

Pinker, of course, knows even better than most of us how young children do in fact develop the language they use – in a later chapter in his book he gives closely observed and delightful examples of the changing language skills of infancy. He argues a comprehensible and skilful case for innate grammatical structure. Mentalese, though, remains mysterious. The only real result achieved by its introduction, so far as I can see, is to push all the problems of relating world and language, all the problems of thinking, one stage further back, so that they are now even less accessible. When we have conscious difficulty thinking in words the problem doesn't usually seem to be one of translation. It is precisely of finding the words to capture something that is *not* in words. Once you've found the words you've got your thought. It's not remotely as if one were moving from some precise formulation in scientific terminology or legal Latin to a vaguer equivalent in the vernacular. It's not even as if one had a suspicion that a precise formulation *must* exist if only one knew it (as one might looking at some unfamiliar insect or architectural feature). The process, when it's slow enough for one to be conscious of (as it often is, thanks to the mild aphasia from which we all suffer so much of the time) is quite the other way round. One is moving from something inchoate, vague, ungraspable, cloudy, something that one can't quite put one's hand on, to something definite and fixed, to some kind of formulation which has already been found for a case which one sees has a certain likeness to this one. What that unsaid something is in itself is unsaid precisely because it is unsayable. Sometimes, it's true, when we can't quite find the right word, we have the feeling that what we are trying to express is something more precise than the expression we have found; at other times, though, we feel that the expression we have found is *too* precise to capture the broadness, even the ambiguity, of what we have in mind.

The only possible attraction of mentalese, so far as I can see, is to serve as an operating language that explains how the human mind could function like a digital computer. The fundamental power of the computer lies in the requirement that it places upon its operators to break down all the information with which it is presented into discrete pieces – a series of irreducibly precise atoms of information upon which precise logical operations will be performed. Natural language, with its endemic haziness and its endlessly shifting colours, its dependence upon context (often non-linguistic), tradition, and allusion, has to be translated into mathematical terms before a computer can do anything with it. (Operating a computer with no operating language at all, as with Wilhelmina and the robin, presents an even less tractable problem.) So, if the mind is a computer, we need to postulate some equivalent, and a hard-edged code that will translate out completely into bytes and algorithms.

I make a gesture with the hands . . . like *that* . . . and you understand. What's that in mentalese? Is it expressed in a movement of mental hands?

I have to decide whether to climb a tree to get a kite that's tangled in its branches. Do I need a set of general principles about trees and kites and human life and the fragility of bones? Or do I look at the tree and imagine myself with my left foot here, my right there . . . and then begin to feel a little exposed and fearful, or able to cope, or recklessly eager for the credit of succeeding, or all three?

We don't manage to think *in spite of* the ambiguity of all our concepts. The ambiguity is of the essence. It's what enables us to extend them to each new candidate. Is this a house or isn't it? Even if we have rules and definitions and illustrations to help us, in the end we have to abandon them, and *decide*, in each particular case, by making some kind of leap of the imagination. Imagination often gets treated as if it were an optional extra in the battery of mental processes. But it is at the heart of seeing and recognising – of all our thinking.

Could we programme a computer to imagine? Well, yes, in the sense that we could instruct it to state that things are other than they are. But when we imagine how things are other than they are we don't mean just *randomly* other. If we imagine that, say, the Russian revolution never happened we don't imagine it being replaced by a packet of cornflakes, or the key of A minor. We mean that things might have been different in some historically interesting way. So we should have to specify in exactly *what* way they are allowed to be different. But then we don't know in *exactly* what way! If we did we shouldn't need to imagine them! So we need another rule to specify in what way the first rule could be modified; and then another to specify in what way *this* rule could be modified; and . . .

But leave aside the streets and *souks*. Doesn't our ability to grasp the structural analogy between different uses of words like 'we', 'you', and 'me' imply the existence within each of us of some kind of pre-existing syntax?

So we are back to the question of an innate grammar, but this time from the point of view of semantics. How, without a grammar waiting for them to take their places, do words like this mean what they mean? How, without a syntactical rule already established in its brain, could a child learn – and learn so early – to call himself 'I' and others 'you', when everyone around him is using these pronouns the other way round? Mother says 'You put your coat on at once or I won't take you to see Granny now.' Why doesn't the naughty child reply 'You *won't* put your coat on! I *will* take you to see Granny!'?

I might try to persuade you that the ambiguous 'you' and 'me' could be translated into perfectly unambiguous plain language as 'the person to whom these words are addressed' and 'the person who is uttering them'. Do I really believe that this is how little Wilhelmina learns to use them? Of course I don't, since she doesn't understand 'addressed' or 'uttered', nor probably even 'person'.

There are other transformations that the poor child has to learn to make.

'Bring your coat *here*, then,' says mother, 'if you want me to help you put it on, because I'm not coming *there*!'

And little Wilhelmina, who doesn't seem to be remarkably precocious in other ways, understands instantly that Mummy's staying 'here' involves her staying *there*, and that it's 'there' which is *here*.

'What's *that* you're holding?' demands her mother in exasperation.

'You mean *this*?' inquires Wilhelmina, looking at the copy of Chomsky she has found.

'Just drop it and put your coat on *at once* because we have to go right *now*.'

'I can't drop it *at once* and we can't go right *now*,' replies the child with sudden reasonableness, 'because *at once* was when you told me to put my coat on before, and *now* was when you wouldn't take me to see Granny then.'

Even if we are enabled to sort these things out so effortlessly because we have some ingenious bit of logical circuitry hard-wired into us, this is really (once again) just pushing the difficulty one stage further back. How did we evolve the circuitry in the first place if it's so counter-intuitive? Were there generations of our forebears who recognised 'I' as identifying whoever claimed it in the conversation first, and as continuing to do so even when a subsequent speaker responded? Was there a particular individual who as a result of some genetic fluke in his cerebral cortex insisted he was 'I', even though his dreaded father had spoken first; and did this early assumption of the verbal crown discourage his enemies in battle and make scores of wide-hipped girls eager to share his bed?

We can explain more simply than this how little Wilhelmina learns the use of 'you' and 'I': by hearing her mother use them, and her father and her grandmother, and by seeing the similarity in the situation of each of them as speaker, and the similarity between their situations and her own. She does it, in other words, by seeing the analogy between one speaker and another, just as later she will learn the use of 'speaker' and 'analogy'.

There is a parallel, for that matter, between the way in which Wilhelmina learns to use personal pronouns and the way in which she learns to move from proper names to common nouns.

Our identification of things in earliest infancy, and our understanding of them, begin long before we have any notion of language, as one or two recurring sensations take shape. If we were to name these as yet unnamed experiences they would perhaps be something like: discomfort . . . comfort . . . And in the first place each of these sensations as it occurs is a unique event, a complete world exclusively the property of its one inhabitant and undifferentiated at that moment from any other world because no other world exists. A *this* which is everything for all time. In the same way that a proper noun designates one individual (if this is a thinkable thought), the *this* is a proper universe.

Until the discomfort is replaced by the comfort . . . the comfort by the discomfort . . . And in our consciousness of this replacement over time of one totality of experience by another is the first beginning of our ability to discriminate the variousness of the world.

Now the discomfort again . . . The comfort again . . . And from the repetition, from the alternation, comes a familiarity, a first inkling that this present world of discomfort has some relation to an earlier world of discomfort, this present world of comfort some relation to an earlier world of comfort. And *this*, our apprehension that one thing can be like another thing, is surely the beginning of recognition.

Now that we have begun to discriminate, we discriminate the texture of our experience a little. The comfort has a softness and liquidity, the discomfort a lack of them. And this softness and liquidity, this lack of them, have some relation to something outside our own control. Softness and liquidity and pinkness . . . Smiling pinkness . . .

Now a smiling pinkness which is not the old familiar smiling pinkness . . .

Now sounds begin to accompany some of these experiences. 'Mummy. Mum-my. It's Mummy. Where's Mummy . . .? Dad. Dad-Dad-Dad. This is Dad. That's not Dad. This is Dad . . . Mouth. *Mouth*. Where's your mouth, Wilhelmina? That's not your mouth, that's your nose. Nose. *Nose*. Where's Wilhelmina's nose . . .?'

All these sounds are proper names. Each of them denotes one particular individual aspect of the infant's world. Until one day, by one manner of means or another, comes the first breakthrough from proper to common. 'Listen, Wilhelmina, where's your nose? Where's Wilhelmina's nose? Good! All right, where's *Mummy's* nose . . .? That's *Wilhelmina's* nose . . .! Yes – except that's *Daddy's* nose! Where's *Mummy's* nose, Wilhelmina . . .?'

Wilhelmina . . . Wilhelmina . . . Who or what is Wilhelmina? The syllables keep recurring, whether Mummy is there or whether it's Daddy, even when there are no noses or mouths or teddies around. 'Hello, Wilhelmina . . . ! Good, Wilhelmina . . .! Naughty Wilhelmina . . . ! Where's Wilhelmina gone . . . ? Oh, *there* she is . . . ! There's Wilhelmina . . . !' Some common denominator evidently lurks in the middle of the Mummies and Daddies and noses, some feature of the world that persists throughout everything.

Proper names are the primary material of language. Mummy. Daddy. Teddy. The *souk* in Jerusalem, the *souk* in Marrakesh. East 18th Street. West 93rd Street. Then Wilhelmina meets Arabella next door, and Arabella also talks about Mummy and Daddy and Teddy. But when Arabella calls to Mummy and puts Teddy to bed, they aren't the Mummy and Teddy Wilhelmina knows, and when Cousin Louisa comes to play it turns out that what Cousin Louisa calls 'Mummy' and 'Teddy' are different again. So the three of them discover, as the day goes by, that 'Mummy' and 'Teddy' are variables with three different values each.

Wilhelmina would like to tell the others that they can safely turn on the television because Mummy's too busy talking to Mummy and Mummy to notice – but it wouldn't make sense, because Arabella's mother and Louisa's mother are not named by the word 'Mummy' when she herself uses it. She would have to say 'the person I call Mummy and the person Arabella calls Mummy and the person Louisa calls Mummy', but it's rather long-winded. So she talks about 'my Mummy' and 'your Mummy' and 'Louisa's Mummy' – and a class of Mummies has been created. 'Mummy' without qualification will continue for her to be a proper noun; but the word has now taken on a secondary role as a common noun, to be understood in each particular case by analogy with that original primary usage.

What would happen if the name Wilhelmina became so overwhelmingly popular that the girl next door and the cousin, and every other little girl that Wilhelmina knows, was also called Wilhelmina? Then 'Wilhelmina' would become for her (and them) a common noun. To find a proper name to designate each of the different Wilhelminas of her acquaintance she would have to qualify the word for each of them, or use it only in an unambiguous context; and in time she would become interested in the differences between Wilhelminas and boys, and in Wilhelminas' studies, and so on. What would it do to her concept of selfhood? Nothing, probably. Each of us is 'I' to him- or herself, after all. 'Wilhelmina' as applied to herself would remain a proper name just as 'Mummy' does within the context of her own family, or 'I' in the context of my own talking and thinking.

Properness and commonness are always dependent upon context, and are often inextricably intertwined. Is 'King Henry' a proper name? Yes, if it's understood to refer to one particular King Henry out of the eight English monarchs who have been so called. There is no way of being 'King Henry', if by 'King Henry' is meant (say) King Henry VII, except by being King Henry VII. But suppose we are surveying Tudor history at large, and we say something like, 'By 1485 yet another King Henry was occupying the throne.' Is it a proper name here or a common noun? You might argue that it is proper, on the grounds that it must refer to one of eight specific individuals. The only way of being a King Henry, in the context of English history, is to be either Henry I, or Henry II, etc. On the other hand you might argue that to be a King Henry is to be a member of a class that remains theoretically open – that any individual can be called a King Henry who has been rightfully crowned king of England and whose name happens to be Henry. Only eight individuals have so far qualified, but there may be more King Henrys to come, or historians may somehow discover another King Henry in the archives. If we take this view, then how shall we know whether a ninth or a tenth individual qualifies as a King Henry? By analogy with one or more of the earlier Henries. Because he has qualifications analogous to theirs for being considered a king of England, and because his name is analogous to theirs.

You protest that calling the qualification of being rightfully crowned King of England *analogous* to the qualification of being rightfully crowned King of England, and the name 'Henry' *analogous* to the name 'Henry' is ridiculous. But identity is a *case* of analogy – the limiting case, where its contingency is finally extinguished. Even here, in any case, the relationship might well stop short of this convenient extreme. Suppose that some future heir to the throne, in deference to demands for a more informal style of royal behaviour, is christened 'Harry'. Suppose that the overlooked earlier king was king not of all England but only of Mercia, and originally christened not Henry but Hendrik. Is he to be regarded as a King Henry? The question is as open as the question of whether he was a king at all, given the dubious circumstances of his coronation by an unfrocked archbishop and the discovery of his secret sex-change.

It's not only nouns that have proper and common usages. Where pronouns denote particular nameable individuals, they are in practice as proper as the name they replace. But an adjective or an adverb can also be proper. You can be very you-ish at times, you know – you can behave very you-ishly. By which I mean you can be very . . . yes, like *that*! Just like you're being at the moment! So how are you being at the moment? I can't really explain. I could say 'difficult to convince, obstinate, capricious, not open to reason . . .' But it doesn't really come at what I mean, any more than an account of your physical and mental qualities is the same as what I mean when I refer to you as 'you'. I can really only say that 'you-ish' means – like *that* – you're doing it again! Can *I* be you-ish? Well, I can certainly be difficult to convince, obstinate, capricious, not open to reason . . . Is this enough to make me you-ish? Well, would it make me *you* if I put on your clothes, lived in your house, did your job, drew your salary, said and thought what you would have said and thought . . . ? Yes, for some purposes, no for others.

Now a man writes to the woman he is in love with: 'You had that wonderful Neapolitan light in your eyes again this evening.' Well, anything, from architecture to ice cream, could be Neapolitan. But no one else could ever have that wonderful Neapolitan light in their eyes, because no one else could have shared what the besotted letter-writer saw when he looked at her on that unforgettable June evening in Naples; and whatever light he saw in another woman's eyes he would never accept that it was the wonderful Neapolitan one. Now she writes to him after that same meeting: 'And when you sweetly insisted on my sharing your umbrella you did it so uggily-puggily . . .' The adverb names one single individual way of sharing an umbrella, which others would find too icky-wicky even to think about.

And if there are proper adverbs then there are proper verbs. When he does things uggily-puggily, she says (even more nauseatingly) that he is *uggy-puggifying*. This couple are plainly going to be continuing their courting in the Valentine communications in the *Guardian*. But when I say quite simply 'I go like this and then like *that*,' and accompany it with what seems to me an entirely idiosyncratic action, I'm also using a proper verb; and if you think that a complete phrase and an accompanying mime don't constitute a verb then let me tell you I have a single word to encapsulate the procedure – I say I am xynophosing – which you will understand perfectly now that I have demonstrated it to you.

Your wife, as irritated as I am by your tiresome you-ishness, makes an impatient gesture that I've seen her make before when confronted with exactly the same aspects of your behaviour. Isn't her gesture a proper one? You catch my eye, as you have on the earlier occasions, and rearrange your features slightly for a moment. Aren't you pulling a proper *face*? Doesn't it, in the context of our friendship and all the previous instances of the same thing happening, refer specifically to this one particular kind of behaviour by this one particular person?

You divorce and marry again. Your new wife finds the same you-ish things to complain about as the first. I realise that you are raising your eyebrows at me in much the same way as before. I understand your intention by analogy with the proper face you used to pull then. Your proper face now, though, is moving towards the common. By the time you get to your fifth wife . . .

You might object that the only use that your first wife's gesture might have had, the only function that the faces you pull could have, would be to express *classes* of feelings, however private those feelings might be – to make an analogy between the feelings being felt by your wife or yourself on this particular occasion and the ones that you each had on some other occasion or occasions – to generalise them. And that even the first time you both behaved like this, you were each identifying a particular class of behaviour that you were beginning to recognise in the other, and a particular class of feelings that you were beginning to recognise yourself as having.

You might make exactly the same point about the most straightforwardly proper of proper nouns. The only point in calling this particular child Wilhelmina is so that you can summon her or identify her or refer to her on another occasion. You might say that this is how 'Wilhelmina' *names* Wilhelmina. It groups together as a class a series of different appearances and references by making clear that they relate to the same object. Just as the light in the beloved's eye comes and goes, just as the particular feelings that your particular qualities induce in your various wives show themselves as the marriage curdles, and are then concealed again while you are courting the next one, so does little Wilhelmina come and go. In *my* experience, at any rate, since I only see her once or twice a week. But then so she does in her mother's experience, since even her mother has to keep taking her eyes off her to work and shop. Wilhelmina vanishes from her *own* sight, for that matter, when she is asleep. For her, too, the name 'Wilhelmina', by giving proper expression to an indefinite series of disparate phenomena, creates them into a single individual. Here, too, even within the naming of this single individual, we depend upon analogy. I observe the crucial similarities between the child I see on Thursday and the child I see the following Tuesday, and I commit myself, by naming them both Wilhelmina and by

intending it as a proper name, to seeing them as a continuously exis-
tent individual.

When I give a name, 'xynophosing', to this thing that I do from time
to time (going like this and then like *that*) I am implying among other
things that the structure of my action, the pattern of it, the logic of it,
something about it, if only the intention behind it, remains recognisably the same from one instance of it to the next.

'Khnaaj!' I cry when I understand this point. This is the exclamation
I utter which expresses the peculiar sensation I have (and whether you
feel something similar I don't know and don't care) when I realise that
I have understood some of the implications of using a proper verb.

It is, I believe, a proper exclamation.

We use the world to model the world.[7] We have to – there is no other
material available to us. And this is the *point* of our modelling – to
link one piece of the world with another. This is how we survive. We
look at the red blobs on the bush in front of us and we see the berries
that eased our hunger yesterday. We look at that uncertain outline in
the darkness and read into it the wolf that sprang at us the day
before.[8]

Now, by an extension of the same process, we look at a flint and we
see the shape of the fist that we use with such difficulty to break shells.
We look at the irregularities in a tree and we see a human face.

Then we look at another flint which is not like a fist at all – and see
in it the fistlike flint that could be bashed out of it by hitting it with
another flint. We look at a piece of tree with no resemblance to a face
– and see the face that could be found in it by shaping it with the flint.

We look at East 18th Street and we see West 93rd Street. Not everything about it, any more than we see in the wooden face the pinkness,
the mobility, and the light in the eyes of a human face. We go further:
we hold up the rue de la Paix and Cable Street, then vague thorough-
fares in industrial wastelands on the outskirts where there are no
buildings and no traffic – none of the things that seemed to link the rue
de la Paix with East 18th Street, except for some notional separation
of public space from the empty lots, some sketchy possibility that
buildings and traffic might one day occupy these dim outlines.

We go further still. We reach the street, we reach the *souk*, we reach
Sydney and the South Pole, and although we do it in some cases on
foot, in some cases by train or plane or dog-sleigh, why we call all

these different actions 'reaching' and why our hearers understand what we say is that we see the structural analogy between one case and the next. And on the chain of analogies goes. We reach, without moving a centimetre, all kinds of other goals and targets in life – things that have no physical location at all. As easily as we perceive the structural similarity between, on the one hand, the walk along the corridor that brings us to the street, and, on the other, the succession of walks, taxi-rides, and flights that brings us to Sydney, so we recognise it again in other sorts of effort we make to achieve goals that have no similarity of any sort to the net at the end of a football field, to targets that have no concentric rings painted on them. And on we go again, to reach things even less like the objectives of journeys – conclusions, verdicts, agreement; boiling point, equilibrium, age sixty-five.

Where does physical analogy stop and metaphor start? – the one moves seamlessly into the other. The concept of reaching that seemed so sharp when it involved the street or the South Pole is already becoming quite extended with even some clearly located physical destinations. At what point do we reach Sydney? As the plane's wheels touch the runway? As we step out of the plane? As we pass Immigration? At some point on the taxi ride into the centre? Sydney suddenly seems to be rather more elusive than unlocated abstract objectives such as boiling point or a verdict. The metaphoricality of reaching an agreement or a conclusion has long since vanished, the daring of the leap from the physical to the abstract long since forgotten. Reaching an understanding of reaching now seems as literal as reaching the street.[9]

The word 'souk' is the key that unlocks access within myself to a possible analogy between various places I have never visited and a place I have. The word 'street' is a key that unlocks an endless interconnecting suite of analogies. The words 'souk' and 'street' are also the keys I turn to unlock an analogy, or a suite of analogies, in you. By using a symbolism we have in common I appeal to your own experience of the world. I tell you something new by referring you to what you know already.

Language is like a signal between two computers. I type the letters 'a' and 'b' and 'c' in London, press the 'send' button, and (after who knows what on the way) your computer types an 'a' and a 'b' and a 'c' on your screen in New York. I have not transferred the actual letters that I had on my screen. All that was actually transferred from my computer to yours was a string of ones and zeros. What I have done is to *evoke* letters of similar structure (though possibly of a different size and in a different font) that were already stored and ready inside your machine. This is what structuralist literary theorists mean when they say that it is not a question of the reader reading the book, but of the book reading the reader.[10]

How did the string of ones and zeros call forth letters? How did the vibrating air between us call forth your experience of streets and *souks*, how did the grains of carbon on the paper in front of you call forth your longings and terrors? In the case of the binary digits, because the circuitry of your computer had been configured precisely so that they would. In the case of the spoken and written words between us, because our ancestors evolved systems of using the existing circuitry in this way, systems that you and I and the author of the book all learned.

4

Likeness

The gift of analogy

I suggested once[1] that a glance at even something as familiar as the back of your hand revealed more than could ever be expressed in language, or any other system of representation. I wonder, now that I think about it again, whether the elusiveness of that small patch of skin isn't even more radical than that. There's not only more there than you can ever express – there's more than you can ever *see*. Look at it now. Abandon all ambitious projects of description. Think merely about the information passing along your optic nerve. It is, as you know, fragmentary and unsystematic. The eye jumps back and forth in the wildest way – it has to in order to function. The image of the hand that the eye transmits is broken down into odd sample scraps, extended not in space but in time. However hard you look, and for however long, the process can never be completed.

You know, in any case, that there is an infinity of detail too small for the eye to resolve. You train a microscope on your hand. The closer you go the longer it takes to see what you see, and the more of the process that remains to be completed (the more, too, your hand changes over time even as the observation continues). You might nourish some hope that when you get close enough to itemise the discrete elementary particles that compose the cells of the flesh you would have reached a point where an inspection could theoretically be completed; but you know that here, at the quantum level, the world will escape again into an unfixability and a wealth of co-existent possibilities even more elusive than the skin you started with.

When you look at this infinity of unresolvable material in front of you, though, before you peer closer and begin to worry about it, you see, quite simply – well, the back of your hand.[2] Where has it come from, this familiar object, that vanished so comprehensively when you examined it more thoroughly? If it didn't come from without then there's only one other place it could have come from, and that's within. The back of your hand was in some sense inside you all the time, waiting to be *evoked*, just as the Chrysler and Arthur Miller were evoked from both of us.[3] And just as Arthur Miller vanishes again when I tell you that I'm not sure about the source after all, and you have to put your 'Arthur Miller' file back on the shelf again, so the back of your hand might disappear if you lost the conceptual framework you are using to reconstruct it. If, for example, you no longer felt the kinaesthetic connections that reassure you that the hand is yours (as with the unshakeable conviction of an early patient of Oliver Sacks that his leg, familiar as its appearance was, was someone else's – the severed leg of a corpse, which the nurses had put into the bed with him as some kind of ghoulish joke, but which had in some bizarre kind of way become attached to him, so that when he hurled it out of bed the rest of him inexplicably tried to follow[4]).

So we have come circling back to perception again, that unavoidable staple of our traffic with the world. What *is* it, to perceive something? It's not, plainly, just having it in front of our open eyes, or even having its image on our retinas. It requires some kind of active participation from us, and what that participation consists in is reading meaning into what we are looking at. That's to say, drawing it into our purposes. And to do this we must in some sense nominate it to ourselves, with or without words. Language, with or without words, is a social act; but once we have learnt it the society in which it is used can be our single self at different moments. Perception is also a social act in this extended sense. It enables us to relate our present experience to our experience in the past and our behaviour in the future.

The act of nomination, wordless or otherwise, that draws an object into being for us, is not the final stage of something that has begun as pure experience. It shapes the experience itself from the beginning, even in the most ambiguous of cases. I span round in terror at the sound of the house falling down behind me, and discovered that it was someone a few feet away blowing his nose. Even in that first instant I had identified the scale of the sound by reference to the perspective of possible sources, all of them remote since I was unaware that there was anyone so close behind me, exactly as you see the distant father as larger than the nearby child. You might argue that nothing can make us see Betelgeuse as larger than the apple in our hand; but this is because we have *never* seen Betelgeuse any closer – we have no larger editions of Betelgeuse in our library to refer to.

Even at the most primitive level of seeing an undefined shape – even when exact terms escape us, but seem to hover just below the horizon of our consciousness – seeing it at all involves an act of recognition, however minimal. We see at any rate the analogy with other undefined shapes, if we are to draw it into the texture of our activities and purposes at all.

Another elusive case: I see someone in the street I know, and cannot remember his name or even any circumstances that would help to identify him (a frequent occurrence). I cannot 'place' him, but even in the first instant of glimpsing him he is more than just a person. This is the trouble. He is . . . *that* person . . . even though I can't fill in the details. I have recognised him as recognisable, even though I can't make any use of that recognition.

Now I see a complete stranger, and even in that first instant I identify him as such. He is someone about whom I know that I know nothing. 'A stranger' is what he is. He has evoked analogies with all the other people I have seen but known that I didn't know.

And now I realise that I was mistaken in both cases. As I get closer to the person I thought I knew but couldn't place, I realise that he is not *that* person at all – he is a stranger after all; while the person I took to be a stranger is someone that I know I know. At once each of them changes. Each of them is as different from his previous self as blowing your nose is from the house collapsing.

Below a certain threshold nothing is evoked, nothing offers itself. You look out of the window of the high-speed train and try to read the name of a passing station. No word forms at all, not even an initial letter. In so far as anything happens at all you experience a blur, evoked from other indistinct and minimal events; with perhaps, in this case, since you know it must be a word, just the suggestion of word-likeness about it.

You might object that, if Merleau-Ponty is right,[5] most objects seem to select themselves. Even when they are unfamiliar they offer themselves up to us as objects because they have some perceived separation from their background, from other objects. They have clear boundaries; they are in motion relative to the world around them; they behave independently. Both the unplaceable acquaintance and the total stranger are already clearly isolable from the street and the other people walking along it. Even the unresolvable name on the passing station is at any rate distinguishable from the fields and distant hills beyond. This is to a large extent because (as I suggested earlier) the world is pre-categorised for us by the iterative nature of biological and chemical reproduction, and by its widespread adaptation to the human purposes that we have imposed upon it already.[6]

Thus do objects offer themselves; and thus, by accepting the offer, do we select them. Sometimes we deselect them by opting to decline the offer, as we do when we see Cygnus – and then realise that we're looking in the wrong direction, at some assortment of stars that probably don't constitute a particular constellation at all. Nations seem to offer themselves as being defined by natural frontiers, or by ethnic and linguistic ones. It's convenient to regard this river as the western boundary and that range of mountains as the southern, and it's going to suit most people's purposes if we draw the eastern boundary along the line of division between Slavonic and Turkic speakers. But whether we *do* accept these apparently obvious frontiers is another matter. Conferences are constantly held, and wars constantly fought, over this very question, and it is the decisions embodied in the treaties that result that in the end define the nation (in international law, if not in the minds of all its citizens).

When scientists talk about objects that exist in space-time (world-lines, as they are sometimes called – spatially three-dimensional objects also extended like filaments in the temporal dimension), it seems for a moment as if this makes them somehow independent of human experience, by freeing them from their imprisonment in the present, the only realm in which the human consciousness operates (or is alleged to operate), so that they extend from the by now uninhabited past into the as yet uninhabited future. But (all questions about the logical status of past and future apart), until this filament has been selected – until it has been detached by a human mind from its fellows, even if this detachment involves an intellectual and abstract projection from what can be presently apprehended – it lies side by side (and end to end) with all the other filaments of the universe in one undifferentiated mass. The course traced by the world-line in the temporal dimension has to be accepted or selected just as much as its extension in the spatial dimensions. Are the world-lines of the various atoms of nitrogen, oxygen, carbon, and hydrogen extinguished as they lock together to form a molecule of trinitrotoluene? Does the world-line of the molecule of trinitrotoluene then begin? And end again as the molecules of trinitrotoluene blow themselves apart into atoms of nitrogen, oxygen, carbon, and hydrogen? Or do the world-lines of the atoms persist throughout? It depends. On what? On us – on whether we are interested in the atoms and their behaviour as atoms, or in the molecules and their behaviour as molecules.

Some things seem to select themselves; some don't. You see a hand, yes, as a hand. But now, as once again you put your eyes close to the back of it, things don't seem nearly so cut-and-dried. Now we have to set to and find features for ourselves as reference points. The highlights and shadows along the tendons; blotches; hairs; the bluish shadows of the veins. As we keep gazing, we begin to note individual cracks and folds in the skin, graduations in the pinkness and smoothness. If this is a landscape that we are going to have to inhabit for some time then we shall no doubt find tinier differentiations still, and eventually they will come to seem as significant as the great rivers and mountain ranges breaking up the land-masses that are our more usual home. The features that propose themselves as features are the ones that we have already explored; the territories that propose themselves as territories are the ones we have already colonised.

We break the world apart as the weather does. The slightest crack in the smooth continuity of the rock, and the frost gets into it and splits it open. The slightest fault in the earth's surface, and the rainwaters draining down from the higher ground scour it out to form a river wide enough to feed thousands – or to drown them – and to divide nation from nation. Any deviation from uniformity, and huge forces waiting for expression seize upon it and magnify it. The smallest difference between one person and another, in ability, accent, or appearance, and both parties dramatise it and mythologise it, until it dwarfs all the ten million things they have in common. A two per cent difference in genetic structure, and we place chimpanzees on the other side of the great divide between mankind and the rest of creation.[7]

Structuralism proposes a digital model of meaning, where simple 'oppositions' (cooked/raw) elegantly reflect the $0/1$, on/off of binary arithmetic. But how often do we perceive things in terms of such simple oppositions? Many objects present themselves with clear outlines, but their qualities and categories shade imperceptibly into each other, and the models we construct for ourselves are more usually analogue than digital. We recognise and identify things because they are like other things, often in indefinite and holistic ways which are very hard to break down into digital components. We locate them on vague continuous scales, where this chair is more rickety than that one, but less rickety than the table. This meteorite may be the largest ever to have hit the earth, but it's not large in comparison with the earth itself, or the solar system. When are you rich? When you are richer than someone else. Richer than whom? The Sultan of Brunei or an Ethiopian peasant? It makes a difference! But your choice of comparison depends entirely upon circumstance – on whether, for instance, it's the Sultan or the peasant who lives next door. There is no one standard example of richness against which all others are measured.

You could be rich by being richer than yourself: richer than when you were young, richer than you feel at three o'clock in the morning, when you wake and know that you will soon be old and unable to earn a living. This is why, once again, there is no absolute sensation of speed – because nothing is changing – only, by comparison between the successive moments of your own experience, of acceleration and deceleration.

So our aspirations in life are often surprisingly modest. We don't want a lot. Just a little more than we have, or a little more than our old friend George, who has a little more than us. And if our modest aspirations are rewarded and we achieve those humble aims it doesn't corrupt us in the least, it doesn't give us grandiose ideas. Our simple hopes in life remain exactly what they were: to get a little more. A little more than we now have, a little more than our friend Charles, who still has a little more than us. Another million pounds would satisfy us entirely. A second executive jet, perhaps. We haven't changed in the least.

To locate things by means of oppositions and comparisons you have to see the analogy between one thing and another – you have to see the way in which both individuals can be fitted on to the same scale. You and I are not *inherently* ranged on a scale of tallness, or ability at playing the bassoon. Someone has to ask the question – and for the question to be raised at all someone has to see us as both having a height, as both being, at any rate potentially, bassoon players. You're sceptical; it seems to you an obviously objective fact that one human being is taller or shorter than the other. So, was Geoffrey Chaucer taller or shorter than Joseph of Arimathea? Well . . . one or the other, certainly, unless they were the same height, and if we saw them side by side we'd know at once. But we *haven't* seen them side by side, and we won't – and we *hadn't* thought of them like that. Is Anna Karenina taller than Elizabeth Bennet? You never *could* see them side by side, and there's no way now in which Tolstoy and Austen could be brought together to decide the matter between them. And we hadn't ever placed them on that scale; we hadn't made an analogy between them as measurable individuals.

When we place objects on a comparative scale it's not for neutral or abstract reasons. It's because the scale serves to relate the objects to our purposes.[8] The more rickety of the two chairs is a worse platform to stand on to make a speech, but a better one than the yet more rickety table beside it. It's also a better platform for a speech than a perfectly well-maintained stage, if the speech is to be made by an aged and infirm dictator. The largest meteorite to hit the earth is at one end of a scale of objects whose size is interesting because upon it depends the amount of damage they do, and the likelihood of their putting an end to the human race and all its relativities. To be richer than me, or than yourself, is only to be remarked upon in the context of what social or practical advantage, or moral disadvantage, money might afford us. No one remarks upon whether a meteorite is less rickety than a chair, or a lion richer than a pig; not unless we have a choice of a meteorite or a chair to balance on, or the prospect of approaching either a lion or a pig for a charitable contribution.

You object that these comparative scales are not so vague.[9] They can all be analysed, just as the structuralists tell us, into sets of simple binary oppositions. A rich man is one who is richer than x, and/or richer than y, and/or . . . They *have* to be so analysed, because the individual comparisons are a series of separate relationships between discrete individuals. A curve on a graph can be constructed digitally on a computer. And precise frontiers often have to be established for a particular purpose between one individual case and the next – you've got into the college and I haven't. But everyone knows that there is an element of abitrariness in this. If it had been a different examiner . . . If we'd got the question on the Thirty Years War instead of the Hundred Years War . . . And a lot of the time we don't want to quantify our comparisons too precisely, or make them too universal. We want to leave them indeterminate and local. You want to be found cleverer than me, certainly, and clever enough to get into the college when I don't. But thereafter you want to be simply one of the cleverest people around, not the 33,007,499th cleverest person in the world.

It would be ridiculous to call an analogue watch really a digital one in disguise simply because in practice we read off the times from it in whole minutes and whole seconds. Just as ridiculous as calling a digital watch an analogue one, simply because the whole-minute and whole-second points in time that it marks are a practical convention for indexing what we all still instinctively feel to be a continuum.

How simple, in any case, are the oppositions of the structuralist? On/off and o/1 are clear-cut and unambiguous, certainly – but only because they are *understood* to be so. We know that we are to ignore small leaks of current, small deviations from the full voltage, and all the infinitely many values that can be written between zero and one. How in practice do we distinguish between the raw and the cooked? By knowing that food in the latter category has been subjected to one of a fairly indefinite range of processes involving the application of heat. The usual marginal cases arise. Has dough that has been left to prove in a warm room been cooked? Meat which has been seared? Milk warmed for the baby's bottle? A baked Alaska? Melted ice . . .?

Do we need to posit some absolute state of rawness and cookedness to settle these questions? Black is obviously separated from white by a continuous gradient of greys. You might say, yes, but there *is* an absolute black (at any rate in theory) – a surface that reflects no light at all – and there *is* in the same way an absolute white – a surface that reflects all the light across the entire spectrum. And you might argue that by black and white in practice we mean surfaces that approach these absolutes.

Is that what we mean? The grubbiest white shirt we might reasonably describe as white by comparison with a grey shirt, or a blue one. The same white shirt we might want to call grey in comparison with a clean one. The clean shirt itself scarcely seems white against the freshly fallen snow. When we talk about a cold day or a cold meal or a cold welcome we don't mean one that has a specific relationship on the thermometer to absolute zero. We mean a day that is colder than some other days, a meal consisting of dishes that have not been heated, a welcome that compares poorly with other welcomes we have known.

But then we use simple geographical polarity in the same way – even though there are clear and precisely located poles. The north of England is north of the south of England because it's closer to the North

Pole. But who needs to refer to the North Pole when he locates Sunderland as being in the north of England? Or to the South Pole, when he goes north from the north of England to the south of Scotland? After all, people lived in the south and travelled to the north long before they understood that the world was a sphere revolving around a polar axis. We still travel east and west, for that matter, without, however far we go in either direction, ever getting any nearer to an East Pole or a West Pole. And *did* travel east and west, even before longitude was invented, and Greenwich established as its starting point.

So the horizon recedes as we move on. Nowadays, we like at moments to believe, everything is permitted. Not at all. But what is not permitted has changed, to include forms of behaviour that no one considered in this light before (discrimination against outsiders or against the less able which before seemed as natural as breathing), or that no one considered at all (strange new sexual outrages). As we permit the old impermissibles so new impermissibles take their place. As we profane the old sacred places so the gods withdraw to new altars.

Now we weary of our travels to the west, since the west seems as far off as ever, and we turn back to the east. We impose ever stricter limits on our freedom, surrender more and more of the world to undiscussable prohibitions and taboos. Does human behaviour become less outrageous? Are God's commands more respected? Not at all. Women are ravished and defiled in their thousands by a man's passing glance. God's holy laws – the nature of which is plainly manifest, surely, to the simplest soul – are wantonly trodden underfoot by every child who chalks a face on the wall.

Our sense of how far east or how far west we are changes from moment to moment, in any case, regardless of these general shifts in perception. One of the apparently simple oppositions at the heart of our experience is between freedom and necessity, between what originates in our choice and what happens regardless of our choice. In fact the opposition is far from simple; the boundary moves perpetually back and forth in our mind, so that there seems to be an element of the determined in even our freest choices, an element of consent in even the harshest compulsion; even, at times, that all our choices are determined, or all our compulsion consented to. Sometimes practical consequences flow from these boundary changes (the function of the law, for example, and the role of therapy). Often, though, we go on puzzling over it quite disinterestedly, because it is another of the dimensions in which we can locate our experience.

It would be simpler to impose general standards of justice and equality on society if there were fewer of these scales along which we perpetually arranged each other. It's difficult to ensure that we treat white and black, or male and female, or clever and stupid, as equals, if we perceive them as distinguishable. Radical reformers, despairing of abolishing the discriminations between groups perceived as different, long to cut the Gordian Knot by obliterating the perception. The programme seems even more hopeless than the one they wish to short-circuit. How can we unsee what we have seen? All we can do is unsay it.

Well, it does happen; unsaying can lead to unseeing. In English-speaking countries we no longer see the inanimate world as being divided into male and female, as in all other European countries – in some of which they also see the universal neuter of the English inanimate world as a *third* possible gender. (Another aspect of the deep grammar snubbed by the Anglo-Saxons!) Though this doesn't mean, so far as I know, that the French have any lingering respect for the nurturing qualities of spoons and forks, or any secret conviction that knives and glasses ought to have the right to be kept in special cupboards of their own, from which spoons and forks are excluded.

Nor do English-speakers any longer distinguish – and this is something that has practical social and political consequences – between two great classes of people who exist in other European countries, the *Thou*'s and the *Ye*'s. In the Romance lands, in the nations where Teutonic, Slavonic, and Finno-Ugrian languages are spoken, these two classes are as separate as night and day, as complexly overlapping and intermingling as rich and poor. And yet English-speakers can't tell them apart. When we learn foreign languages we laboriously memorise various rules that have been distilled for us from the observation of local practice.[10] We learn, as it were, that people in Rolls-Royces are in general to be thought of as rich, and that people in clapped-out Ford Transits are by and large to be called poor. But we find it very difficult to see, as the locals immediately do, that some of the people in Rolls-Royces are really poor people who happen to have money, and that some of the people in Transits are really rich people who happen not to. In Sweden, in a fit of leftist moral absolutism in the 1970s, people abolished the distinction between *ni* and *du* (the Swedish *vous* and *tu*), by appointing everyone *du*. This was a strange way round to do it, since the former *ni* class, the people you don't know so well, in a population of 8,462,000, necessarily outnumber the *du* class by about 8,461,950. Do you think *ni* wasn't still written all over a new *du*'s face, as livid and unremovable as the scarlet letter? How long, for that matter, did it take in England before the *Thou*'s looked absolutely indistinguishable from the *Ye*'s?

Even if it were possible, though, to remove all the remaining perceptions of difference, what would the result be? We can think about the world at all only through oppositions and polarities and longitudes – and we can't stop thinking, because no thinking means no eating. If we extinguished the existing oppositions wouldn't we have to find others? And wouldn't precisely the same political difficulties arise with these? You might think that one of the consequences of our apparently wholesome failure to distinguish the *Thou*'s from the *Ye*'s in Anglo-Saxon countries is our pathological sensitivity to other social distinctions: in Britain to the obsessively fine gradations of class that seem to elude other Europeans; in America to the brutal divide between perceived success and failure, which is always, in Britain, mercifully a little blurred by class, and in other European countries no less mercifully a little confused by the counterweave of *Ye* and *Thou*.

The established usages of language are part of the standing timber in the great forest that surrounds us. They seem to have been there for ever, all these great oaks – not only the words and syntax of language, but all the thoughts that have already been thought, all the concepts that have already been understood, all the objects that have already been discriminated. And yet the real life of the forest is out of sight once again: the fetching of the buried minerals and underground water to the surface, and their transformation by the energy of the sun overhead; the fetching of the not-yet-said and not-yet-understood to the surface, and their transformation by the energy of our needs and wishes; the bringing of the visible forest into being.

This is why youth usually has more savour than middle age (foul as well as sweet) – because the possibilities of the imagined future are more engaging and painful than the actualities of the realised past. There is a somewhat similar difference between the satisfyingly well-organised world inhabited by the critic and the art-lover on the one hand and the perpetually chaotic world inhabited by the artist on the other. The first is a forest of finished works. The second is rough ground littered with unresolved problems and baffled attempts, split acorns and scrawny seedlings.

There are plenty of experts who will tell you how to write a novel or a play, because the world is full of novels and plays. A good critic can see how they've been composed, and a good teacher can tell you how to do likewise. But then they've all been written already. The problem confronting someone who actually writes novels and plays is a different one: not how to write *a* novel, or *a* play, but how to write *this* novel or *this* play – the one that hasn't been written, the one he's writing now; and until he's done it nobody knows. Least of all the writer himself.

(One of the difficulties in planning a story is how to make some note of what you've glimpsed as about to happen. Everything in a story is perpetually changing – and this unceasing onward movement is its essence. It's very difficult to remember the shifting complexities you have in mind if you don't note them down – but as soon as you do, the life and movement seem to drain out of them.)

You wave as you turn the corner at the end of the street, and once again you're gone. Of course, I could run after you – and there you'd be. I could phone you, and hear your voice. I could nominate other particular points of view, and use tools to access them: set up a mirror at the corner, and follow your progress until you have dwindled into the distance; install surveillance cameras monitoring your every move. I could give substance to my night-thoughts about all the births, deaths, and tortures occurring *at this very moment* by setting up more cameras in various delivery rooms, police barracks, and hospices. I could have a great many security cameras running. However many I set up, though, I could not – even in principle, if each was a discrete object in space – have enough to see everything everywhere. And the more cameras I had running, the less attention I could give to any one of them. I could keep you in my world only until I had to turn away from you to look at the delivery of the baby, when you would cease to exist for me just as surely as you did when you turned that first blind corner. The baby would cease to be as soon as I turned to deal with the cry of pain from the torture chamber. I could not in practice confer my personal objective reality upon any more virtual events than I could real ones.

A large part of human intellectual endeavour is undertaken to extend the limits of this tiny world of mine; the very existence of the restriction is what stimulates us to such ingenuity in circumventing it. Language and mathematics were no doubt first invented to label and order the world in front of our eyes. But, once Abel has established that this is a sheep and *this* is a sheep, he can say 'sheep' and point into the distance even when no sheep is present, and thereby allude to the existence of sheep elsewhere. Once he has got the sheep/finger analogy going in order to count the sheep in the fold in front of him, he can go on bending fingers down to count the sheep hidden on the other side of the hill, or the number he *will* have after the next lambing. Once Cain has painted a portrait of Abel from the life he can paint him again from memory or imagination, to console himself for his absence after he has mysteriously gone missing.

God has the advantage of us here. He sees all things – and with no need for mirrors, snapshots, or banks of security monitors. His eye is everywhere simultaneously; he sees everything from all points of view at once. It's difficult to imagine what the image on his retina is like as he surveys the totality of his creation. Or perhaps it's not so difficult, when you think about it. Imagine one small and simple part of it – a table, let's say. What would its image be like, if it had been perceived by an eye which is both everywhere around it and everywhere upon its surface and everywhere inside it?

Three-dimensional, presumably. Solid. Of the same texture and consistency as the table it represents (it's not just a mere *visual* image!). If it is truly accurate (and it is) then it is the same size. What is it like seen from inside? Unilluminated, certainly, but that needn't be a problem for God. Endowed with a solidity and texture which are now all-engulfing, and a sense of the physical contiguity of cell with cell, molecule with molecule . . .

The complex reality of this image can be summed up in one simple word; it's a table.

The image of the universe in God's eye, in other words, is a universe. *Another* universe, then, as solid as this one, but . . . somewhere else? Difficult to imagine where, if the universe is as universal as it sounds. In any case, since God's eye is everywhere, it is not only here, but, if there *is* a somewhere else, then it is somewhere else as well. So this second universe, too, must be represented in the same way by a third, and this third by a fourth . . .

This infinite regress can be replaced by a tidier solution if we accept that God's image of the universe is co-terminous with the universe itself. This is the limiting case of representation, the uboama where meaning is extinguished in generality.

I lift my eyes from the inexhaustible possibilities offered by the few square inches of skin on the back of my hand – and there you are in front of me again, back from your travels. And what I'm looking at now is not just a patch of your skin but you as a whole person. At once the inexhaustibility multiplies itself – multiplies itself inexhaustibly.

The sheer blinding complexity of you! The sheer variety of contexts in which you can take your place – physical, biological, intellectual, emotional, historical, social, political . . . ! The sheer number of different ways of coming at you – snapshots, X-rays, medical tests, brain scans, friendly conversation, interrogation, exam results, psychiatric reports, old love-letters, references from employers, sporting records, graphological analyses, shared experience . . . ! It's like looking up at the sky on a starry night. You've been one of my closest friends for forty years, and I'm no closer to grasping you than I ever was.

Your complexity is not the only difficulty – there is a *logical* problem, too, in comprehending something at least as complex as myself. Can you pack a car into a car, or put a matchbox into a matchbox?

And then there's another logical difficulty. I know that, even as I am looking at you, you are looking at the world around you. Even while I am examining your old diaries and talking to your friends about you, you are writing more diary entries and making new friends. You are changing in front of my eyes! You are also looking at me as I look at you, wondering what I'm up to, trying to make out what sort of person I am. I am a part of your world just as much as you are of mine. Each of us is an actor in the drama. I smile – you smile back. You frown – I smile again. Each of us perceives not only the other but himself; perceives the other perceiving himself; perceives himself perceiving the other perceiving himself . . .

The immediately objective world around me is like the electromagnetic field around a charged body. But you, too, have your field. You are a particle in my field, I am a particle in yours. The electrodynamics are complex.

Like me, I know, you are an endlessly flowing and shifting river of perceptions and feelings, an ever-springing source of decisions and initiatives. How can I ever begin to lay hold of this subjectiveness?

The only way I can do it is to take a small selection of the available evidence, and from this manufacture a character – some kind of stable entity, with a personality and an identity that have something of the physically tangible about them. We are like fetishists, who simplify the objects of their desire by reducing the intolerably complex whole to a pair of shoes or a piece of underwear.

I locate your qualities by the same kind of relativities and oppositions by which I locate physical objects. I categorise you as intelligent, because you're more intelligent than me, handsome, because you're more handsome than our friend Charles. Even classes of thirty and nations of fifty million can be broken down into a series of simple binary oppositions, just as the structuralists say. Vicky is clever and Diane isn't. Or they are in the eyes of Louise, because when she comes nineteenth in the class examination list at the end of term she is one below Vicky and one above Diane. But being nineteenth also means being thirteen below Daisy and five above Holly, nineteen below Lindsay and eleven above Dawn, and different selections of these oppositions are important to Louise in different contexts. Her sense of inferiority to Vicky and superiority to Diane, immediately above and below her, might be much sharper than her feelings about Lindsay and Dawn, because everyone knows that Lindsay is absolutely brilliant and that Dawn is as thick as the lid of her desk – they're a maharajah and an untouchable whose differences in station are largely irrelevant to Louise's own experience, somewhere in the middle. A relief, then, for her to get on to the netball field, and be the seventh person to be chosen when Vicky is the eighth. Poor swotty Lindsay is the very last to be picked, but then she always is, and Louise has gone off with her friends long before they get to her. Nor does Louise or anyone else at school notice when a few years later Lindsay comes first in the national Scrabble championships.

Of whose varying positions in the shifting hierarchies of her friends is Louise most conscious? Her own. Whom does she feel most intimately to be defined by the admiration, scorn, jealousy, and indifference of the people around her? Herself. Among the people she has to characterise she herself is the foremost. She does it in precisely the same way as she does with Vicky and Diane – relative to them, and everyone else, by this primary opposition of me/them, me/her, me/him.

Now she leaves all the shifting relativities of school and college behind her and moves in with Henry. So now the pair-bond, me/Henry, becomes the most defining opposition. Among Henry's characteristics relative to his friends is his notorious profligacy, which is to say that he is more profligate than Albert and Dale and Spencer. Now he is more profligate than Louise, too. His profligacy doesn't colour Louise's perception merely of *him*, but of *herself*, too. It makes her see herself as thrifty (not how she's ever thought of herself before), even as mean. More – it actually *makes* her mean, because if one of them is spending the money the other has to be saving it, or very soon there won't be any money left for the other to spend. This is why the pair-bond becomes such a straitjacket, if it's not offset by other relations offering different relativities. It may be why in the end she takes off with Spencer – because she's sick of being forced by Henry to play the skinflint. She's attracted by Spencer's carefulness with money not only because it shows her up as generous by comparison, but because it does actually liberate her from the constraint she has laboured under, and enables her to become, well, a bit of a spendthrift.

What happens, though, when Spencer and Henry suddenly find themselves side by side at the bar one day, and are forced by the awkwardness of the situation to have a drink together? We may get another surprise. It may be careful Spencer and not reckless Henry who insists on buying the drinks, and generously offers to let Louise go off with Henry to his mother's birthday party, because this is a third and different opposition, in which the participants are not constrained to the same relative roles at all. Spencer, as the winner in the contest, can *afford* to be generous, after all – is *obliged* by the terms of this new relationship to show himself in that light.

You look up and see the aircraft, lit by the sun, silver in the morning haze, and your spirits lift as your imagination flies up to join the soaring pilot. But you have the better part of the deal, because you can see that ghostly silver plane, and the pilot can't. The best he can do is to imagine you on the ground below, looking up and envying him.

Recognition, categorisation, and perception are indissolubly bound up with each other; and what the whole complex depends upon is this strange, elusive ability that we have to see an analogy. You might say that even a plant's purely biochemical capacity to distinguish the nutrients it needs is a kind of low-level version of this; in order to absorb nitrogen it must be able to recognise a nitrogen molecule – it must be able to register the analogy between one nitrogen molecule and the next. There is probably a continuum of analogical capacity, from the purely on/off mechanical function of the lettuce, up through the lower ranks of the animal kingdom, to the lion, which sees the analogy not only between antelope and antelope, but also between antelope and ibex, and ibex and gazelle – and sees it whether its prey is lying down or standing up, walking or running, head on or sideways on, in full view or half hidden. The scale continues upwards through chimpanzees, which are able to see analogies between their own behaviour and the behaviour of others, and to express it by mimicry. They can even see analogies with the behaviour of different species – they have been filmed imitating a man blowing through a blade of grass.

And yet this basic animal capacity has proved disconcertingly hard to reproduce in a digital computer. The computer has to be told *precisely* what features count as identifying an antelope as an antelope, *precisely* what variations and transmutations of these features are allowable. I, unlike the computer, can glide effortlessly, without any instruction, from the simple four-legged table in the kitchen to the one-legged occasional table in the drawing-room, and on, through every conceivable variant of the class, built and unbuilt, to metaphorical extensions of the idea: the water table and the multiplication table; the table (and here comes a new extension of the concept, never seen before, but instantly comprehensible) formed by this paragraph, with its spread of sentences laid out on the white cloth of the page for your delectation. A computer programmer can write a complex algorithm which might just about see the analogy between the image of you caught on a security camera and the image of you held in police records. I, on the other hand, can usually see the analogy between you today and you on all the other occasions I've seen you – and do it instantaneously, even when your face is distorted by emotion, or half in shadow. I can do it when it has been transformed by age, or represented in oils and marble. I can locate the traces of you in other members of your family. I can see something familiar about you, even when your face is hidden, in your gestures and the way you walk.

I can go on, and see the likeness between you and a cat. A cat? Certainly – look at the wary way in which you pad around some appetising new idea I've put on the plate in front of you! And between you and a dog. A dog as well? Yes – think of the simple-minded delight with which you wag your tail as you go bounding off into the thickets after some idiotic hare you've started! You don't have paws or a tail, and couldn't possibly pad around ideas or bound after them even if you did. But I can kind of . . . *see* . . . the paws and the tail, the padding and the bounding, just as I can kind of see the hare you've started, and the wild goose that the hare is going to turn out to be.

You protest. Incoherently. 'But even if you . . . I mean, look . . . How could I . . . ? I can't . . . Not simultaneously . . .' And at once, without any special effort or skill, almost without noticing that I'm doing it, I read a *sense* into this confusion of incompleteness, vagueness, and syntactical error. I don't have to correct it in my head – I don't even notice that there was anything astray. I *understand* you to be telling me that, even if I see you as a cat or a dog, I can't be seeing you as both at the same time. Isn't it as if I'm seeing the analogy between your incoherent verbal fragments and a coherent proposition, rather as I see the analogy between a random scattering of stars in the night sky and the figure of a hunter? 'Yes,' I respond at once with my professionally polished articulacy, 'I think . . . I mean, I don't mean . . . I don't actually . . . Well . . .' And on we go, each of us taking the intended sense from what was said, just as if it had actually been expressed, reading into the muddle the message that we expect – or, even if not the one that we expect, one that fits the context, that relates to our purposes and interests.[11]

There's no limit to the volume of analogical traffic that our thinking can bear, to the layers of sense that can be read into something, any more than there is a limit (another analogy!) to the volume of electromagnetic traffic that space can bear, as waves travelling from any point in the universe to any other weave through each other without interference or overload.

Our skill at analogy is tested to destruction in the pun, where the coincidental use of homonyms to denote unrelated objects invites us to imagine the possibility of something in common between (for instance) a glass container and the partial openness of a door, or the Christmas dinner in front of us and a country in the Middle East. At this point our almost endlessly elastic ability is finally stretched beyond its limit, and we are amused by our helplessness to focus any single common object for the term, rather as we are when our almost endlessly adaptable ability to move around on our two feet is finally stretched beyond its limit by drunkenness or a pitching deck.

There's nothing mystical or ineffable in all this. One day, no doubt, neurologists will begin to piece together the mechanics involved in seeing an analogy. Then, perhaps, we may be able to program a computer to do the same. What all this surely suggests, though, once again, is that our thinking is *not* binary in structure, even if it is built out of binary events such as the firing or non-firing of a synapse, the opening or closing of a logic gate. The discrete elements, like atoms, compose wholes with qualities that have nothing in common with the qualities of the atoms. After all, we already use the binary code in electronic devices to represent the continuities of the world's visible appearance and sound without its inclining us to believe that the world itself consists of digits, any more than our ability to represent it in the silver salts of a photographic emulsion inclines us to believe that the world is composed of silver salts, or even that it can be resolved, anywhere except in a photograph, into the same distribution of particles as the one that composes the photographic emulsion.

Analogy is a simpler concept than the structures of invisible rules that Chomsky and others postulate, and it's also logically anterior to them. Without the idea of analogy it's impossible to explain how things are categorised so as to fit the rules. Give me any set of rules for deciding whether something is to be categorised as a cat, and I will think of an entity which fails the rules, but which we might in certain circumstances want to call a cat. Analogy offers some explanation, as rules do not, of how categories are extended to colonise the unknown. You need the idea of analogy to understand how it's possible to have rules at all – and how the rules themselves can be constructively subverted.

So, if analogy is a simple concept, it begs a simple question: what is it?

Well, what is blue? What explanation can I give, except to show you various things that are blue, and hope that you see the analogy between them? What explanation can I give of analogy in its turn, except to show you various forms of analogy, and hope that you see the analogy between one analogy and another?

There comes a point in our search to understand any concept when we can analyse no further, when all we can do is – well, see analogies. And when we've grasped the analogy we've gone as far as we can go.

And yet . . . there's still something eluding us here. Seeing blue, in the sense of *understanding* that it's blue, that it relates to other examples of the same thing, isn't a simple passive event that just happens to happen to me, any more than seeing a bus and recognising it as a bus. Nor is grasping an analogy, in any way that makes it possible to build on what I've grasped. They are both examples of thinking a thought.

So how do I think a thought? Or, even more simply, how does a thought get thought?

V: Homewards

Off-Line

Thinking of nothing; idle thoughts; dreams

Let's try to reduce thinking to its simplest level. Forget for the moment all the difficulties of understanding how thinking relates to the outside world (even if it's only to the floaters behind my own eyelids), or the problems of capturing the thoughts I think in words. I'm going simply to lie down in a darkened room and think about what's going on right here inside my own head, without regard to anything else. Or rather, not even think *about* it, since introspection is so notoriously difficult – simply *think* it. No – something even less active than that. I'm going to let any thoughts that happen to occur think themselves. I'm going to try to describe only what is immediately, totally, and intimately present to me.

So . . . Darkness. Quietness . . .

In the darkness, though, *something*. What is it? Some ghostly sense of something that's not blackness. Some sort of light, then? No, not light. A ghost of light. An allusion to light. Not an image of some sort? Not an image of any sort. A pattern, then? There's no pattern in it. Just a kind of presence – flickering, mercurial, gone. No, there all the time. There, certainly.

Something indefinite, like mist? Not really, if only because mist is visible. So this is invisible? Neither visible nor invisible. More like a very faint flickering gaseous electric field in a failed discharge tube, or the place where it was after the tube has been switched off. Simply . . . *there*.

Where? *There* – in front of my eyes.

But it's pitch-dark! So it's something in my mind? Not at all – it's something in front of my eyes. Right . . . *there*.

And I can give not the slightest account of it. A somewhat discouraging start to the exercise. Scarcely of much significance, though. It's a vanishingly marginal phenomenon. I should be little the poorer if it had never happened to me, and I doubt whether I should notice if it never happened to me again. I shouldn't even have noticed that it *was* happening if I hadn't been on the lookout.

Already I've ceased to notice it . . . And now here's a thought wandering into my head of its own account – a sharp recollection that I haven't written a letter I should have written. Ignore it – worries are far too complex to consider for our purposes . . . Nothing, nothing . . . And now – yes! Someone I once knew slightly thirty years ago has for some reason come sauntering into my mind. No, or I'll get sidetracked into thinking about exactly what his features are, etc. . . . Now something soft . . . red . . . made of cotton . . . A skirt, and emerging from it a suntanned bare knee, remarkably smooth and warm to the touch . . . No – no distractions. Now a feeling that I should think of some other way of going about this whole business . . . No – no calls for decisions – that's another problem altogether . . .

Now, here's something more like it. Something very simple. The sound of an oboe. Not a complex tune but the single sustained A sounded by the oboe at the beginning of a concert for the rest of the orchestra to tune to. Good. There it is . . .

Well, I've described it, as I set out to do, in the sense of describing the phenomenon that it represents. But what I *haven't* described is the thought itself. What do I mean, when I say it's *there*? What *is* it exactly that's there? The thought of the taste of potatoes and custard, I argued earlier,[1] wasn't *complete* until it had been identified in words, or by means of some other symbolic device. What we're trying to do here, though, is to dig back to locate the incomplete raw material of such a thought.

Do I mean that I have going through my head the experience that I would have if I were actually hearing someone play an A on an oboe? No, what I have in my head is some kind of representation of the experience. A representation actually pitched at A, like a recording, with something inside my head actually vibrating at 440 cycles per second? Of course not – nothing's vibrating! So in what sense is the note A? Well, in the sense that it seems about right for A. *About* right? – Yes. I don't have perfect pitch. In any case the question of whether it's an accurate representation of A is not at issue. I'm thinking about the experience itself, not about what it represents. So how do I know it's an A? – Because A is what the oboe plays for the orchestra to tune to, and this is what I specified I was thinking about. Should I recognise it as A if I hadn't specified it as A? – It's not a question of recognising it, since I *did* specify it. It's A because I say it's A!

But I'm certain it's the sound of an oboe, and not a clarinet? – Absolutely certain. What makes me so certain? Well, I could talk about its plangency . . . its acidity . . . But I don't need to. I can hear it perfectly clearly! Well, not *hear* it. But there it is. And now I'm thinking about the sound of a clarinet, for comparison. There's the sound of the clarinet . . . and now there's the sound of the oboe. I can think about the ways in which the two sounds differ. So *how* do they differ? Well, the oboe sounds like *this* . . . and the clarinet like *that*. I realise that I'm not making myself entirely clear to you. But then I'm not trying to make myself clear to you. I'm not talking to you, for once. I'm talking to myself. Think about an oboe and a clarinet yourself, if you want to. To you, for all I know, they may both sound indistinguishable from a foghorn.

In any case I can think about the A of the oboe, and think about it usefully, think about it purposively, without imagining or remembering it. I can think of how it always induces in me a pleasurable feeling of excitement and anticipation. When I think this I don't go through the full performance of remembering the sound, and then another complete dress rehearsal of the excitement and anticipation, and then a third of recalling some regular connection between the two. But then nor (in the first place) do I think about it in words. What I have in mind is not the assertion, 'The A of the oboe always induces in me a pleasurable feeling of excitement and anticipation.' This is what I say to you (or to myself) after the event, to fix something much more elusive (as with the taste of the potatoes), which involves neither words nor images.

What is this elusive event, which is so easy and natural to experience, and so difficult to capture? It seems to me that I make a kind of . . . passing mental reference to the sound, then a passing reference to the feeling, and that I kind of . . . associate the two. But then I associate them because they are (it seems to me) already associated!

I shouldn't like to have to offer this account in a court of law, to hostile counsel.

Sometimes, it's true, I do feel that I think in words. It sometimes does seem to me that the thought forms itself, inside my head, at the very moment of its birth, in some kind of quite coherent verbal form. It is the choice of those particular words, and their falling into place in that particular construction, that actually constitutes the thought. But this makes the event not clearer but more obscure! Where have these words come from? How have they (as it seems to me) chosen themselves? What is it that makes the proposition forming itself inside my head a real expression of something, rather than an arbitrary assemblage of well-formed formulae? Even if introspection is impossible, and I can't watch any one particular proposition take shape, how can I not look back with hindsight and give some account of it? If I had ever fought in a battle I don't suppose I should have been able to give you any account of it while I was fighting it, or even of my concomitant feelings of terror, despair, hope, and pain; but that doesn't mean that I couldn't have given you some kind of historical account of both battle and feelings after the event. The formation of that last sentence, however, is still, even now, after the event, as opaque to me as the formation of this present one. I can't explain it to you – I can't explain it to myself.

I try thinking a different sort of thought – one that involves not slippery sounds and ambiguous visual images, but entirely abstract notions, expressed entirely in mathematical notation. I multiply fifty-four by forty-seven – a calculation somewhere on the outer limits of my mental arithmetic . . . And, yes, once again I'm imagining. I'm picturing the sum being done on paper, with the 8 from the first 4×7 being written down below the line, and the little 2 carried into the space above the tens column . . . The source of these pictures is not abstract, though. They represent propositions from the multiplication tables, already formulated in symbols, just as the written figures themselves would as I put them down on paper.

I imagine something that has no connection, however tenuous, with any kind of objective correlative. I imagine that . . . I'm playing an oboe myself, now, in spite of the fact that I'm not, that I've never played an oboe, that I couldn't begin even to get a sound out of one. And at once, effortlessly, some kind of visual ghost of the instrument springs into being, and is in some kind of relational ghost of contact with some kind of kinaesthetic ghost of lips and fingers. These spirits have been summoned up not by mysterious forces within me, but by me myself. By the director of the whole enterprise, quite openly and straightforwardly.

But, as a source of oboes and lips and a sustained A, *I* am just as obscure as every other provenance.

So what we're talking about is sounds and images that are not in any normal sense sounds, images, or sensations, and that involve, so far as one can tell, none of the mental or neurological events that would be involved in the experiencing of actual sounds, images, and sensations. Now I go further, and imagine things that *could not* be sounds and images in any normal sense, impossible things that *could not* involve any of the mental or neurological events involved in the experiencing of actual ones. This oboe I'm playing is made not of ebony or silver, but of hopes and fears and good intentions. And playing not A, but sausages, say, and the cube roots of negative numbers.

A slight pause here, certainly, while I work on this. Dimly, though, uncertainly, I begin to imagine . . . *something*. Something not quite visualisable or hearable, but which is, in some way, an oboe made of hopes and fears and good intentions, with, emerging from its bell, the plangent tones of sausages, the acid harmonics trailed by the cube roots of negative numbers. How do I identify the material it's made of? How do I recognise the note? Well, how did I identify the ebony, how did I recognise the A? I knew they were what they were because I had specified them. My specification came into being as a constituent of the universe, whether it specified anything or not – whether it specified anything capable of being specified – just as Canute's command to the tide did, even though it could never have had any effect upon the tide. (The vacuousness of the command was after all what he was attempting to demonstrate, but its significance as an utterance is demonstrated by the fact that we've been talking about it ever since.)

How do I think of something that *can* take real form – and that can't take real form *until* it's been thought of? How, in other words, do I think of an answer to a simple open-ended problem in the real world, where no multiple-choice selection of answers has been provided by the management? I've got to buy a birthday present for an old friend, and I have no idea what. I go through the day with my mind a blank. I lie awake at night – nothing comes to me. And then, next morning . . . yes, sitting in my head as innocently as a child who's slipped into class late while the teacher's back was turned, an idea. But how this idea formed and got itself thought, simple and concrete though it is – of that I have *no* idea. There was an input in the shape of a problem. There were various constraints to bear in mind, various ineffectual efforts to remember what my friend might have expressed an interest in, what I might have noticed in the shops. Then other thoughts took their place and the whole thing was forgotten for the moment. And then there was an idea.

As an account of its provenance this is about as illuminating as explaining how to make a pizza by saying you phone a pizza delivery service and one gets delivered.

Well, you can imagine the kind of thing that goes on at the pizza parlour to connect the two events, even if you don't know exactly how it's done. You can at least feel confident that there is some fairly clear and definite chain of cause and effect involved. But you have to have a considerable faith in determinism – a *blind* faith – to take on trust the existence of some coherent causal chain between the problem of my friend's birthday present and the well-chosen lithograph that represented its solution.

Defeated in my efforts to examine what's going on in a factual way, I try another approach. I write a story about it. I invent a character who is trying to think of a birthday present for his old friend. I provide an account of various experiences he has had during his life: of his friend's character and of my protagonist's understanding or misunderstanding of it – of the whole tangled history of their relationship and the choices they have made that express it . . . It's all very coherent and convincing, so that when, at the end of the story, he finally gives his dear friend not a valuable first edition or a well-reviewed recording of the *St Matthew Passion* but a lithograph with a particular pattern of orange and blue spirals you feel, yes, I understand every labyrinthine twist and turn of the thinking that led him to this surprising solution. But where did *I* get all this from when I wrote the story? All these labyrinthine twists and turns? This challenging but tasteful piece of artwork? I've no more idea than I had before!

You look wise; you've read books about this kind of thing. These ideas, you tell me solemnly, come out of my unconscious. They are the product of my suppressed fears and forbidden appetites. They are connected with experiences in my early childhood that I have laid away and concealed from myself. Oh, sure. But as a practical explanation of their origins you might as well say that they came from fairyland.

No, because their source could be confirmed by independent investigation. Under expert psychoanalytic guidance I might well recall an incident in which orange and blue, or spiral shapes, had performed some relevant symbolic function. Might I also discover the sources of all the apparently arbitrary examples I have hit upon in the course of this discussion? Why I suggested thinking of an oboe rather than a bassoon or a tuba, for example. Why the imaginary oboe was to be made of cube roots rather than cosines or council by-laws. Why, for that matter, the alternatives I have just suggested were a bassoon and a tuba, cosines and by-laws, rather than a mouth-organ and a game-lan, religious convictions and twinges of neuralgia . . . Why these alternatives to the alternatives were a mouth-organ, etc. . . .

Might I also discover why I had chosen this particular way of presenting the argument? Why I had chosen those particular words? Why I was now choosing *these* particular words . . .?

Fairyland is beginning to seem a rather more hopeful place to look for the source of my ideas than this country of endlessly receding planes where every surface you touch turns out to be merely the ghost of something else – of something that is a ghost in its turn.

Now, tired of thinking up practical solutions to practical problems, and of imagining the unimaginable, I let my mind wander. I allow myself to daydream a little. An agreeable experience. Somehow (it now seems to me) the task I have set myself, whatever it is, is getting itself done – is done already – has brought me success and repose, the regard and affection of the world . . . So what does this experience – a perfectly familiar one of mine, perhaps also of yours – actually consist of? However indefinable the act of imagining something has turned out to be, am I at any rate imagining definite things? Particular words of praise? Particular sensations of pleasure? Particular expressions of regard and affection?

Well, yes, to some extent. Particular words do come into my mind, do repeat themselves, do edit themselves until they are even more pleasing. Particular smiles and expressions on particular faces come and go, come and go again. Particular sensations – an excitement in the pit of the stomach, for instance – announce themselves. But for the most part, I think – though of course it slips away out of the corner of my eye once again even as I turn to look – the words, the looks, the feelings are *not* specific. I am in a shifting world of agreeable and reas-suring . . . *presences*. They hang in the air like the presence of royalty at a social occasion, changing everything even though you never quite manage to set eyes on the royal party yourself.

Now the daydreams fade. For some reason I have begun to think of all my failures and shortcomings, of all my anxieties, of all the things that still have to be done, still have to be decided. Do I have a sequence

of clear images of failure and problems and looming disaster? Well . . .
possibly . . . Then the images swirl up like the pack of cards at the end
of *Alice*, into a panic as formless and distressing as the daydream was
formless and reassuring.

There are good reasons for me to be reassured and hopeful. There
are good reasons for me to be fearful and distressed. Some last shreds
of purposiveness and relevance cling to these inward events. But the
forms that these feelings take, the actual sequence of experiences
through which my mind passes, are as indeterminate as the paths of
electrons around the nucleus.

And then, sooner or later, all this considering and imagining and daydreaming and panicking ceases, and I begin to glide downwards into
the borderlands of sleep. Now thoughts drift into my head entirely at
random, unshaped by even a general sense of gratification or anxiety.
Perhaps memories, perhaps half-memories, perhaps imaginings . . .
Too fugitive to give any account of . . .

Until at last I am completely asleep. 'I lose consciousness', as the
phrase goes, I 'become unconscious'.

Or do I?

I lose consciousness of the external world, certainly. I lose the thread
of inward experience in any form that retains the least hint of serving
some intentional end. But often, of course, I do continue to experience
something in the form of dreams. My consciousness persists, in spite
of my losing it. Dreams happen to me no less than my imagining of the
sound of an oboe happens to me. They can be as vivid and engulfing
as the most vivid and engulfing of waking experiences. *More* vivid and
engulfing; how often, except in waking from dreams, do we find ourselves in the freezing paralysis of horror?

When you first begin to think about them, dreams seem to have a lot
in common with our waking experiences. They often make allusion to
real events, in the recent or the remote past. They often, like daydreams and panics, embody familar emotions and anxieties. Very
occasionally they even seem to present us with practical solutions to
real problems, as mysteriously but as helpfully as the solutions that
offer themselves in daylight (as, most famously, when the nineteenth-
century German chemist August Kekule discovered the ring structure
of the benzene molecule by dreaming of a snake biting its tail).

They also, however, allude to purely fictitious events, and offer a fraudulent familiarity with emotions and anxieties completely alien to us. They offer solutions, sometimes apparently detailed and compelling ones, that turn out to have no relevance to our problems when we try to examine them in the light of day, and often not even a coherently recoverable form. Often, too, these solutions are to problems that have themselves no existence in the waking world. The scenes and situations for books and plays that I have dreamt! And that in the morning have faded, together with the books and plays themselves, to nothing but a faint golden afterglow.

The first difficulty in thinking about dreams (as opposed to having them) is that our waking access to them is largely arbitrary. The dreams that we recall, according to scientists whose speciality is the research of sleep, are the ones during which we wake. They can assert this because dreaming seems to have an objective, externally observable correlative – a certain form of rapid eye movement that recurs intermittently throughout the night. At any rate, sleepers wakened during these periods always report dreams, whereas sleepers who are wakened during the intermediate periods often report that they have had no dreams, even though they have previously had periods of rapid eye movement.[2]

(This leaves open the existential status of dreams that we *don't* recall – for example, the ones that are assumed to have accompanied the rapid eye movement during which the sleeper was *not* wakened. A similar problem arises with the objects of all forgotten experience, but where the source was external there is at any rate the theoretical possibility of its being reconstructed from the record or recollected by others. But of a piece of experience that has no foothold in the mind of the subject after it has occurred, and no foothold in the record or the minds of others, either, what can one say?)

The next difficulty in accessing a dream, even if you feel you can recall it clearly, is giving any account of it, because as soon as you try you can feel it slip away through your fingers. It happens even when you try to get it straight in your own mind – even when you try to recollect it without the intervention of any words at all. The very act of giving the details any kind of coherent form, even if that form is one of wordless recollection, falsifies them, because coherent form is something that they do not possess. Worse, even as the details vanish, one by one, so the whole essence of the dream leaks away with them, in the way that the magical lustre of the underwater stone dries and vanishes when you bring it to the surface. The same kind of loss occurs when you try to give an account of waking experience, too, particularly of feelings, or of complex events, or to fix it in your memory. But with the dream the loss seems catastrophic and total.

We do not allow ourselves to be deterred, of course, by these apparently fatal difficulties. We go on trying to recall and recapture our dreams. I'm going to tell you some of mine before the chapter is through. If I gave you half a chance you would start telling me yours. One of the characteristics of dreams is their air of importance. They seem to require us to take note of them, to recount them, to make sense of them. We are readily persuaded that they are messages from God, or predictions of the future; that they give us access to an inner self, to a storehouse of memories we thought we had lost. We want to accommodate them to the meaningful, just as we do with everything else that happens to us. We want to believe that they are part of the great chain of cause and effect through which we make sense of the world, that they are the output, however transformed and of whatever significance, of some input.

And of course there *is* an input from our conscious experience. The characters and events in dreams, like the characters and events in stories, do contain elements that we recognise have been drawn from our waking life. Even the elements that we *don't* recognise, even the material of the most outlandish fantasies, must have been taken in from the external world in some form or another, through our eyes and ears, through our nose and tongue and fingertips. If we had had no experience of the outside world then none of this inward experience could exist either.

You might say, for that matter, that everything in a painting or a poem could be accounted for in terms of input. The caverns measureless to man are Wookey Hole, etc. But Wookey Hole is not a cavern measureless to man until it has passed through Coleridge. Even if you take a straightforward description of some actual aspect of the world, a picture or a passage which you might say is attempting to be all input, that input still doesn't account for *everything*. Or even for very much. There's Henry VIII, planted solidly on his two great hams, gazing head-on at the painter. Now here he is in Holbein's almost photographic representation of him, planted solidly on his two great hams, gazing head-on at us. Input – output. King in – portrait out, as straightforward as cow in – hamburger out. And yet you can't help

feeling, both in the case of the Holbein and the hamburger, that everything of interest is really subsumed in that dash written on the page
between input and output.

Coleridge, Holbein – these are elevated examples of the great transformation process. About the processes of art and literature one can
imagine anything. But the same is true in every kind of invention. The
two essential inputs of a mousetrap are, on the one hand, the mice eating you out of house and home, and on the other, the odd lengths of
wood, pipes, rusty iron bars, and coils of wire in the lumber-room; but
there is no mousetrap until the mind of man has brought the two into
relationship to each other. Here are the brochures for Ischia, Tunisia,
the Costa del Sol, and the islands of Greece; and your summer holiday
is invented wholly out of them; but until you have selected and
wrought there is no summer holiday.

What inventory of input, likewise, what catalogue of conscious
experience, could begin to account for *everything* in a dream? All
right, the seven lean kine represent seven lean years (never mind the
magical precognition – think of it as a straightforward economic forecast), the flight of stairs represents sexual intercourse . . . But what
about the way the stairs rise out of some kind of industrial workshop,
and pass through something that seems to be an operating theatre?
What about the expression on the face of the sixth lean cow, as it turns
to look at you with a sorrowful smile, just so?

What seems to be important to Pharoah in the dream as he dreams
it may well be not the number of kine, or their implications for
Egyptian commodity futures, or the key they offer to his neurotic
obsession with funeral monuments, or even their derivation from
some particular cows he once saw in a field near Cairo, but the way
in which that sixth cow turns and looks at him with a sorrowful
smile, a way in which no cow or any other creature ever has in life.
Just . . . yes – *so* . . .

I say that the cow turns and looks at him just *so* . . . And already I've
gone wrong, because just *so* is exactly how things in dreams *don't*
look. Nothing in a dream is just so – not the actions of the participants, or their identities, not the time or the location. Even when my
three children, who have such an overwhelmingly definite and individual presence in my waking thoughts, visit me in my dreams, they
change their form, and even their number. They exchange part of their

identity with other people of the same names who mean nothing to me.

Everything in a dream is in continuous flux. Even before I get to the depredations caused by the attempt to recount or recall it – even at the time, during the course of the dream itself (if I can somehow imagine myself back to this, perhaps in the most general way, without any attempt at specifics), almost its only stable characteristic is a kind of perpetual deliquescence. If it's difficult to describe the back of your hand in waking life, imagine trying to do it in a dream. Even as you looked it would turn out to be the back of someone else's hand, and not the back of their hand, but the back of their head, in fact not the back of anything at all, but a small pet animal such as a crab, looking up at you with a plaintive expression, or rather nothing to do with a living creature, but the surface of the Kara-Kum Desert as a mediaeval army crosses it on its way to play golf in Palm Springs . . .

No – the difficulty would be even more fundamental than this. The details would lack not just any continuity of identity from one moment to the next – they would lack any definite form even at each single moment. The creatures and objects that seem to come before you in a dream have no explorable extension. The problem with the back of your hand in waking life is that its details elude you as soon as they escape from the tiny focus of foveal vision; the determinability of the back of your hand as an object is reduced to a sense of its potential determinability, the feeling that every detail *could* be determined if you directed your attention to it. In a dream, however, even this potential determinability has disappeared. If you look at the back of your hand in a dream, even in the moment before it changes into a plaintive crab, you won't quite be able to. There will be no features to which to direct your attention, unless you consciously invent them. There will be no sense of even an ungraspable definiteness. It will be more as if you simply *know* it's the back of your hand – as if the object itself were somehow just out of your field of vision, its actual nature suggested by supposition from your knowledge of its identity.

Sometimes, too, it's not just the detail that shifts but the whole mood and sense of the dream, with unexplained abruptness. I dreamt that I was leaving the house on a sunny morning, in a genial mood, putting on a crocheted trilby hat that I found slightly comic. But as I closed the kitchen door behind me it wasn't a sunny morning at all – it was pitch-dark. As I blindly edged up the steps to street level I realised that there was some terrible unknown creature squeezing up against me in the blackness, and I awoke in horror. From sunny morning to black midnight, from cheerfulness to blind terror, in the blink of an eye, without the ghost of a narrative connection.

All the same, in so far as you can say anything about it at all, a dream, like a story, is something complete unto itself, not verified or justified or given meaning by any causal relationship with the world. Whatever goes in as input, something happens to it that transforms it and makes it what it is. Something or someone takes over.

The question is: who or what takes over?

The answer's obvious: *I* take over. The dream, after all, is being performed inside a theatre owned and managed by myself, and I surely have a considerable say in the way things are done here.

At once, though, a difficulty arises. I am also an actor in this theatre. I am in amongst the action, just as I am in life. I am one of the characters, limited to a particular point of view, and while I'm down there on the stage I can't easily take an overview of the script, or keep an eye on the box-office returns. I often seem to know no more about what's going on in my own theatre than the rest of the cast do. In fact I often seem to know a lot less than I do in the waking world. Things that I know perfectly well in daylight I don't know at all when I'm dreaming. I also entirely fail to notice the existence or significance of a lot of things that seem obvious as soon as I think back on them in the morning.

I dreamt that a producer told me he had persuaded Roger Moore to play a part in one of my plays. (This was a literally theatrical dream.) I told the producer that I didn't know what Roger Moore had been in before – that I had never even heard of him. And I was speaking the truth; in the dream I didn't and hadn't. The producer expressed amazement and amusement that I could be so out of touch. And then he told me that Roger Moore had played James Bond in some of the Bond films. The producer whom I was dreaming – whom I had made up, and educated, and taught all he knew – was aware of this; but until he told me, I, in my own dream, was not.

Another dream: I was trying to make a presented documentary about Cambridge. I wandered round the town, unable to think of anything to say about it, and met Alan Bennett, who was also making a presented documentary on the same subject. He explained to me, in a characteristically self-deprecating way, that he was using Cambridge as a metaphor of national life, so that the first court of each college would serve as a symbol of the political world, 'etcetera' with the other courts. I was jealous that he'd managed to think of this. But of course *I'd* thought of it, and supplied it to him, even though I couldn't think of it for myself.

There are parallels with this in the waking world. The failure to know what one knows, or to appreciate the obvious significance of things, is often a characteristic of panic. Once you've got it into your head that the house is on fire the smell of burning persists, however calm and quiet it all seems. Once you've turned your brother, in a single terrifying first glance, into a cardboard and cotton-wool Father Christmas, it's difficult to turn him back into flesh and blood, however much like flesh and blood he subsequently appears to be.

The story, like the dream, as any novelist or playwright will tell you, develops a logic of its own that begins to generate events of its own accord. The author finds himself in the same position as his characters, swept along by events he can no longer quite control. The same thing happens in small children's games of (so-called) make-believe, which change their form with bewildering speed as new possibilities emerge that change the situation, even the nature of the game. 'You be the Daddy who's living in the wigwam, and I'll be the lady who comes in to buy things . . . No, *I* know! I'm a princess, and you're a Mummy who comes to my birthday party . . .'

The different players in the game often battle with each other for control over the game's content and conventions. The author, too, sometimes finds himself struggling to regain control of his own story. Some stories fight back at the author; some games defeat all their participants. The same sort of struggle occasionally goes on between the dreamer and the dream he is participating in.

I dreamt that the Chairman of the Board (*what* Board? – I've no idea) expressed impatience as once again I took his summing-up as the starting point for further discussion. He made me feel that I was a newcomer to corporate life who didn't understand the conventions by which boards of management operated. But, reflecting on the dream afterwards, what surprised me was that this was not the end of the story. I fought back – was conscious of fighting back – by suggesting forcefully that the Chairman was trying to cut the discussion off before the matter had been properly thrashed out. The Chairman was inside my dream, and I could, as the author, have crushed him from above. But instead I was in there with him, as a character in the story, on the same logical level as himself, defending myself in exactly the same way as any of the other characters might have done, responding to events, as in life, trying to exercise some control over them. I wasn't a passive perceptual machine; wasn't a vessel through which forces not myself (from the unconscious, or from the memory-bank) passed. I was a person in the same sense as I am a person in my waking life.

Yes, I was more than a character. I was not being written or manipulated by anyone – not even by myself. I was an autonomous person. But, autonomous as I was, I was still located within the story, not outside it or above it. I was bound by the story's terms and conventions just as we are bound by the terms and conventions of our waking world.

The Chairman seemed to be just as autonomous himself, just as much a person. From my observation, at any rate, he gave no sign of being supplied with his actions and attitudes and dialogue by some higher authority – certainly not by me. Which may suggest merely the subtlety and skill with which I was supplying them. A subtlety so profound that I could detect no sign of my own authorship, even from my point of view as author.

Something similar could have happened if the dream had been a story. In my outline for the scene the Chairman of the Board is scheduled to win his confrontation; in the finished text he loses it. I may consciously have changed his mind as I worked, and reorganised the plot. But the Chairman may have lost in spite of my intention. As I write the scene, the dissenting director argues his corner with a vehemence and determination that I hadn't foreseen. No doubt I'm somehow in there with him, just as I was in my own character in the dream. In the case of the story I still know I'm the author, even if I'm losing the fight. I still – regrettably – have all the labour of invention and writing to perform, however much the character is jogging my elbow. Dreamer and author alike, however, have become entangled in their own creations. Although the author began by telling the story, and the dreamer by dreaming the dream, the story has ended up by telling the teller, the dream by dreaming the dreamer.

Or has this transfer already happened, in some sense, at an earlier stage? When I as author *plan* the Chairman's triumph (whether in my written notes or inside my head) I'm already telling the story, and the same considerations already apply. How do I decide, as I write my notes, that the Chairman is to win the debate? How do I know that this is what I want to happen? Because, in one way or another, swiftly or slowly, easily or reluctantly, the story comes to me, because it tells itself to me. I'm free to accept or reject it, of course. I'm pleased with having found the story of the Chairman's winning, because it contrasts interestingly with everything that the story up to this point has made us expect. On the other hand it's going to make things difficult in the next chapter. As I ponder this an alternative story begins to tell itself to me. I see the roof collapsing and falling on him in the very moment of his triumph . . . Absurd; I put it out of my mind . . . Except that I can see his hand in its pinstripe cuff, projecting from the rubble . . .

In stories and in dreams, as in perception, something offers itself that we then take hold of and transform. The same is true in every form of invention, even the most modest. Where are you to go on holiday? How is the kitchen to be arranged? Some kind of notion separates itself from its background, from the noise of considerations and arguments and other possibilities, and moves towards you. It seems to relate to your purposes. Something about it attracts you. And you select it. You name it to yourself. There is a dialogue, once again, as in perception, a traffic between you and the world.

Or between you and yourself.

Sometimes one is present in one's dreams not only as a character, but as the controller of the dream in person, manipulating events, as if Hitchcock, making his habitual brief appearance in one of his own films, had continued to call for rewrites and design changes even while he was on camera.

I dreamt that I was driving through France, on holiday with my family, in a Citroën grand piano. Actually from what little I could see of it under the dust-sheets it looked like a Challen; but I thought of it as a Citroën. We kept being forced to stop because one of the shock-absorbers was breaking down. Or rather it broke down once, and I told the man at the garage in Amiens that it had broken down many times before (which is exactly how one might establish the repetition of an event in a film).

But what struck me was this: that I, as author-producer of the dream, was conscious of struggling to make the dream allow us to continue the journey, of trying to make sure that it had an upbeat tone and a happy ending, the way that we in the Front Office wanted.

Who was thwarting my wholesome proposals? Who or what was I struggling *with*? I was struggling with the dream!

To win my battle I made the garage-man at Amiens discover, in a last-minute script change, that enough of the other shock-absorbers were working for us to continue. And indeed the scriptwriters, after their grudging acceptance of the demands of management, began to warm to the idea. They had the garage-man go further. He rocked the iron frame of the piano gently back and forth, demonstrating that it still had plenty of flexibility in its mountings.[3]

Sometimes we intervene in a dream much more dramatically still. We hate it so much that we march on to the studio floor and close the whole production down unfinished. It takes a real effort of will, as it did for Samson when he stretched out his arms and shattered the Temple around him. As it does in the waking world when you force yourself to break out of the dialogue you are locked into, to abrogate the convention you have been bound by; when you walk out of the argument, the job, the marriage; when you finally say the thing that could never be said.

The philosophers of the past who worried about how we knew whether we were awake or merely dreaming might have adopted the deliquescence of dreams as a practical test. If you've managed to sustain a coherent anxiety about whether you're dreaming or not for long enough to seem to have written it down then you're almost certainly awake.

Do we in fact ever seriously wonder about this, outside the pages of ancient philosophy textbooks? There are sometimes moments of confusion immediately upon awakening, certainly, when the anxieties of the dream trail on, and it takes time to recognise that the projects and problems filling one's head have no reality in the waking world. In the delirium induced by fever, too, it's hard to be sure of the existential status of one's surroundings – perhaps also in the strange remoteness that can overtake you when you're on the point of fainting. I can't recall ever feeling, when I was in good health, that I might be dreaming, though people in stories seem to, at moments of great joy, or when events take a turn which seems difficult to credit. It might be more to the point if the suspicion came to them that they were in a story.

The only time I can remember the question ever seriously occurring to me was when I was not awake but indeed dreaming (though I didn't manage to commit my doubts to paper). I was approaching Amsterdam by car, and I suspected that it must be a dream because I had started arriving as soon as the intention of going to Amsterdam had been formed. But it was all so intensely real and familiar that I thought, No, this can't be a dream – this must really be happening. And I remembered from previous visits exactly how to get to my destination. It involved going right through the great central square, surrounded by high Gothic buildings, to a road leading out of town on the other side. Then I saw that the country started immediately beyond, and I realised that it must be a dream after all, because it's only in dreams that the country starts in the middle of the city.4 When I awoke I realised that the city had borne not the slightest resemblance to Amsterdam. It had been a fictitious city, with perhaps touches of Delft and Haarlem, and a central square loosely based – in the way that fictitious locations in novels are on real ones – on the Grand' Place in Brussels.

The only entirely definite details that I recall ever retrieving from a dream have been linguistic – particular words that have survived into the waking world with remarkable integrity and unambiguity. I dreamt of reading a review of one of my plays in the *New Statesman*. It was unfavourable, but moderate in tone. One sentence I recall word for word: 'The arrival of the wife in an eighteenth-century emu-feather hat is the first contact with any external reality.'

Perhaps words can sometimes survive intact because, like numbers, they are the coinage that we have minted to make the evanescent products of our experience bankable for ourselves and marketable to others. They are intended, like silver dollars and copper pennies, to survive being passed from hand to hand, from seller to buyer, from long storage in dark vaults to the brightness of the shop. Not even the most vivid experience, even in the waking world, is entirely definite until we have found linguistic expression for it. And even as we do, some of the brightness and immediacy goes out of it. Its particularity begins to fade. When we want to retrieve the experience now we find only the words, and we have to reconstruct the experience from them, as if we had sold our home-made cheese to make money, and were now using the money to buy factory-made cheese from the supermarket. Since the words are coins – parts of the common language, not proper names in a private language of our own – what they purchase is no longer the exact experiences for which we sold them, but slightly more generalised versions: something closer to the mean of all the particular experiences put into circulation by all the various traders.

How do I know, in a dream, that I'm in Amsterdam? Well, how do I know that I'm in Amsterdam when I actually am? Do I look around and recognise characteristic details and famous landmarks? No, I know I'm in Amsterdam because . . . because I know I'm in Amsterdam! Because I bought a plane ticket for Amsterdam, and got on a flight that was announced as being for Amsterdam. Or because I followed the map and the roadside postings. Now that I know I'm in Amsterdam I look around, and yes, everything fits, it all looks most Amsterdam-like. And one of the things that colours everything I see, that helps to make it look Amsterdam-like, is – the knowledge that I'm in Amsterdam.

We fill in the nature of the sunset, the nature of the back of our hand (whether in the dream or in the waking world), from our knowledge that this *is* a sunset, this *is* the back of our hand.

Then something undermines that knowledge. I realise that I have become a little confused about time zones during the course of a long and complicated flight. This is the dawn, not the sunset. At once the whole nature of what lies before my eyes changes. Or, as we look at the characteristic view of the Amsterdam canals, we become uneasily aware of a dustcart going by with the Delft city arms on the side. The profoundly unsettling possibility opens that we have somehow got off the train at the wrong station. Now the canal doesn't look Amsterdam-like at all . . .

There's a wonderful moment like this in Alan Bennett's play *The Old Country*. Its middle-class English characters are living, at the height of the Cold War, in a villa that is evidently somewhere in the wooded outer suburbs of London. Halfway through the first act it dawns on you, from a reference in the conversation, that these aren't the outskirts of London at all – they're the outskirts of Moscow, and the English characters are former spies who have taken refuge there. In front of your eyes the whole set seems to go through a transformation scene as elaborate and expensive as anything brought about by the Fairy Godmother in a Christmas pantomime. The light becomes cold and alien. Defects in the plumbing silently announce themselves.

Strange edible fungi spring up unseen among the trees. There is an unsmelt whiff of low-octane Soviet petrol from an unseen, unheard, offstage road.⁵

Another dream. The doorbell rings, and when I open the door there's a nun, standing with her face turned away from me, and saying that she's collecting for charity. I ask her, with my usual caution, which charity – but am already realising that there's something more astray here than a distasteful cause – that the voice I heard was not a woman's at all, but a man's, disguised, and that the supposed nun is already pushing his way past me into the house.

Now I, as a character participating in this dream, didn't know, when I heard the doorbell, that there was going to be a nun on the doorstep, let alone a false nun. But I am participating not only as a character, and for that matter not only as the audience, but as its author (I *must* be, because there's no one else involved in providing this particular entertainment). So why did I, as author, introduce the doorbell, unless I had someone in mind to be on the doorstep – unless, in other words, I had the next event in the story already prepared in my imagination, but without my being aware of it? Did I, as the author of the dream, know what I, as the participant in the dream, didn't know?

Or is the making of a dream like the telling of a story by a storyteller who works from hand to mouth – from event to event, almost from word to word, as the story goes on? But even here, surely, there must be some anticipation, some foreknowledge. The storyteller must know, as the door opens, that *someone* will be waiting; must know, as he describes how the nun's face is hidden because it's turned away behind her cowl, that there's something to hide.

Another dream of sudden terror. I was walking across a small dark courtyard at night to the brightly lit street beyond (Marylebone Road), when with terrifying suddenness everything changed. The darkness was filled with noise and fear and falling. I realised that I had been attacked by some hidden assailant. The whole event seemed to have come out of nowhere, as if I had been struck by lightning. But perhaps I was in some sense the author even of this thunderbolt, because now I think I recall an almost subliminal thought that flashed upon me, in the brief moment when I was walking through the darkness – how this was exactly the kind of place in which some assailant might possibly be lurking. Perhaps in that very thought I was already plotting the unforeseen attack.

Or perhaps, in that very thought, the unforeseen attack was plotting itself.

This is the heart of the mystery, this is what is so difficult to examine: the moment when the action takes shape inside your head. Often at this very point, in fact, just as the plot delivers the great scene, the dream seems to lose interest in itself and peter out. The important events that have been set in motion never quite materialise. We fall from the high building – but never quite hit the ground. In the brief instant of terror before I awoke, in that dark courtyard off the Marylebone Road, I had time to note, with surprise and interest, that in the midst of all this overwhelming mass of sensation I couldn't locate any actual physical pain.

Sometimes, when you wake from a dream, the dream goes on dreaming itself in your waking mind. Voices continue to speak – more coherently now, in a way that suggests they have a bearing on the normal course of life. But still they speak fictions. Still they're voices that don't belong to you, speaking messages in whose form and content you have no part.

And yes – for me, at any rate, these afterdreams are mostly couched in language, and spoken by various voices. But then so is a great deal of my thinking. Sentences form and utter themselves inside my head. They do not proceed from me. Often the sentences follow each other in a way that suggests not a monologue by a single voice, but a dialogue between various voices. And when, instead of dreaming the incident of the false nun on the doorstep, or the assailant lurking in the darkness of the courtyard, I write it as a story, the unexpected emerges almost as suddenly as in the dream. I am almost as surprised in writing such an incident as I am in dreaming it – by the violence and suddenness of the attack, by the words that the assailant chooses and the expression on his face, by the shocked extremity of the victim's response. They all leapt out of the shadows as inexplicably as my dream assailant.

Isn't there, at the point of origin of *all* our actions, *all* our thoughts, some such element of autonomy? The moment of creation always occurs a fraction of a second before the conscious mind discovers the created material, takes it over, and organises it into coherence.

To Descartes the '*cogito*' seemed apodictic – so much so that it could serve as the foundation stone for the construction of a world.[6] But the word begs all the questions we have been looking at; the first-person construction conceals more than it reveals. Even if it's certain, from the act of doubting (which is what constitutes the thinking in *cogito*), that thinking of some sort is occurring, the authorship of that thinking remains an open question. My thoughts think themselves, and from that thinking the author has to be constructed, as dubious an entity as the argument suggests that everything else might be. The life of the family is the life of its members, but the lives its members lead are their own. Without the family – no members. Without the members – no family.

What would it be like if it were otherwise? What would it be to feel that the thoughts inside my head – or the words issuing from my mouth, for that matter – were being consciously crafted by a single central authority which in some way represented the whole of my being? Well, who would it be who was feeling this? The same central authority, the same whole person? As if the Government really did speak as one – and also listened to itself and was convinced and wrote the leading articles passing favourable judgement on its total unanimity.

Am I really 'I' through to the very core of my being? Or do I, at my very heart, incorporate active originating forces which are not 'I' at all?

An obvious question: *why* do we dream?

The answer to most apparently teleological questions about the natural world has to be sought not in objectives but in origins. We and our fellow creatures behave in the way we do because it has proved a relatively successful policy in supporting our survival and the propagation of our genes. So how has dreaming evolved? What evolutionary advantage could it ever possibly have given us? You might think that it would have had, if anything, the opposite effect – that sleeping gazelles *not* dreaming of green grasslands (or plaintive blue cows) would be more likely to hear the lion approaching, and that couples who never had cause to bore each other by recounting their dreams over breakfast would be more likely to stay together long enough to breed and raise their young.

Psychoanalysts see dreams as the working out of some inner conflict. And one can imagine how they might be used as texts into which such interpretations could be read. But then random ink blots can be used in the same way without anyone, even a psychoanalyst, thinking that the meaning read into it by the patient was unconsciously put there by the person who scattered the ink. You might think, likewise, that when a dream seems to offer a practical solution to a conscious problem it's because the dreamer reads a solution arrived at by other means, even if unconscious ones, into the ambiguous material on offer. What's remarkable about Kekule's celebrated adventure in dreamland is not that his unconscious supplied the image of the snake to indicate the ring structure of the benzene molecule but that his conscious mind saw the molecular structure in the snake. The solution might equally well have been released by a circle of tea-leaves in the bottom of a cup, or the sight of a children's merry-go-round. For that matter he might have understood the dream of the snake as an inspiration for recycling, or for a belief in reincarnation – or as a sign that he should give up going endlessly round and round in his mind trying to find the structure of the benzene molecule.

Psychologists have proposed various mechanistic explanations, usually by analogy with computer practice. They have suggested that it is a kind of mental housekeeping, in which the brain sorts the day's information, deleting unwanted files and backing up others. Some elements in dreams, it's true, do seem to relate to the experiences of the day just ended. Most (of mine, at any rate) don't. They relate, if to any experience at all, to events in the remote past. Usually they seem more like pure fiction. And far from suggesting any parallel with orderly filing, dreaming seems much more like the breaking open of files, both familiar and unfamiliar, and the chaotic scattering of their contents.

In any case, even if some cerebral housework were in progress, why should it have to be conscious? Shouldn't we get a better night's sleep, and thereby work harder and afford better nutrition and beget more children, if the brain got on with the clearing up behind closed doors, as it does so much else? We don't have to dream about it at night to keep the blood going round, and the food going through, and the kidneys functioning.

What about other species, who can't recount their nocturnal adventures to us (or each other) in the morning? Do we have to suppose that dogs and cats need to dream to get their brains clear? Or do we suppose that they manage without it? Dogs stir in their sleep sometimes, from some indeterminable inner cause. They start awake for reasons which seem unlocatable in the external world. You can imagine that some kind of experience is going on. Do fish dream, though? Do spiders? Now the question begins to seem not so much unanswerable as unaskable.

Perhaps dreaming has no function. If it really does have no bearing on whether we live or die, or whether we mate or fail to, perhaps it's not subject to the pressures of selection. Its functioning seems to be largely random. Perhaps its origins were, too – a random series of mutations in the brains of individuals whose genes were dominant for quite other reasons.

It's difficult not to see elements of the random and the indeterminate in all our experience, most striking in dreams but also present, as we have seen, in waking perception. There is a parallel here with the indeterminacy of particles. (And the indeterminacy of experience prevents it shaping our knowledge and behaviour no more than the indeterminacy of particles prevents them determining physical structure and chemical behaviour.)

Then again, maybe there is an element of pure randomness in the apparently unlocatable deciding force at the heart of dreaming (and of all our thinking). Maybe this generative principle, which is both I and not-I, stems from the quantum behaviour of individual particles in the system, deflecting and shaping the throughput of information from external sources as fundamentally and massively as the random release of the beta-particle affects the state of Schrödinger's cat.[7]

One way or another, we have lost the deterministic thread of the universe right here, inside ourselves.

An odd minor footnote to all this. Or perhaps not so minor.

Sometimes, when I 'enjoy a good night's sleep', I mean that I enjoy a night of sound and dreamless sleep, or at any rate one with no dreams that I recall in the morning. How can I *enjoy* something of which I am entirely unconscious? Do I mean in the abstract sense of enjoying a privilege? Or do I mean that it induces a mood of well-being that I enjoy when I wake? I don't think either. There is something actively enjoyable about deep and dreamless sleep itself – something *deeply* enjoyable. I wake with a feeling of *having* enjoyed it.

This on the face of it is a contradiction in terms. Perhaps I'm simply projecting later feelings back upon the past. Or do I, even in deep and dreamless sleep, have some faint residual consciousness? As elusive as the faint light in front of my eyes in the darkness? After all, one can be awoken from sleep by external sounds; things do 'penetrate one's consciousness'. One retains in one's unconsciousness at any rate a potential ability to experience.

The enjoyment of dreamless sleep is the limiting case of consciousness – the boundary at which all connection with the external world, however indirect, and every last shred of causality, is finally extinguished. Pure experience.

Home Address

And who lives there

Whichever way we go, waking or sleeping, we come back to ourselves – you to yourself, me to mine – the monarchs of all we survey, whether we like it or not.

A sovereign of the old school I had always felt myself to be, benevolent but absolute, the source of all the edicts that constitute the fabric of the court and its business, the master of my own revels. Now it has pleased me to command this inquiry into my own authority, however, I discover that I am not an absolute ruler after all. I am a mere constitutional fiction, a face on the postage stamps, a signature at the bottom of decrees written by unidentified powers behind the throne over which I have no control. I am manipulated by competing factions about whose divisions and debates I am kept ignorant. Even my private entertainments are devised for me by invisible courtiers working in parts of the palace that I have never entered, and could never find my way to. (And where did the idea for this inquiry come from? What dark purpose – and whose – was it intended to serve?) All I can ever know about this world behind my back is its product – the words I utter on its behalf, the mysterious masques I find myself watching.

Just a moment, though. These never knowable sources, these forever mysterious chambers, begin to sound remarkably like the ineffable world of things-in-themselves which was supposed to be the source of the phenomena that are present to our senses. I must surely be a little positivistic here once again. I must limit myself to what is observable – if only by myself. What I am, observably, is simply this: the product itself, the stream of emerging thoughts and decisions, dreams and sensations – and that is all there is to be said about it.

So when I say 'I think', 'I feel', 'I have decided', am I really saying 'thought is being thought', 'feeling is being felt', 'a decision has taken itself'? When I say (and think) that I am writing these words, is this to be understood as meaning simply that these words are being written?

I surely mean more by 'I' than merely the thinking of the present thought, the feeling of the present feeling, the writing of the present words. I mean to assert a continuity at least with some other thinking. With what other thinking? – With the other thinking that is going on and has gone on at this address. And not just with the other thinking, but with everything else that has gone on at this same address, all the feeling and deciding and imagining as well.

So what address *is* this, if not 'I'? Could it be the address written by you as 'you'? (Looking at me from the outside you have no reason to question the suspect metaphysical origins of my acts and utterances. You can use the word in all innocence.) Could it be simply my name, which can be used just as innocently by anybody in the world who wants to reach me? Could I learn to use these addresses myself, when I think about my own thinking? 'The sequence of thoughts and actions and experiences that you call "you"'? 'The sequence . . . generally known as "Michael Frayn"'?

However I write the address, what is it that situates all these different enterprises at it, and identifies them as activities of the same group of companies? What differentiates the complex of thinking and feeling that we're talking about here from your thinking and feeling, and all the other thinking and feeling that are going on in the world?

It has a single address, surely, because there is a single something that it all is. After all, the processes of the group are not entirely random. There is a connection, however intermittent and unreliable, between the input of observations and considerations and the output of communications and decisions. There is some character and pattern to the process, some organising principle. Some sort of *control*, however haphazard and inconsistent, is being exercised. The firm that feels the punch on the nose can hardly be separated as a legal entity from the firm that tries to punch you back. It's also closely allied with the company that sees the connection between the two and contemplates the difficulty of establishing exactly what that connection is.

Your home has a single address, but many rooms. Which of the rooms is *really* home to you?

The living-room? The kitchen? The bedroom? All of them, surely – or any of them. If the builders are working in the living-room you can always take refuge in the kitchen, and if the kitchen is flooded as well you can withdraw to the bedroom, the attic, the garden shed, the bathroom, the airing cupboard, without ever entirely losing the sense of being at home. Your home and your self are like holograms, where each part of the picture contains the whole.[1]

I know *something* about my inward workings – I'm not anaesthetised, or lobotomised. But what exactly do I know? Descartes knew beyond doubt that he was thinking. That is, even the act of questioning his own existence (which is what was at issue) itself *constituted* thinking. So, for that matter, did the act of questioning whether he was thinking or not.

The anxiety about one's own existence now seems difficult to recreate, but the certainty that I am having some kind of experience, however elusive its nature, however transitory, however uncertain its relationship with a shared objective reality, still seems like a bedrock. Whatever I know or don't know about my inmost nature, or the way in which I take the decisions that define me, I know that some sort of thinking is going on. What sort of thinking? Whatever sort of thinking is involved in producing the conclusions that I'm giving you an account of now. I know this simply because I'm giving an account of them – if not to you, since by this time you may well have ceased to listen, then at least to myself.

Even if I'm giving the account purely to myself, though, I'm giving it in the language that I learnt (and in the process adapted slightly) by speaking to you, and listening to you. In the language that we both learnt together (and adapted a little between us) by speaking and listening to various other people, who learnt it (and adapted it) in their turn from the long procession of generations before them who had learnt it, and adapted and adapted and adapted it to their ever-changing needs and purposes.

Do I know anything about myself that you *don't* know – that you *can't* know, however closely you attend to what I'm saying, and for that matter to what I do, not to mention the expressions that cross my face and the way that I stand, sit, walk, etc.? Of course. I know (for instance) about the pain I have that comes and goes in my back. Not just *that* I have it. You know *that* I have it almost as well as I do, since I've complained about it to you so often. What *I* know, and what I can never make *you* know, however much I groan, however graphic the objective analogies that I find for you, is that when it comes it feels . . . like *that*!

Like what? I can't tell myself any more than I can tell you. But I don't need to tell myself, since I can feel it. Though now it's ceased and I can't feel it any longer I can't really recapture it. It's as closed to me as it is to you.

I also know that I have a certain feeling in my stomach, my shoulders, arms, legs, head . . . What – pains everywhere? No, not pains. Simply the sensations in each part of my body of its being there, the elusive background aliveness that differentiates the parts of *my* body from *yours*, or from dead meat. These feelings are like the 'atmos' that film sound-recordists are always careful to make tracks of – the sound of the location when no one is speaking and nothing is happening, over which the voice and effects tracks will later be laid, and without which they would sound inert and unreal. I'm not actively aware of these feelings (any more than I am of the atmos) until I turn my attention towards them. Or until they cease, as when the dentist anaesthetises my lower jaw, and it seems to have been replaced by a wooden prosthesis which is no more a part of me than the broken leg that Oliver Sacks' patient tried to throw out of his bed.

Why, if all these shifting lights are so inexpressible, to others and even to ourselves, do we struggle so obstinately to find some expression for the self that they seem to illuminate? Because we want to give others (and ourselves) some *reference* to them, so that they can place their own experiences (and so that we can place ours) in a context. We have a historical purpose in mind, too; to leave some mark, for others and ourselves, of our passage through the world; to celebrate it.

If I speak about myself in the language I have learnt by communicating with you, can I discover more about myself by analogy with you? Can I read myself into you?

I look at you – at your face, and at the way it changes as you become animated or listless, angry or amused, as you take the point of what I'm saying, or fail to. I listen to what you say in your turn. And yes, what I'm observing is not just a succession of disconnected events. It is a coherent organic whole. I am looking at a person.

So now I try to construct myself from the outside in the same way. I look at myself in the mirror, and I see . . . yes, a person again. One with a very limited range of behaviour, it's true, forever restricted to two activities – to putting on a performance for the benefit of the mirror and to watching it. What is the performance? I think it's the performance of being himself. *Him*self, not *my*self? Yes, the himself who is being looked at. It can't be *myself* I'm looking at; there must be more to me than *that*! A self of some sort, though? A complete person, with feelings and a history, not just an arrangement of eyes, nose, and mouth, and a few slight conventional modifications of their contours? – An attempt at a self, at any rate, a superficial sketch of a self. And an attempt involves an intention; making a sketch requires the transformation of input experience into expressive output. *Something's* going on at the back of all this!

I step closer to the mirror, too close to see the expression on my reflection's face, and look into its eyes. *My* eyes, as others see them. So now I'm not watching the performance – I'm watching the watching. And what I'm seeing is not just a pair of eyes, but a pair of eyes engaged in a particular activity that implies intention. To see what I'm seeing I must have some sense of that intention, and of the train of thinking that led to its formulation. I feel I have a glimpse into some kind of inner world.

I put another mirror behind me and look at the reflection of my reflection. Now I see myself going away from myself, out into a world beyond. I can imagine that the unseen expressions on my face have become indicative of rather more than just performing and watching. On the other hand, I seem even less like my *self* than I was before. If it weren't for the familiar bald patch on the back of my head I'm not sure I'd even know who it was.

I forget all about my experiment, and go to a party where I suddenly catch sight of myself in a mirror unawares, talking to other people, and for a moment see myself as they see me. A person, yes. A person, certainly. But not me! Surely not! A grotesque caricature of me, a coarsely satirical rendering, entirely removed from what it feels like being me from the inside. So far removed that for a moment he continues to behave quite unselfconsciously, because for a moment I don't even recognise him. And now he realises he's being watched he gets a grip on himself, and begins to put on a more acceptable performance – or at any rate one that I find more familiar from his appearances in the mirror at home.

There was a man in college with me who looked like this same reflection of myself caught unawares. Each time I saw him I went through that same disconcerting moment of delay in realising that it was me – then a second even more disconcerting moment of realising that it *wasn't* – that it was once again this man who had devoted his life to being a slightly exaggerated reflection of me. Not once in three years did I ever bring myself to exchange a word with him, or even to find out his name. I wonder if I had the same effect on him as he had on me. Perhaps he never saw the likeness at all.

So, yes, I plainly do have some idea of myself as a person, after all, if I react so sharply when I think I see myself misrepresented. In rather the same way I find it difficult to explain what something I've written is about – until some editor, interviewer, or critic attempts an explanation of his own, and at once I know that at any rate it's not *that*.

We need to be *objectivised* in some way, even if it's inadequately or inaccurately, to make ourselves visible to ourselves. The inadequacy and inaccuracy might even help. You sometimes feel that you catch a glimpse of yourself, even without the aid of mirrors or doubles, when you're ill, particularly with a fever – precisely because you're *not yourself*. It's as if your everyday pair of spectacles, which you never notice because the world appears through them as it always does, has been replaced by a pair of which you can't help being aware because the world seems strangely remote through them, tinted the wrong colour, arranged in a strained and unfamiliar perspective.

In some moods we're tempted to believe that there are laws of psychology and sociology by which we're located with the same comforting objectivity as physical objects are by physical laws (though in *my* case, of course, unlike yours, I can slip free of the bonds in an instant whenever I choose, like a prince putting off his beggar's disguise). Or we see ourselves as the object of other people's perceptions. Society's; the audience's. Through the eyes of someone who loves us. Even of someone who hates us. We transfer the burden of perception to others, and in their universe acquire the clear outlines of ourselves that we find it so difficult to glimpse in our own.

All these attempts to grasp the elusive phenomenon of the self depend upon an even more elusive phenomenon at the heart of it, the living nerve of the self: consciousness. Without invoking what I *feel*, what I *think*, what I *understand*, and so on, there is no way of characterising the events of which I am composed, even the glimpses of myself from outside as the subject of photographs and psychological laws and your gaze, and there is no way of linking them into the continuous something that I feel myself to be.

About consciousness much has been said, and not a word of it that told us anything that we didn't already know perfectly well from our own lifelong experience, which is nothing. We can't even say what *sort* of thing it is. It lurks like an outlaw on the boundary between philosophy and science, and continually slips away across the frontier to elude capture by the forces of law and order from either side. Or perhaps (another metaphor from the world of criminal mystery) like some kind of subversive organisation that has its hand in every branch of the traffic between us and the world – and that conceals itself in the classic way, as soon as detectives look into its affairs, by infiltrating the police department and taking over the investigation.

There have been brisk attempts to identify it as a purely and objectively neurological phenomenon,[2] and there have been heroically elaborate and sophisticated ones to do the same.[3] On the other hand a lot of philosophers (and scientists) continue to treat mind, which is the domain of consciousness, as some kind of different stuff from the material things that make up the rest of the universe, quite outside the scope of the scientific laws and theories that describe the latter,[4] and most neurologists recognise that, although consciousness is plainly altogether dependent upon the mechanics of the brain, no account of the mechanics begins to explain what it is to feel and be aware.[5]

The difficulty is not just an abstract one. How can we give even an external, objective account of human behaviour without the concept of a conscious self? Some people have evidently found it possible to imagine that there are human beings in the world – perhaps all human beings in the world apart from themselves – who are zombies, creatures that behave like ourselves but that have none of the feelings we ourselves have, nothing of our internal world. These creatures know to take their hand out of the fire even though they feel no pain, to say 'I was moved by the naked intensity of feeling in the slow movement,' even though they have no capacity to be moved. But the zombie's inability to experience anything surely makes zombiedom inexplicable. *Why* would a zombie say it found something moving? *Why* would it bother to remove its hand from the fire?

Well, I'm no zombie, and you tell me you're not, either. But we have both on occasion claimed to have been moved by something when really we didn't have any feelings about it at all. We have both seen doctors and dentists laboriously shielding themselves from X-rays even though they can't feel them. But we do these things because we have *other* feelings, or an anticipation of them. We want to enjoy being regarded as emotionally responsive. The dentist wants to avoid the suffering and the shortening of his life that exposure to radiation will produce later. No such motivation for the zombie. No feelings – no reason for action – no action. Human affect is integral to our functioning.

And if you tell me that pain and pleasure, shame and pride, pity and obligation, and all the other feelings that seem to motivate our behaviour, are in fact irrelevant, because everything we do is really a series of conditioned reflexes and learned responses, to which our feelings are merely a functionally irrelevant accompaniment, then tell me as well why such useless distractions from the business of surviving and propagating should ever have been selected by evolution. In any case, how are our reflexes conditioned? How are our responses learned? Through punishment by pain and reward by pleasure.

Even if we leave human behaviour aside, how can we give any account of the objective world around us, now that relativity and indeterminacy have made the description of the objective world dependent upon the notional possibility of an observer, without a notionally conscious self? Science, it turns out, is in this respect simply an extension of all the other means we have found of representing the world. As with pictures and narratives, the world is unimaginable without the focus of a viewpoint.[6] To understand the world in any way whatsoever – scientifically, historically, artistically, anecdotally, imaginatively – we find ourselves compelled to assume a potential point of convergence from which everything is viewed, measured, and recounted. About the entity that defines this point we know only that it is by definition invisible to us. And since the instruments of science, of logic and mathematics, and of art, are all products of this invisible entity, they will not serve to represent or explain it to us.

Anything in the world, or out of it, can be perceived or thought about, or both, and represented in our various codes. The only thing that systematically eludes us, whichever way we turn, is the something upon which everything else depends. The conscious subject that gives meaning to the objective universe cannot give meaning to itself. Without it nothing can be understood; about it nothing can be said.

'The point of convergence . . .' And already we've gone wrong. The self isn't a point, in any remotely geometrical sense, but a complex organisation; and the consciousness that it generates, and by which it is animated, is a complex phenomenon. The phrase also suggests something passive, but our role as a conscious observer is a highly active one. Observation is not a matter of sitting in a trance. It involves physically directing the gaze, perhaps also straining the ears, sniffing the air, reaching out with the hands, putting things into the mouth. It requires us to direct our attention mentally, to interrogate the world for particular purposes, and to make sense of the answers we get.[7]

The self is an active participant in the world in other ways, too. Not only do I mentally shape what I see but I shape it physically, by the actions I decide upon and initiate. Even if you regard the history of the universe as in principle entirely determined, and established from the beginning by the unexaminable disposition of the infinitely dense point at the beginning of the Big Bang, you have to accept an infinity of later interventions along the causal road by the systematically random behaviour of particles at the quantum level. To give a complete account of how the world got to be the way it is now, though, you must also include the modifications made in the last few moments of cosmological time by the behaviour of human beings and other animals. And if you think that even this is determined, that even this is part of the great causal chain, then you are saying that our perceptual participation must be determined along with everything else. We have been inexorably shaped to see this thing as a dachshund and that as the Risorgimento; to understand our centrality; even to understand that we are determined – and to understand that we understand this . . . We have no choice but to see ourselves as having no choice. This is the kind of circularity involved in religious faith: I believe because I believe.

The active function of the self in shaping the world compounds the problems of reflexivity that arise in introspection, because one of the elements of the world that the self attempts to shape is precisely itself. This was the other problem to which I thought philosophy might give me an answer when I first began to study it: how could one ever change oneself (as it seemed to me people sometimes did) if it was the self to be changed that had to do the changing?

I'm not sure that my studies ever enlightened me very much on this point. If the self is a palace, though, with all its courtiers and counsellors, rather than just the single figure in the throne-room, it's not so surprising if reforms of court procedure, and even palace revolutions, do sometimes occur.

*

Another effect of being ill, which strikes even deeper than the effect it has upon your perception of the world: the will itself seems to be undermined. You lie on the bed and feel not only that your body won't respond to the commands you give it, but that your very power to issue commands has ebbed away.

Yet I can imagine that I might choose to overcome the feeling. If there were some compelling reason for action then my will might – would – bestir itself. I should get up from my sickbed and walk. Some even more inward powers of will would emerge to overrule the ones that I feel to be so attenuated.

I put this to the test once, without a compelling reason, simply out of intellectual curiosity. On the fourth day of a fever, as I lay in profound lassitude, unable, as it seemed to me, to move arm or leg, or even to wish to, I simply . . . got up . . . and tried an experimental press-up. To my surprise I found I could do not just one, but the usual twenty – and that it seemed to require no more effort than it ever did. Perhaps it was even a little easier, as if I had become a less significant object to raise from the floor.

On the other hand, when I was taken ill in the theatre once, and helped out semi-conscious, I was quite unable to direct my body where to go, or keep it upright unaided. I *did* walk, though, with assistance, to where I wanted to go. Was it my willpower, my inner self, driving the defective machine forward? It didn't feel like that, in this case. It felt as if the machine itself was carrying on as best it could, quite regardless of me – that I had no influence over it at all, no will to operate it.

Once I fell in the shower. My soapy feet were suddenly away with the migrating birds, and I became a mere unsupported object in the earth's gravitational field. Having fallen I remained an object, with strange cries of distress emerging from me. Did the cries express pain? Astonishment? Dismay? The wish to be helped? I couldn't say what they expressed. I was in pain, certainly, and astonished, and dismayed, and anxious to be helped. But the cries uttered themselves, in the same way that the electronic cries of a car alarm utter themselves; a purely mechanical reaction programmed, in the case of the car, by the manufacturers, in my case by evolution, to alert passers-by. Could I have *not* uttered them? If, for instance, there had been a danger that they would have given away my position and helplessness to an enemy? I don't know. Possibly, if I had had the danger firmly enough in mind, and a moment to think about it. Or possibly not, until it was too late.

And I lay there, at the bottom of the bath, inert and inanimate. There was no physical reason why I couldn't get out of the bath – I'd broken a few ribs, but not my leg. It was as if I couldn't get my willpower into play. I couldn't will myself to will.

The ever-elusive self has an outward manifestation, rather in the way that an underground organisation has a political wing that represents its interests in the daylight world. The outward manifestation of the self is the personality, an objective if shifting assemblage of character-istics that seems to make a recognisable whole. It is something that we feel we can become familiar with and describe, certainly in the case of others and even to some extent of ourselves. We believe that we can recognise regularities in the behaviour that we observe, and we per-haps also try to make some estimate of the motivation that produces it. The personality seems to offer some basis for predicting future behaviour.

Looking at ourselves, even if not at others, we feel that we have some intimation of the relationship between personality and self – between how we outwardly seem and what we inwardly are. The nature of this relationship, though, is as difficult to pin down as the nature of the relationship between the anonymous gunmen and the public figures who are alleged to have no organisational ties with them, nor any responsibility for their actions, but who apparently know their minds and have some influence over their policy.

Psychologists and psychiatrists, in their efforts to systematise their insights, and to give them the coherence of the physical sciences, have tend-ed to treat the personality as an exclusively objective phenomenon, which can be fully located in the causal production line, where it has been pressed out by genetic inheritance acting upon local circumstance in the way that a Ford car is pressed out by machine tools acting upon steel sheets and rods. Something like this does happen with our phys-ical characteristics. Although the permutation of our genes is not fully determinate it is probabilistic, and the only other factors involved are the equally objective (if equally unpredictable) ones of accidental dam-age and random genetic error.

But do you feel that the differences between your personality and mine are really no more profound than the differences between the colour of our eyes and the length of our noses? We're not all *that* different facially – we might even get mistaken for each other by someone who doesn't know us very well. But our personalities are completely different! You're . . . how can I describe you, in a word? Yes – you're *you*! And I'm *me*! Do you feel your you-ness was simply dumped upon you like the slightly comic shape of your ears? No, you feel it's in some way expressive of you as you are in your inmost core. You take some personal credit for your charming insouciance, courage, and generosity, just as you acknowledge some personal responsibility for your unpunctuality, thoughtlessness, and unreliability with money. You feel, as once again you miss the vital meeting, that you yourself *chose* to behave once again with your characteristic insouciance; that you, no less characteristically, *failed* once again to choose to set out in time; that, in some sense, for worse as well as for better, you *made yourself* as you appear to the world.

You relate to your personality rather like an actor to the character he is playing – or, better, like an actor in a Mike Leigh play, whose part isn't written by an outside author, but developed by the actor himself out of some elusive combination of the director's suggestions, his own researches into the world portrayed in the play, the possibilities of his own nature, and his own fantasy. Some puppeteers I once saw in Prague offered another striking model of the relationship between self and personality. Each of them, making no attempt to conceal himself from view, held out a life-size puppet in front of him like a kind of human shield, and worked it so that it played its part in the common story, where it fought its corner against the puppets facing it. And, curiously, what you found yourself looking at was not the puppeteers whose skill animated these inanimate objects, and whose imagination and personality shaped their behaviour, but the inanimate objects through which these powers were expressed.

Look at a small child playing. Not for one moment are you tempted to believe that this is simply a selection of genetic traits interacting with the wind and the weather. You see at once that there is another element involved as well – the child itself. From the moment that the child comes into the world it is a principal in the action, just as its parents are, a player in the game. Human beings are irreducibly subjects as well as objects. They don't just summate genetic influences – they integrate and transcend them. They don't just suffer external events – they *use* them, they draw them into the game. They are supplied with the material, certainly, just as they are with the box of bricks; but out of that material they construct themselves, in the same way that they construct their own idiosyncratic artefacts from the bricks (and not only from the bricks, probably, but from the box itself, and the instruction book, and various other uncovenanted materials they have seen a use for).

Imagine that your mother had conceived you a year later than she in fact did. How do you think of this counterfactual child? Do you think simply: 'Oh, then I should be a year younger than I am. I should have many different characteristics, of course, because I should be a different permutation from the same two stocks of genetic material, but then again I should probably have been brought up in almost exactly the same circumstances . . .' Or do you think: 'Well, in that case there would be a different person in the world, and I shouldn't exist'?

Well, do you regard your younger brother, who actually *was* conceived a year later, as in any sense whatsoever yourself, or as any sort of variant or Doppelgänger of yourself? Do you regard even your identical twin, born at the same time from exactly the same genetic material and brought up precisely alike, whose every thought and feeling you seem to know before he expresses them, as in any sense whatsoever your own self?

This sense of the self comes very early in life. Even a baby plainly sees itself as the author of its own actions. How else would it ever learn to mimic the expressions of others, when it can see them only from the outside, and can organise its own expressions only from what it feels on the inside? How else would it make that strange analogical leap from the 'I' designated by you when you talk about yourself to the 'I' designated by itself, a creature who is 'you' and 'he' or 'she' to everyone else?

Even the shape of your ears depends partly upon the self, when you come to think about it. Not *your* self, of course – your mother's self and your father's. Your particular set of genes was assembled only because your mother decided, for good reasons or bad or none, to switch from mathematics to physics, a subject being read by a certain young man who later became your father, and who, after long struggles with himself or on a sudden impulse, summoned up the courage to talk to her after the Diffusion Theory class; only, for that matter, because *her* mother decided to . . . and because *his* mother thought that she would . . . ; because *her* father chose to . . . *his* father tried to . . . their grandparents agreed to, refused to, insisted on, thought they would . . . ; because Eve decided to offer Adam the apple, because Adam decided to try it.

But then even the Ford production line, when you look at the process as a whole, is not really the mechanistic ballet of robots that it appears to be. Those jigs and moulds didn't shape themselves. The great machine was set in motion by designers working from a combination of the marketing department's estimates of public taste and their own fantasy, by managers trying to guess the curve of demand and the impact of the Government's fiscal policies.

What should I think if I looked in the mirror and saw an unfamiliar face looking back at me? Should I feel that I was a different *self*?

A great many things about my face have changed during our long relationship without my feeling that my *self* had in any way changed. But – suddenly – a total stranger gazing back! Someone profoundly alien in every way! Someone apparently recognisable, even – Josef Stalin, say, or my maternal grandmother. A terrible shock, of course, and one that might quickly make me start to feel rather differently about myself (as about many other things). But whose would be the feelings that changed? They would be mine, evidently, as I have described the situation. To be shocked I should still have to be me.

What should I feel if it was not just my appearance that changed but my behaviour, and in a way that suggested some even more profound physical alteration: an alteration to my brain? I discover to my amazement that I am picking up a bassoon, an instrument that I have never even taken in my hands before, and am playing the first movement of the Mozart bassoon concerto. I explain to my astonished friends how much more astonished I myself feel at this development – and find that my explanation is emerging in fluent Hungarian. Then I recall learning the instrument while I was growing up in Budapest . . . as the daughter of a famous Hungarian entomologist . . . just before the First World War . . .

Now I don't know what to say – except that for this situation to retain any last fingerhold upon the conceivable I must still have some sense of myself as being myself. Of course, we can easily imagine a woman who plays the bassoon, and who recalls growing up in Budapest – if this person is not also encumbered with an awareness of being a man who cannot play the bassoon, and who grew up in the suburbs of London.

What, in the end, gives the self its continuing identity, its perseverance in time as the element that relates the various inputs and outputs of the mind to each other in some coherent way? Locke suggested that it was precisely what we jettisoned within the thought-experiment above – memory. But the human memory, even outside implausible thought-experiments, is as fickle as water. Of course, you can imagine using it as a practical test, in one of those rare cases where someone claims to be identical with a person who vanished a long time ago. What was the colour of our old bedroom curtains? What did I once whisper to you, and to no one else? Even as a practical test, though, it couldn't be conclusive. If the claimant answered every question correctly he might still have been coached by someone else (the real candidate, perhaps).[8] If he failed to answer the questions he might simply have forgotten, or remembered things differently. My own identity would have become dubious long ago if it depended on memory. Old men forget, and old women, too. Everything, sometimes: their name, their history, the appearance of everyone they have ever known. Have they ceased to be who they were? They also remember things that couldn't possibly have happened to them.[9] Have they at this point become someone else entirely?

The answer might depend upon the use to which it was being put. Even the loss of a leg might completely change your identity as far as the captain of the athletics team is concerned. Now some neurological disaster devastates your whole temperament and character. If your loyal friends insist it's still you, in spite of your treating them as hostile strangers, isn't it largely for old times' sake? Or from the force of habit? Or in the way in which we agree to accept that a few picturesque fragments of masonry surrounded by well-tended lawns and well-maintained public toilets are identical with the muddy, fever-ridden, excrement-odoured, starvation-haunted killing machine that had once stood on the site when it was a working mediaeval castle?

What makes Tompkins and Son, Candlestickmakers, still the same firm, when Tompkins and his son are both long dead, the original staff have all been made redundant, and the firm has ceased to manufacture candlesticks and moved into polystyrene picnicware? Firms and nations are like ropes – bundles of threads overlapping each other. The

Celts are marginalised by the Saxons, the Saxons are adulterated by the Danes, the Saxon/Danes are overlaid by the Normans, the Norman/Saxon/Danes are flavoured by the Dutch and the Huguenots, by Italians in the catering trade, by White Russians and Polish Jews and West Indians and Pakistanis . . . Every aspect of life is changed by technological and political developments. How much do the builders of the new Wembley Stadium have in common with the builders of Stonehenge, except perhaps the weather they work in? No – even the weather has changed.

The continuity of the individual is similar. Every cell of your body is replaced. The dark hair you are born with grows blond, grows mousy, grows long, grows short, turns grey, turns blond again, falls out. Teeth come and go, and so do knowledge, muscular strength, common sense, courage, and lovability. You look back in old age upon the child whose name and DNA you share, and yes, it's you, of course it's you; and at the same time it's not, it's someone into whose skin and mind you can scarcely begin to think yourself.

Is it for one moment imaginable, though, that I might have some sense of the self that is totally independent of the body – including even the brain? It evidently is to some people. They retain a feeling that some sort of mental entity might exist without any physical stuff at all to support it. Even after the brain has entirely ceased to function – even after every last shred of brain tissue has been removed from the skull by worms or flames – they believe (passionately believe, half-believe, fearfully suspect, almost entirely disbelieve but not quite) that some kind of essential something would remain.

What could it possibly be like, this something? Its nature can be arrived at reductively if you think of death not as some mysterious event beyond our understanding but as the sum of a series of somewhat less dramatically transformative events with which our living experience has made us all too familiar: strokes. In a stroke a small part of the brain dies, and a function is lost. Speech, perhaps, or any of the perceptual abilities catalogued by Oliver Sacks,[10] most of them so taken for granted that they're unnoticed until they're gone. One of the five senses in its entirety. The memory. Any continuing sense of one's own identity. Now add up every possible variety of strokes, every possible variety of functions lost . . .

Is it possible to understand this (as we all surely do in the depths of our hearts, however much we may insist to the contrary), and yet maintain the venerable idea, which some philosophers still seem to find intelligible and even self-evident, that mind is a different *substance* from the substance that makes up the material world?

What place does this different kind of stuff have in the history of the universe? What are its origins? Are we to suppose that it was not subject to the physical forces that shaped material substances? Did it precede the Big Bang, as some kind of as yet undifferentiated non-physical something? Or did it somehow emerge – in the last few moments of cosmological time – in step with the various forms of life? Did the early forms of mind somehow mirror the relative simplicity of the early creatures? Did this mental stuff evolve from species to species, starting with something suitably primitive for bacteria, and mutating step by step to achieve primate richness and complexity?

How did it evolve? What random non-physical events produced changes in it for success to nurture? By what non-physical mechanism were these mutations passed on from generation to generation? Or did they survive and transmit themselves not through any characteristic of mental substance itself, but through the evolutionary mechanics of the physical organisms with which they were (non-physically) associated? Are we to think that there simply happened to be some kind of regular coincidental reappearance in each new generation of the same level of mental substance as was (non-physically) associated with the previous generation?

Well, there must be a whole series of intangible links between the two forms of substance if we are to feel a (mental) pain when someone sticks a pin in our (physical) backside, and if the (mental) impulse to remove our (physical) backside from the pin is to take effect. I suppose the idea of some kind of existential coordination between the two substances, whereby mental substance evolves in step with physical, and comes into existence in each individual conscient creature more or less gradually from the moment of conception onwards, then ceases at the moment of that creature's physical death (or doesn't), is no odder than any of the other supposed links.

What do we get in return for all these difficulties, though? What does the notion of mental substance actually elucidate? Even the familiar notion of physical substance has turned out to be remarkably elusive. The material world, we now understand, is not quite as solid and coherent – as *substantial* – as it sounds. It's a succession of events;

a concatenation of probabilities; a series of endlessly changing values for such abstract qualities as mass, energy, charge, and spin – values which are never to be precisely and completely determinable. This is tenuous enough. What the proponents of mental substance are trying to tell us is that consciousness, mind, self, is not a stuff even in this sense – not even a cloud of indeterminate probabilities. So what residue of substantiality is there left for us to get hold of? If material substance is as insubstantial as it is, there doesn't seem to be even a metaphorical handhold left for the notion of mental substance to cling on to. To talk about mental substance doesn't offer any graspable parallel – doesn't *explain* anything. It's like writing Terra Incognita on the map over an unexplored area – and then coming to believe that, because it has a name written in the same typeface as Outer Mongolia or Uttar Pradesh, Terra Incognita must be a geo-political entity of the same sort.

If we reject the idea of mental substance, do we have to suppose that consciousness is subsumed under the category of material substance? Is the self somehow a flow of electrons, a smear of co-existent possible states? Or could we simply forget the whole question of substance? Not everything has to be a substance! No one – not dualists nor materialists nor idealists – thinks that eating breakfast, for example, is a substance. Cornflakes are physical stuff, certainly, and so are spoons, and so is the mouth and the digestive system, and the whole performance of eating breakfast is plainly and unmysteriously stuff acting on stuff. But the *eating* isn't stuff – it's the event that occurs when various sorts of stuff act upon each other in certain ways. There's no mystery about it, because there's no reason to suppose that an event is any kind of object.

Isn't it the same with consciousness? For all the conceptual difficulties we have with it, it's plainly more like an event than a thing. Or, more accurately, perhaps, a series of events (just as eating breakfast is, after all), that occur when some sorts of extremely complex stuff act upon each other in certain extremely complex ways. It evolved in step with neural development for the same reason that eating breakfast evolved in step with the human digestive system, the invention of spoons, and the manufacture of cornflakes.

The centrality of the self is embodied in one of the most remarkable inventions of the Renaissance: perspective. Into every representation of the world there was introduced the metaphysically as well as artistically profound idea of the vanishing point, beyond which the world ceased to exist, because it was beyond the range of the observer's senses. You might see this new expression of an individual viewpoint in the doctrines of the Reformation, that gave each of us unmediated access to God and the responsibility to judge his own actions by the light of his own conscience. Once the idea had been discovered it could never be entirely forgotten again. When you look at the magnificence of the great Baroque churches, built by the Counter-Reformation quite deliberately to dwarf the individual and his conscience, and to re-establish a theocentric universe in which the existence or non-existence of man was only a secondary consideration, you see that they have absorbed the very relativism that they are designed to suppress. They have incorporated perspective into the architecture itself. The false cupolas and the heavens glimpsed beyond them have to be viewed from a particular spot – they are directed towards an individual human eye. The focus of metaphysics has shifted decisively from producer to consumer, just as the focus of the economy would in the nineteenth and twentieth centuries.

Now our thoughts have moved on, and suggested a possible qualification to our humanism. The runaway success of one of our own tools, the digital computer, and its ability to replicate certain aspects of our thinking, has compelled us to wonder whether it could in principle ever go further, and replicate human consciousness: whether it could ever become a self in the way that each of us feels he is. Early enthusiasts for artificial intelligence seemed to take it for granted (and some still do) that computers would become conscious simply by virtue of becoming larger, and having more capacity, or by becoming somehow more complex. If billions of elements got interconnected . . . with loops that somehow enabled one element to monitor what was going on in another element . . .

There's an immediate and obvious difficulty here, though. In the case of human beings, consciousness is plainly inseparable from the body and its functions. We can give the computer ways of taking in information – cameras, microphones, etc., which have at any rate some plausible similarity to their human equivalents – and programs for processing the information. But why *should* it take in information, why *should* it process it, if it doesn't have a reason to, other than the instructions we have written into it? If it doesn't, as we do, feel the necessity to eat and excrete, to find allies and flee enemies, to avoid injury and premature death, to reproduce and protect offspring until they can protect themselves? We can give it a voice, and robot arms and the means of locomotion, but *why* should it talk, why should it think, why should it do *anything*? Why should it *want* to do anything? *How* should it want to? Why should it be interested in anything? Why should it care about history, if it had no idea of what it was to have antecedents? Or politics, if it wasn't part of a group of creatures that had to find ways of living together? Or geography, if it didn't know what it was like to go from place to place, to find mountains impassable, to know cold and heat and the difficulty of scratching a living from the earth's crust? Why should it solve equations if there weren't some reason why it needed to know the answer? Why should it bother even to drink beer and eat crisps and watch television?

To make it function autonomously we should have to begin by giving it a context of purposes of its own. What might these purposes be? Ones parallel to ours, presumably. We should have to involve it in the rough and tumble of the world – push it out of the nest and leave it to seek out its own fuel, to discover how to improvise solutions to problems that we had never foreseen, to get together with other similar machines to perform cooperative tasks. We should have to instruct it to seek to be well fed but to avoid over-eating; to respond in a variety of ways to threats; to seek to continue in being and to avoid destruction.

If it's to adopt these purposes as its own, though, and not simply pursue them because they're *our* purposes for it, we should have to equip it with feelings. We should have to enable it, at the most elemental level, to feel pain and pleasure, to like and dislike, to prefer one thing to another. (And, if it's really to be a sovereign self in the sense that a human being is, we should also have to give it the liberty to be perverse – to choose pain rather than pleasure, to seek what it disliked rather than what it liked. Now we need to write more instructions . . .)

How could we set it up so that it had feelings? Nothing comes to mind. Well, how were *we* set up to have feelings? How were other animals? By adaptive selection, as with all our capacities. Our forebears happened upon whatever neural machinery is involved bit by bit, and the tools it provided us with have proved successful. To put it at its most elemental level: we have evolved a desire to survive and propagate, and we have evolved it because a desire to survive and propagate has proved an effective mechanism for enabling us to do just that. How can a machine be endowed with the same device without undergoing an interlocking set of events as long and complex as the set that has shaped our species? Or, at any rate, without a neurology copied from the complexity of animal neurology?

I'm not trying to suggest that there's anything *mystical* about human consciousness, anything that couldn't in theory be reproduced artificially. It's just that nothing that has emerged in computer technology so far has the slightest resemblance to it. The best chance would seem to be to imitate the physical structure of the human body and brain by genetic means – to construct a genome out of suitable biochemicals, which would direct cell growth biologically, and might happen to replicate whatever neural structure it is that produces consciousness in us and other animals.[11]

Another question now arises, even about this procedure. What would be the *point* of it? We can and do achieve the same results by the very simple and inexpensive procedures we already have, and on a massive scale (probably more massive already than the planet can support). Your parents, with no training in molecular biology or artificial intelligence, produced a most excellent example of a conscious machine by these means. As a machine, it's true, you have your drawbacks. Cheap as you were to produce, you're very expensive to run. You're cocooned in rights and privileges that sharply restrict your employability for many tasks, and you malfunction as a result of all kinds of moral and temperamental problems that are difficult for a service engineer to fix. You see why I might be tempted to give you the push in favour of a robot. The trouble is, though, that if the robot has to have the same feelings as you in order to function, then it will presumably have to be treated with the same expensive and time-consuming circumspection.

Would the question ever arise, in practice, as to whether the sophisticated machine we had built was conscious or not? If it did everything we required of it? Reminded you of my birthday, suggested a tactful present, found out where to buy it, and got to the shop before it closed, in spite of a Tube strike and an unforeseen Bank Holiday? Would it even occur to you to wonder whether it really had any genuine interest in your needs and wishes, or even any desire to earn its keep?

Well, if it really *didn't* matter to you, if you really *were* as indifferent to its motivation as you are to the motivation of your electric coffee-grinder, then you've made absolutely plain what you consider the answer to the question is. You're as confident of its lack of feeling as you are of the coffee-grinder's. If you're dealing with a sentient creature like yourself, even a servant or a slave, the question of his feelings is *always* a consideration. You have to make some estimate, at the very least, of his readiness to serve you; of his thresholds of pain and fear; of the chances of revolt; of his quiescence or discontent; of the *sincerity* of his attitude, or the lack of it.

Much stress is laid by proponents of artificial intelligence on the supposed *creativity* of machines – on their ability, when suitably programmed and commanded, to draw pictures, compose music, and write stories – and to produce novel and surprising results. This is a much more modest programme than the ones we have been looking at. Here we are concerned merely with variations on a theme, produced in accordance with a given set of rules. Creativity[12] is supposed to be suggested by the unforeseen nature of some of these developments. But surprisingness in itself is no indication of autonomy or originality. Earthquakes surprise us, and produce entirely unforeseen patterns of destruction, without this being found a sign of creative intelligence.

So, you are presented with a plausible picture, or a coherent story. Now you are told that they have been produced by a computer. Does your attitude to them remain the same? Well, suppose you're in prison and you hear tapping. You discern certain regularities which seem to suggest a code – and you begin to work out the message that it conveys. Now you discover that the tapping is caused by rainwater dripping inside the wall. Would you be inclined to think that the rain had somehow generated a message which it hadn't understood? You have to believe, with even the trashiest writers, that they are offering you some genuine personal vision of the world, however trite, that they are in some sense transmuting their own experience and appealing to yours. If you begin to suspect that the author is simply manipulating a formula then the situation changes at once. I suppose you might try to close your mind to the knowledge, or take an interest in what the originators of the formula thought would appeal to you. In the same way, I suppose, if you suspected that the provenance of the work was an insentient machine, you might take an interest in the minds and experiences of the programmers who had written the instructions.

Now the situation changes – the machine produces a poem about its own experiences as a machine. It talks about the feeling of being switched off at night, or of suffering a glitch in a programme. From the poem it is apparent that it doesn't like glitches, but that it finds being switched off a restful prospect . . .

Oh, come on! One way or another it's been *told* to behave like this! A machine is a tool; a tool is an instrument for carrying out its designer's purposes. Real communication is the expression of free choice. You seem to want something that is simultaneously both subservient and sovereign, like a slave ordered to be a king – not to *play* a king but to *be* one, to command the chamberlain who is commanding *him*.

Let us imagine, nevertheless, that in spite of every logical and conceptual difficulty I have designed a machine that will write plays, and do it without hope of a royalty. Writing a play is only half the transaction, though; the other half is watching it – and a terrible chore that is for a human audience, in the case of my playwriting machine's apprentice works. So my next contribution to the world is to design a machine that will *watch* the plays that the first machine has written. Rows and rows of my new machines sit laughing sympathetically at even the feeblest jokes, wiping furtive tears from their visual sensors at the kitschiest of happy endings. Their heartwires are manifestly plucked, their motherboards wrung. They insist on recounting the plot to us over dinner afterwards . . .

What's so ridiculous about this? When we hear a flock of starlings chattering away to each other we accept that they have a genuine life and world of their own. – Yes, but starlings are like us! They have been formed by mutation and survival – they have our kind of aetiology and mortality. Even with starlings, in any case, we don't imagine that they are communicating, as they chatter, all that much about the meaning of their lives.

You suggest that my mockery may be a little premature. You remind me that there are already machines that have sufficient linguistic capacity to *translate* from one language to another – a rather high-order activity that certainly taxes *my* abilities. Well, translating machines translate in the sense that Searle's famous Chinese Room speaks Chinese. The Chinese Room is operated by a man who understands no Chinese, but who takes in questions in Chinese and hands out answers in Chinese, to the Chinese-speaking customers' satisfaction, because he has a set of rules that tell him which answer to give to which question. Presumably the staff of telephone chatlines work on a similar basis; without any of the feelings that sexual overtures are intended to arouse they make responses that some code of practice, composed by the managers or remembered from their own private experience or learnt from watching pornographic films, suggests to be appropriate.

But these simulacra only work because someone who actually does understand Chinese has already marked which questions should elicit which answers, and because there is real sexual arousal whose expression can be mimicked. The actual work of translation in a translating machine has already been done by the programmer who wrote the rules. What the programmer had to do is what any human translator (or the author of the dictionary he consults) has to do: go back to the situation in which the words and phrases of the source language would have been used, and to start again, from the beginning of the linguistic process, to find the words and phrases that represent that situation in the target language.

With familiar expressions, no doubt, the paths have long since been beaten. The programmer composes the rule 'Replace *artichaut* with *artichoke*' without needing to summon up memories of the actual vegetable to which both words refer. But now he comes to 'Replace *il a un coeur d'artichaut*, with . . .' At this point, probably, he has to pause, and recall first in what situations a French-speaker says that a man has an artichoke's heart, and then what an English-speaker says when he wants to describe the same situation. What qualifications would a machine have to have to do likewise, and actually translate the expression without the benefit of a rule already written by the programmer? It would have to have had similar experiences. It would have to have:

observed the behaviour of an emotionally promiscuous man;
heard how an indefinite range of native French speakers sometimes describe it;
observed the multiplicity of leaves clustering around the centre of an artichoke;
seen the analogy between the two;
heard what an indefinite range of native English-speakers have said when describing the same sort of man.

Actual translation, in other words, requires all the functions of consciousness for which it is so difficult to conjecture any mechanical equivalent. Indeed it requires more than this. For the machine to have done all the things listed above, without input from a programmer, it would have to have done quite a lot of living. It would have to have knocked around a fair amount with both French speakers and English speakers. It would have to be familiar with eating artichokes. And *why*, without orders from a programmer, would it have knocked around the world in this way? *Why* would it have eaten artichokes? Only because, once again, it had feelings: an *interest* in France, the French language, human nature, food.

The elegant digitality of the computer is seductive, and we are easily persuaded that human mental processes work in some similar way, so that they are all analysable into series of discrete steps, very suitable for embodiment in rules. The circuitry of the brain may well be entirely digital – but, even if it is, most of the thinking that we do with it, and all of the feeling, is in broad-brushed analogues. Do we need to feel that weather patterns are really digital because they can be modelled by a digital computer? Or that music is because it can be reproduced in digital recordings? Or that a moving object is really not moving at all, but occupying a sequence of twenty-four varying static positions per second?

Do we need to suppose that the interest we take in the weather is digital, or the delight in the music? Or that the hunger driving the lion's analysis of the gazelle's movement runs at twenty-four frames per second?

We come back once again to the unresolvable paradox at the heart of all philosophy – perhaps of all human thought. The universe plainly exists independently of human consciousness; but what can ever be said about it that has not been mediated though that consciousness? What can ever be wordlessly *seen* of it that is not dependent upon the existence of a single viewpoint from which to see it? What can be understood of it without the scale and context of human purposes, or the instruments of human thought? And if nothing can be said about it, nothing seen, nothing understood, what kind of a universe would it be? We have to imagine something like the situation at the beginning of Genesis, before the first day of Creation – something without form and void. But the difficulty goes deeper than this. Even to say that it is without form and void – even to describe it as *something* – to imply that it so much as *exists* – is to resort to language and to all the powers of human consciousness.

This has bizarre historical implications, among other things. Am I saying that the universe didn't really begin when cosmologists say that it did, fourteen billion years ago, but only a few hundred thousand, as homo sapiens began to evolve? Of course not! We give it its existence and its form retrospectively, by projecting our thoughts back upon it, exactly as we project them back upon the vanished events of yesterday, or forward to the still unsubstantiated events of tomorrow, or for that matter outwards to all the parts of the universe that are beyond our immediate purview, from the inside of the house next door to the remotest galaxies. But the inexpressible question returns: what was it like at the time? What would it be like if we were not here?

Even if the physical universe came into being with a bang, human consciousness plainly didn't. It evolved slowly, and the support it gives to the physical universe with it. The unknowable black holes and the ambiguous sub-atomic particles first began to take shape in our fore-bears' dim animal intimations of the eatable and the mateable. No – before that. In the blind nosings of the worm, in the first responses of primitive organisms to heat and light. The very first faint antecedent of a conceived universe was the primal effect of matter-as-energy on energy-as-matter.

In Genesis, as in modern accounts of our antecedents, man is a late arrival on the scene. Not until the sixth day of the week does he appear, with all his feelings and laws and words, to make sense of everything. However, if you understand by 'God' man's projection of his own con-sciousness and intelligence back to the beginning of things, then I'm making a somewhat similar point to Moses about the emergence of the world through our conception of it. The events of Day One, in this sense, were made possible only gradually and retrospectively by the emergence of the physical environment and living systems on Days Two to Five; until at last, on Day Six, the best part of fourteen billion years after the beginning of that long week, the arrival of man and his dominion finally brought the long darkness of Day Zero to an end.

We have to recognise that even consciousness, which sets us apart from the world (and, in its human form, reinforced by language and other forms of symbolic representation, from the animal species closest to us), is itself an emergent feature of that world, a developed instrument in our traffic with it, a product of it. It creates a world within which, and upon which, we can live. In this, our most inward and intimately human aspect, which gives us our one claim to uniqueness in the universe, and a kind of divinity, we are bound in every particular to the contingent world. Even those workings of the body to which we are privy are part of that world. Even the workings of the mind that we struggle with such indifferent success to understand.

'The mystical thing', said Wittgenstein, 'is not how the world is, but that it is.' Maybe, but the two are indissolubly bound up together. The fact of its existence is no more and no less than the fact of how it is. To be is to be thus and so (or if not thus and so, then so and thus). Not to be thus and so (or so and thus) is not to be. The truly mystical thing, in the sense of being unexaminable, is our consciousness, and the standing of the world in relation to it; and that is indissolubly a part of how things are in fact disposed. We are contingent in our very essence.

Conspectus

So every path eventually leads us back to where we started: to you looking out this way and that at the world around you, to me doing the same; up at the stars, down at the backs of our hands; at each other; to the two of us talking together about what we can see.

Even the most abstract and apparently impersonal of scientific and mathematical ideas turn out to be describing not a universe in which we are irrelevant bystanders, but simply what we should observe if we looked, and compared what we saw with various other things we could see. However hard we try to seize the complexity of the world, it systematically eludes our grasp. However precisely we try to focus on the details of that complexity, it remains systematically indeterminate. However far and wide we try to reach, however far forward and back in time, everything but one small patch at one instant of time remains forever beyond our immediate reach.

And yet the world is shaped by the traffic that passes between us and it in that one single shifting instant. So are we. From that tiny blur of contact we have constructed the universe and ourselves. When we turn our gaze inwards, though, and try to catch a glimpse of how we do it, it seems even more elusive than the raw material we are working upon.

This is what it comes down to in the end: the world has no form or substance without you and me to provide them, and you and I have no form or substance without the world to provide them in its turn. We are supporting the globe on our shoulders, like Atlas – and we are standing on the globe that we are supporting . . .

Soon I shall close my eyes not temporarily and experimentally, but permanently and in earnest. I shall cease to keep my eye on things – cease even to dream them. So it will be up to you to keep the whole performance going. And there will come a time, improbable as it seems to you now, when even you will be forced to retire from the job. Others will come, and go, until eventually there will be no one left to take over the business. What will happen to the great mutual balancing act, as the last man on earth finally closes his eyes, and there is suddenly no here, no there, no anywhere? No *is*, no *was*, no *will be*?

Nothing will happen. We both know that the universe will go on exactly as before. It will be affected by our departure no more than it was by our presence. What were we, after all? Merely a few fleeting eddies on the surface of the ocean.

The paradox remains. We have not even begun to resolve it.

All we can do is what human beings have always done since they first had words enough to think about it: start again and have another try.

Look up at the stars on a calm, clear night . . .

Notes

Prospectus

1 John Gribbin, in *The Birth of Time* (London, 1999) suggests fifteen billion years, but in 2003 this was refined as a result of NASA's Wilkinson microwave probe to 13.7 billion. (Here and throughout the book, 'billion' is used in the sense that it is understood in the United States, and increasingly worldwide: to mean a thousand million. 'Trillion', correspondingly, is used to mean a million million.)

2 *Are* we objectively at the centre of the universe? See note 11 to chapter I/4.

3 Murray Gell-Mann, *The Quark and the Jaguar* (1994); Brian Greene, *The Elegant Universe* (1999). For the most brilliantly clear and graphic account of the vastness of the universe and the smallness of particles, and of the sheer discrepancy between both and our own dimensions, see the opening chapters of Bill Bryson's wonderful survey of scientific understanding, *A Short History of Nearly Everything* (2003).

4 Carl Haub, 'How Many People Have Ever Lived on Earth?', posted on the website of the Population Reference Bureau, at http://www.prb.org/Template.cfm?Section=PRB&template=/ContentManagement/ContentDisplay.cfm&ContentID=7421

5 This is an example of the so-called 'anthropic principle', a term coined in 1974 by the cosmologist Brandon Carter for a concept around which a whole philosophical literature has since accumulated. Nick Bostrom, in his book *Anthropic Bias: Observation Selection Effects in Science and Philosophy* (2002), deprecates the word 'anthropic', and relates the principle to a general class of cases where the result of an observation is predetermined by the circumstances in which the observation is made. You can estimate what percentage of the population now use mobile phones by asking a random sample of citizens – but if the way you ask them is by phoning them on their mobiles then this is going to pre-empt the result. Ask what the odds are against your winning the lottery before the draw is made, and the answer may be ten million to one. Ask again from the vantage point of having won, though, and the question is otiose. It no longer has a function.

Bostrom traces the idea back through Kant and Hume to Bacon, whose witty account of the self-validating evidence for miraculous preservation he quotes:

It was a good answer that was made by one who when they showed him hanging in a temple a picture of those who had paid their vows as having escaped shipwreck, and who would have him say whether he did not now acknowledge the power of the gods, – 'Aye,' asked he again, 'but where are they painted that were drowned after their vows?' And such is the way of all superstition, whether in astrology, dreams, omens, divine judgments, or the like; wherein men, having a delight in such vanities, mark the events where they are fulfilled, but when they fail, though this happens much oftener, neglect and pass them by.

6 Richard Dawkins, *A Devil's Chaplain* (2003), p. 6.

7 Whether this is what philosophy really is, or should be, is of course itself a philosophical question, and our conception of it has altered radically over the years. Once upon a time there were philosophers who aspired to construct systems of thought that embraced everything. This changed, at any rate in the Anglo-Saxon world, and philosophy began to be seen as having a more modest aim – as being an attempt to address a series of *problems* (mostly, it was held, created by philosophers themselves). This was the view of philosophy which had become more or less taken for granted in British universities when I studied the subject.

It's since been much attacked, and has fallen out of fashion. Although I can't help remaining sympathetic to the modesty and openness of this approach, I've begun to wonder about those 'problems' myself. There are plenty of problems, of course, in making sense of the world and our place in it; but is it really the particular problems themselves that drive us to philosophy? Isn't it more a general feeling of things being problematic? Or something even vaguer and more general than that – sheer ordinary curiosity about the world and our place in it? We want (impossibly) to get behind the scenes, to open up the computer and take a look inside. We know that the colourful display of easily comprehensible symbols on the screen depends upon an unseen complexity, upon a hierarchy of programs written on top of programs – Word upon Windows upon DOS upon Basic upon machine code upon hexadecimal upon binary. We have an uneasy feeling that we are being treated like children – that we ought to be able to track the display back to that fundamental alternation of ones and zeros.

What's structurally interesting, though, is not the ones and zeros themselves but the *fact* that from nothing but ones and zeros all the complexity that confronts us on the screen can be developed. This is the *story*. And this is the real point of philosophy, surely, however you practise it: that it tells a general story about the world, just as science and religions do (even if the story is a modest one of particular problems raised and discussed) – that it sets a tone, casts a certain slant of light upon the world. And if you're going to tell a story, perhaps there's a case for telling a long one, in the old-fashioned way, a huge sprawling novel with larger-than-life characters and melodramatic situations that takes on the whole of human experience. Perhaps there's a case for big metaphysical constructions. Not that this is what I'm proposing here.

I/I Traffic

1 John C. Taylor, *Hidden Unity in Nature's Laws* (2001), p. 141.

2 The grounds of indeterminacy are themselves curiously indeterminate. You can read quite different accounts of it in different sources – so different that it's difficult to see that the same principle is under discussion. Even Werner Heisenberg and Niels Bohr offered radically different versions of it when Heisenberg first postulated it and his former mentor took it up. Here's my own amateur derivation of a version of it (as related to position/ velocity) that seems to me to apply even in classical mechanics. The velocity of an object is a function of (at least) two positions in space and two in time. But its position in space can be precisely determined only at a point where there is no change in its position in time, and its position in time can be precisely determined only at a moment when there is no change in its position in space. In other words, either position can be precisely determined only if the velocity of the object is zero. But this is impossible to specify, because to measure its velocity as zero means having determined two positions in space (even though in this case they are the same position), and two in time; and these four positions can be determined, as we have seen, only when its velocity has been measured (as zero).

Of course, we can specify two mathematical points in time and two mathematical points in space. But now we are talking about a theoretical world, a constructed world of abstractions, not the real world of real measurements.

3 If a farmer has made a path, he is able to saunter easily up and down it. That is what the path was made for. But the work of making the path was not a process of sauntering easily, but one of marking the ground, digging, fetching loads of gravel, rolling, and draining. He dug and rolled where there was yet no path, so that he might in the end have a path on which he could saunter without any more digging or rolling. Similarly a person who has a theory can, among other things, expound to himself, or the world, the whole theory, or any part of it; he can, so to speak, saunter in prose from any part to any other part of it. But the work of building the theory was a job of making paths where as yet there were none.

Gilbert Ryle, *The Concept of Mind* (1949; Penguin, 1990), p. 272.

4 And in modern quantum theory the level of complexity in the universe is dramatically raised yet again by *entanglement* – the continuing communion of particles in a joint quantum state even when they are physically separated:

Take a detector that has detected a particle that was in an entangled state. Suppose it is struck by an air molecule or a photon from the cosmic microwave background. The molecule (or photon) may bounce off, leaving the detector in a different quantum state. Then the detector and the receding molecule (or photon) are in an even more complicated entangled state. Very quickly, a sort of web of entanglement is spreading out into space.

Taylor, op. cit., p. 252.

5 Another way of putting it: 'Human experience is the construction of reality, not a property of a physical world that imparts the same experience to everyone who encounters it.' Sandra Scarr, psychologist and researcher in twin studies, quoted in Lawrence Wright, *Twins* (1997), p. 121.

6 This dialogue that I detect us maintaining with the world sounds suspiciously Kantian. Could the elusive things that offer themselves in the traffic be the strangely reticent *Dinge an sich*, the things-in-themselves that Kant sees as somehow underlying the phenomenal world of our perceptions, yet as having no causal connection with it? Not at all. The things I'm talking about are physical objects, as fully located in the physical world as anything ever can be, if that world exists (as in one sense it plainly does) independently of us. The question at issue is how we grasp that world.

The hypothesis of things-in-themselves merely compounds the problems of phenomenalism. If the attempt to dissolve the world into elements of our experience is unreal, as I am suggesting, then the predication of an unreachable world out there beyond the appearances doesn't restore the objectivity of the world – it simply renders it incomprehensible.

I can't help feeling that any plausibility that things-in-themselves ever had depended, like the plausibility of sense-data, on an artificially simplified picture of the world we are dealing with. The perceptual experiences analysed by phenomenalists always seemed to be restricted to seeing pennies – large, old pennies, at that period – which were neatly resolved into elliptical brown sense-data. Any superficial charm that this procedure has fades when you consider some of the other objects there are to be analysed, such as the formless muddle of putty and tangled string at the back of that conceptually challenging drawer in the kitchen. No handy adjectives like 'brown' or 'elliptical' come to mind to characterise the sense-data we are experiencing when we look at this, only depressingly circular expressions such as 'old-putty-like', or ostensive hand-wavings such as 'coloured like *this* and like *this* and like *this* . . .', all of which seem to require the existence of some external object whose shape and colour can be indicated to others before the internal experience into which that object is supposed to be dissolved can be named. (The same is true of brown ellipses, of course, but their geometrical regularity and smoothness makes them slip past our conceptual guard more easily.) You can't help suspecting that the perceptual thing that Kant had in mind when he believed he could imagine a thing-in-itself lurking behind it was also something reassuringly simple in colour and shape. A pfennig, in all probability. And if a pfennig can be related to a pfennig-in-itself, then I suppose there's no theoretical reason why that indescribable mess at the back of the drawer shouldn't have an indescribable-mess-at-the-back-of-the-drawer *an sich* to keep it company. All the same, there's something about the idea that would surely make even Kant's heart sink a little.

Since things-in-themselves exist in a world beyond space and time then old Mrs Tompkins across the street is supported by a Mrs Tompkins who is not old at all (or young, either), and there are spaceless Spice Girls who remain forever unheard-of, forever famous, and forever forgotten again.

Or do old Spice-Girls-CDs-in-themselves eventually go to a muncipal-incinerator-in-itself, and turn into a little heap of ash-in-itself, a puff of carbon-dioxide-in-itself? Does this lofty realm beyond causality, while remaining completely unaffected by the succession of cause and effect in the phenomenal world, somehow causelessly reflect it?

Then again, how is old/young Mrs-Tompkins-in-herself to be discriminated, even notionally, from a pot-of-tangerine-yoghurt-in-itself? The identities of their physical counterparts are part of the perceptual world just as surely as the space that they occupy, or the time through which they occupy it, or the causality that fills the pot with yoghurt and empties it again, that puts the wrinkles in Mrs Tompkins's face and in the end stops the breath in her lungs. Mrs Tompkins and the yoghurt bring us their discriminability in the physical world; we, as our contribution, seize hold of that discriminability and do discriminate them. The deal is struck in our world. So, just as things-in-themselves are not extended in time and space, so (if one may make so bold as to say anything about them at all) they remain an undifferentiated, solid, fused, homogenous mass, a single all-embracing *Ding an sich*.

Schopenhauer raises a similar objection to Kant – and accepts the homogeneous mass as the solution to the problem. Plurality, he says, is an attribute, and since only the phenomenal world can have attributes, plurality is possible only in the phenomenal, not the noumenal. So the noumenal is indeed one single undifferentiated and undifferentiatable thing-in-itself. The great inaccessible junkyard postulated by Kant has in Schopenhauer undergone meltdown. It's difficult enough to think what's happening in Kant's noumenal world, when with our phenomenal teeth we crack a phenomenal nut – whether we are to suppose teeth-in-themselves cracking a nut-in-itself into various pieces-of-nut-in-themselves, with various traces-of-dental-enamel-in-themselves attached? But if teeth and walnut are somehow the phenomenal manifestations not of various teeth-in-themselves and of a walnut-in-itself, but of one single teeth-and-walnut complex, what's happening in the noumenal world when our little piece of phenomenal cracking and tooth damage occur? Nothing, even more plainly than in the Kantian world. Nothing is getting cracked into pieces. Nothing is flaking off. Nothing is happening. The noumenal world is plainly unchanging from the beginning of time to the end of it.

A peaceful world. A rather dull one, though, perhaps.

7 E.g. M. Merleau-Ponty, *Phenomenology of Perception* (1945, 1958), p. 18: 'Our perceptual field is made up of "things" and "spaces between things".'

8 More on this in Chapter I/4.

9 Taylor, op. cit., p. 290: 'We are always ignorant about the detailed microscopic state of a macroscopic lump of matter. What we know about it is generally of a statistical nature. Statements about its temperature, pressure, magnetism and so on, are statements about average properties . . .'

Taylor says there is one exception to this. 'If we could get the lump of matter to the absolute zero of temperature, it would (in principle) be in a single quantum state: the state of minimum energy.'

The astrophysicist Subrahmanyan Chandrasekhar suggests another exception: a black hole (an object so dense, and consequently with such an intense gravitational field, that no light or other signal can escape from it). Unlike ordinary physical objects, which 'are governed by a variety of forces, derived from a variety of approximations to a variety of physical theories . . . the only elements in the construction of black holes are our basic concepts of space and time. They are thus, almost by definition, the most perfect macroscopic objects there are in the universe.' (Quoted from Chandrasekhar, *Truth and Beauty*, in op. cit., p. 359.)

10 It might be noted that perspective involves not one single viewpoint but two – one located in each of our two eyes. Any real knowledge of the world in fact involves a great many more than two viewpoints – it requires an indefinite series of them extending over time. Our information about the depth and development of the world is stored in the differences between these viewpoints (rather as electrical energy is an expression of the difference in potential between two poles).

I/2 The Laws of Nature

1 Kepler had introduced his laws of planetary motion, and Descartes his laws of nature, in the early part of the century, but see page 55.

2 Newton himself, though, says James Gleick in his biography (Gleick, *Isaac Newton*, 2003), 'never succumbed to this fantasy of pure order and perfect determinism. Continuing to calculate where calculation was impossible, he saw ahead to the chaos that could emerge in the interactions of many bodies, rather than just two or three.' Each successive orbit of a planet, for example, 'is dependent upon the combined motions of all the planets, not to mention their actions upon each other. Unless I am much mistaken, it would exceed the force of human wit to consider so many causes of motion at the same time, and to define the motions by exact laws which would allow of an easy calculation.'

3 The quantum of heat energy was introduced by Max Planck. It was Einstein who understood that light, too, was composed of quanta (photons), and Bohr who saw that an atom of matter would also change its state in discrete steps as the quantal packets of energy were absorbed or given out, and electrons 'jumped' seamlessly from one orbit to another.

4 This of course is the Uncertainty Principle, introduced by Heisenberg in 1927. Wolfgang Pauli, to whom Heisenberg first propounded the idea, had in fact hit upon the first notion of this a few months earlier:

> When electrons approach each other in a collision or in an atom, their quantum behaviour comes into play and manifests a 'dark point': 'The p's [momentum] must be assumed to be *controlled* [by observation or calculation], the q's as *uncontrolled*. That means one can always only calculate the probabilities of definite changes of the p's for given initial values and *averaged over all possible values of the q's*.'

Thus one cannot speak of a definite 'path' of the particle, nor, wrote Pauli, 'may one inquire simultaneously about p and q. In other words, of the two conjugate variables p and q, a probable value for the change in momentum (or velocity . . .) occured simultaneously with an indeterminate value for the position.'

Heisenberg was '*very* enthusiastic about Pauli's letter' . . .

David C. Cassidy, *Uncertainty, the Life and Science of Werner Heisenberg* (1992), pp. 232–3.

5 It also means, at least in theory, that if we know that an object is at rest we don't know exactly where it is; and that if we're sure we *do* then we can't be sure that it's at rest.

6 Even though this unambiguous object may now have the ambiguities of its history somehow inscribed upon it, as in the observable and unambiguous interference patterns produced by particles which have followed unobserved an ambiguous path – the ambiguity of which would have vanished if we had attempted to observe it; and in the phenomenon of entanglement (discovered much later), where the state of a particle changes in concert with changes in the state of another particle with which it was involved in the past, but is no longer.

7 The uncertainty relations apply *in principle* also to much bigger, macroscopic, objects . . . If you walk down the street and I want to measure your position at a given time with the same accuracy 10^{-8} cm [the size of a hydrogen atom], then, with due regard to the uncertainty relations, I can in principle still measure your velocity to an accuracy of a billionth of a billionth of a mile per second.

Abraham Pais: *Niels Bohr's Times* (1991), pp. 307–8.

8 Or so it seemed to the pioneers of quantum mechanics. More on this later.

All this of course has obvious philosophical implications, as well as scientific ones. It's difficult to work out to what extent these implications were registered by the professional philosophers of the period. The logical positivists were highly deferential to scientific theory, and the members of the Vienna Circle, where logical positivism was first introduced, were greatly interested in the work being done by Niels Bohr, Werner Heisenberg and others in Copenhagen. Their fundamental belief that the meaning of a proposition was the means of its verification seems to parallel Bohr's insistence that the result of any observation had to include reference to the circumstances in which the observation was made. But the general tenor of positivism's impatience with any language that doesn't engage snip-snap with hard-edged reality is quite out of keeping with Bohr and Heisenberg. After all, one of the founding texts of the Vienna Circle was Wittgenstein's *Tractatus* (even though Wittgenstein himself deprecated any connection with the positivists), which argues that the world consists of 'facts' – unambiguous, self-contained states of affairs mapped precisely by the series of elementary propositions into which language is supposedly analysable. These 'facts' are very unlike the incomplete states of affairs suggested by uncertainty, and the propositions that so totally express them very unlike the probabilities of quantum mechanics.

It's true that Bohr's ideas foreshadow Wittgenstein's later philosophy, developed during the thirties and set forth in his posthumous *Philosophical*

Investigations (1953). Wittgenstein's later conception of language, not as a precise picture of the world, but as a set of tools for getting things communicated between one person and another, and of meaning as being equivalent to use, is very much in the spirit of Bohr's view of scientific experiment. Then again, Bohr's recognition of the impossibility of both thinking and at the same time thinking about thinking is echoed in Wittgenstein's treatment of the 'other minds' problem. Philosophers have traditionally cast doubt on one's ability to know that other people have consciousness in the same way that one knows it of oneself. They have seen a fundamental contrast in type between the second-hand evidence one has of other people's thinking and feeling and the direct knowledge one has of one's own. Wittgenstein undermined this contrast by discounting the specialness of the access that one has to one's own thinking, and brought out the parallels with the external, objective ways in which one knows about other people's.

No one, though, so far as I know, has ever demonstrated that Wittgenstein was aware of Bohr's work, and many philosophers since have written in a way that implies a definite and unambiguously expressible world that surely vanished in March 1927 (when Heisenberg published his uncertainty paper).

9 Feynman, *The Character of Physical Law* (1965), p. 58. He takes a particularly dim view of the efforts of philosophers in this direction, e.g., 'We have learned from much experience that all philosophical intuitions about what nature is going to do fail' (ibid., p. 53).

10 On the other hand (and, as we shall see, this is an argument with many other hands – though not one where many hands make light work):

> Physics is not mathematics, and mathematics is not physics. One helps the other. But in physics you have to have an understanding of the connection of words with the real world. It is necessary at the end to translate what you have figured out into English, into the world, into the blocks of copper and glass that you are going to do the experiments with. Only in this way can you find out whether the consequences are true. This is a problem which is not a problem of mathematics at all.
>
> Feynman, ibid., p. 55.

11 Max Planck, who first introduced the notion of the quantum, didn't (understandably enough) take in the shocking implications of what he had begun. He shared the assumption of his predecessors that the laws of nature are external and objective. 'There is a real world,' he said, 'independent of our senses; the laws of nature were not invented by man, but forced upon him by the natural world. They are the expression of a rational world order' (Max Planck, *The Philosophy of Science*, quoted in Max Perutz: I Wish I Made You Angry Earlier, 1998).

Einstein, who first saw that if energy was quantal then matter must also be, and who first understood that all <u>measurement was relative,</u> insisted like Planck on the 'objective reality' of the universe, which these new laws would respect and express. But on the nature of the laws he parted company from Planck. He saw them not as parts of the external reality which they attempted to codify, but as 'f

> The enormous practical success of [Newton's] th
> prevented him and the physicists of the eighteent
> centuries from recognising the fictitious characte
> system . . . The fictitious character of the principle
> by the fact that it is possible to exhibit two essenti
> [Newtonian mechanics and general relativistic mecl
> which in its consequences leads to a large measure o
> experience.

Einstein, *On the Method of Theoretical Phy₋₋₋* (1933), p. 460.

By 'fictions' he meant, says his biographer Abraham Pais, 'free inventions of the human mind' (Pais, *Niels Bohr's Times* [1991], p. 460). These inventions were not, as earlier physicists had supposed, derived purely from experimental evidence. Nor were they justified purely, as more pragmatically minded scientists later suggested, by their utility in explaining further evidence. They were constructions that reached beyond the phenomena to express some kind of deeper harmony, and their justification was essentially an aesthetic one, unconnected with any utility that they might turn out to have.

Even Bohr and Heisenberg were a little slow at first to take in the full paradoxicality of the quantum mechanics they had developed. Bohr, in some of his work, seems to accept that a measuring instrument is itself a potential observer. Heisenberg continued to talk about the orbits of electrons, even in the act of demonstrating that they were indeterminable. It wasn't long, however, before they accepted what Bohr called 'the fundamental limitation . . . of the objective existence of the phenomena independent of the means of their observation' (Bohr, *Light and Life* [1932], p. 315). Bohr came to believe that the conceptual apparatus of physics had to be seen not as part of the natural world, but as a human artefact. 'There is no quantum world,' he is reported to have been in the habit of saying. 'There is only an abstract quantum physical description. It is wrong to think that the task of physics is to find out how nature is. Physics concerns what we can say about nature.'*

---

* A footnote to the note above:

Did Niels Bohr actually make the remark attributed to him above about 'a quantum' world'? ('There is no quantum world. There is only an abstract quantum physical description. It is wrong to think that the task of physics is to find out how nature is. Physics concerns what we can say about nature.') The remark is much quoted, and its source is always given as a contribution by Aage Petersen, one of Bohr's closest colleagues, to the *Bulletin of Atomic Science*, September 1963. I asked Finn Aaserud, the Director of the Niels Bohr Archive in Copenhagen, if he could give me the original reference in Bohr's own work. The question turned out to be more interesting than I had realised, and Aaserud responded by sending me a learned and delightful article on the subject by the American physicist N David Mermin (*Physics Today*, February 2004).

Mermin reports nearly three hundred hits on Google for the remark, ninety per cent of them treating it as a direct quotation from Bohr. But this, as Mermin points out, is not what Petersen claims in his article. He was generalising what Bohr 'would answer' when asked about the relations of quantum mechanics and a supposed quantum world. The

...born, another of the founding fathers of quantum mechanics, saw ...elativism and indeterminacy of the new physics as a liberation:

> Ideas such as absolute certitude, absolute exactness, final truth, etc., are figments of the imagination which should not be admissible in any field of science . . . This *loosening of thinking* seems to me to be the greatest blessing which modern science has given to us. For the belief in a single truth and being the possessor thereof is the root cause of all evil in this world.
>
> Max Born, 'Symbol and reality', in *Physics in My Generation*, 1969.

This early view of quantum mechanics came to be known as 'the Copenhagen Interpretation', because it was introduced by Niels Bohr in Copenhagen and a group of physicists who at various times worked with him there, in particular Heisenberg and Pauli (though see note 11 to this chapter). Einstein, in spite of his belief that scientific laws were fictions, remained notoriously resistant to its inherent probabalism. He remained a determinist at heart, and maintained until the end of his life that the new laws (brilliantly predictive as he agreed them to be) were 'incomplete' – a temporary development that would turn out to be merely the limiting case of some more profound laws, rather as Newtonian mechanics might be seen as the limiting case of relativity for inertially homogeneous systems. His jokes about what '*der Herrgott*' would or would not permit the universe to be like (as in '*Der Herrgott werfelt nicht*' – 'The Lord God does not play dice') suggest some kind of moral or aesthetic considerations independent of human purposes, and some kind of freestanding structure with a definite form.

Some physicists now agree with Einstein about the unsatisfactoriness of quantum theory. The mathematician Roger Penrose also regards it, in its present state, as 'provisional'.

Ilya Prigogine, one of the pioneers in the study of unstable physical systems, sees science rather similarly in some ways – not exactly as an invention, but as 'a dialogue between mankind and nature' (Prigogine, *The End of Uncertainty* [1996 in French; 1997 in English], p. 157). He believes, though, that 'the laws of physics, as formulated in the traditional way, describe an idealised, stable world that is quite different from the unstable, evolving

---

closest equivalent that Mermin can find in Bohr's written works is:

> Indeed from our present standpoint, physics is to be regarded not so much as the study of something a priori given, but as the development of methods for ordering and surveying human experience.
>
> *Essays 1958–62 on Atomic Physics and Human Knowledge*, 1987.

Mermin believes that this can be interpreted to mean much the same, and it certainly seems to bear out the part of the original which is relevant to my purposes in this section, the claim that the concern of physics is not how the world is but what we can say about it. He records that Victor Weisskopf, who had worked with Bohr, refused to believe that he can ever have said anything like Petersen's version, while Rudolf Peierls, who had also worked with him, found it on the contrary exactly the kind of thing that he loved to remark.

world in which we live' (ibid., p. 26), and on the question of probability he takes exactly the opposite view to Einstein. All physics relating to the real world, he insists, deals more or less exclusively in probabilities. The zoologist Peter Medawar also sees scientific reasoning as 'a kind of dialogue', but as one 'between the possible and actual, between what might be and what is in fact the case' (Peter Medawar, *Induction and Intuition in Scientific Thought*, 1969).

The biologist Jacques Monod, even though he believes that the biosphere is not deducible from first principles and is therefore essentially unpredictable, describes the basic 'strategy' of science as the discovery of 'invariants':

> Every law of physics, as for that matter every mathematical development, specifies some invariant relation; science's fundamental statements are expressed as universal 'conservation principles'. It is easily seen, in any example one may like to choose, that it is in fact impossible to analyse any phenomenon in terms other than those of the invariants that are conserved through it.
>
> Jacques Monod, *Chance and Necessity* (1972), p. 98; and for the unpredictability of biological phenomena see ibid., p. 49.

The mathematician and physicist John von Neumann, briskly pragmatic, thinks of science as a matter of models rather than theories. Scientific method, he says, is 'primarily opportunistic' – quite how 'utterly opportunistic' few people outside the sciences appreciate. It must be repeated again and again, he says, that 'the sciences do not try to explain, they hardly even try to interpret, they mainly make models. By a model is meant a mathematical construct which, with the addition of certain verbal interpretations, describes observed phenomena. The justification of such a mathematical construct is solely and precisely that it is expected to work' (John von Neumann, 'Method in the Physical Sciences', in *Collected Works* vol. VI). When he says that a construct is expected to 'work', he explains, he means 'correctly to describe phenomena from a reasonably wide area'. He also thinks that a model should satisfy 'certain aesthetic criteria', by which he means that 'in relation to how much it describes, it must be rather simple.'

Mermin wittily concludes, echoing the Petersen version:

There is no Copenhagen interpretation of quantum mechanics. There is only a range of quantum physical positions. Some are held by Weisskopf's Bohr and some by Peierls's Bohr. There are even positions held by my own Bohr, who, unlike the other two Bohrs, is not constrained by my ever having actually met the man, except for a remote sighting in 1957 from the back row of an enormous auditorium where he spoke for an hour, inaudibly. My Bohr is rather similar to, but considerably more cautious than, Petersen's Bohr . . .

Mermin says that he hopes his article 'will serve to restore the unfortunately vanishing distinction between Petersen's Bohr and what we might, for clarity, better off calling Bohr's Bohr'.

He goes on, in the same paper, to stress the value of vagueness here, as exemplified by the use of the word 'rather'. It all depends upon the context:

One cannot tell exactly how 'simple' simple is . . . Simplicity is largely a matter of historical background, of previous conditioning, of antecedents, of customary procedures, and it is very much a function of what is explained by it. If the amount of material which is unambiguously explained – that is, explained with no added interpretation or commentaries – is extremely extensive, if it is also very heterogeneous, if one has clearly explained a large number of things in very different areas, then one will accept a good deal of complication and a good deal of deviation from stylistic beauty. If, on the other hand, only relatively little has been explained, one will absolutely insist that it should at least be done by very simple and direct means.

The two most influential philosophers of science in the twentieth century, Karl Popper and Thomas Kuhn, agree that scientific laws and theories are human constructs. According to Popper: 'Science is not a system of certain, or well-established statements, nor is it a system which steadily advances toward a state of finality' (Karl Popper, *The Logic of Scientific Discovery* [1935], p. 278). We don't wait passively for repetitions to impress or impose regularities upon us, 'we actively try to impose regularities upon the world. We try to discover similarities in it, and to interpret it in terms of laws invented by us. Without waiting for premises we jump to conclusions. These may have to be discarded later, should observation show that they are wrong.' This he describes as 'a theory of trial and error – of *conjectures and refutations*'. Scientific theories, he says, are not 'the digest of observations', but 'inventions – conjectures boldly put forward for trial, to be eliminated if they clashed with observations; with observations which were rarely accidental but as a rule undertaken with the definite intention of testing a thory by obtaining, if possible, a decisive refutation' (ibid., p. 60).

Kuhn takes an even more pragmatic view. Far from searching out counterinstances, he says, most scientists most of the time are so constrained by the current way of thinking that they tend not even to notice them – and if they do, they ignore them. It's only when the evidence for which the current theory offers no adequate explanation begins to accumulate and to create problems that a new theory has to be constructed. The analogy is with politics, and the perpetual alternation of stasis and revolutionary change; between the maintenance of society's hard-won forms and institutions on the one hand and their renewal on the other.

Kuhn agrees with Popper that there is no final state of affairs waiting to be discovered. There is progress in science, but,

we may . . . have to relinquish the notion, explicit or implicit, that changes of paradigm carry scientists and those who learn from them closer and closer to the truth . . . Nothing that has been or will be said makes it a process of evolution toward anything, any more than biological evolution is . . . We are all deeply accustomed to seeing science as the one enterprise

434

that draws constantly nearer to some goal set by nature in advance. But need there be any such goal . . . ? Does it really help to imagine that there is some one full, objective, true account of nature and that the proper measure of scientific achievement is the extent to which it brings us closer to that ultimate goal?

Thomas S. Kuhn, *The Structure of Scientific Revolutions* (1962), p. 170.

Kuhn thinks that his readers may be disturbed by this lack of a final destination for science. He points out that it was the same thing that most disturbed many of Darwin's contemporaries about biological evolution – its overturning of the assumption that it was directed towards some goal, some finally perfect form (ibid., pp. 171–2). Kuhn has been much attacked for 'relativism' – abandoning the notion of some theoretically accessible absolute standard of truth against which scientific theories should be measured. He agrees that there is 'no theory-independent way to reconstruct phrases like "really there"', and that 'the notion of a match between the ontology of a theory and its "real" counterpoint in nature now seems to me illusive in principle' (ibid., p. 206).

The physicist Steven Weinberg describes this as 'wormwood' to scientists like himself, who think that 'the task of science is to bring us closer and closer to objective truth', and 'delicious to those who take a more skeptical view of the pretensions of science' (Weinberg, 'The revolution that didn't happen', in the *New York Review of Books*, 8 October 1998). He is particularly hostile to Kuhn's belief that scientific theories can be properly understood, and therefore judged, only within the context of the particular 'paradigm' (Kuhn's term for a particular system and climate of thought and practice) of which they form part; this, he thinks, implies that these theories are 'not privileged over other ways of looking at the world, such as shamanism or astrology or creationism. If the transition from one paradigm to another cannot be judged by any external standard, then perhaps it is culture rather than nature that dictates the content of scientific theories.'

This seems to me a false dichotomy. Cultures and sciences alike (and all other human activities) are responses to 'nature' and are aspects of our interaction with it. They stand, fall, and evolve according to the same kind of criteria: how well they function, how widely they serve our needs, how coherent they are with the full range of our experience and understanding.

Richard Feynman, who has little interest in the philosophical niceties, shifts his position on the nature of scientific laws disconcertingly. Sometimes he seems to expect some kind of absolute truth to stand revealed at the last (and even speculates that this won't be in mathematical form). What bothers him, he says, is the inexhaustible complexity of the universe, and the impossibility of ever producing any complete account of even the smallest region of space and time according to the laws as we currently understand them:

How can all that be going on in that tiny space? Why should it take an

435

infinite amount of logic to figure out what one tiny piece of space/time is
going to do? So I have made the hypothesis that ultimately physics will
not require a mathematical statement, that in the end the machinery will
be revealed, and the laws will turn out to be simple.

Feynman, *The Character of Physical Law* (1965), pp. 57–8.

Elsewhere, though, in a passage written only a few years later, he seems to
be suggesting something much closer to Popper's and Kuhn's view. Not only
are the experiments on which the laws are based 'always inaccurate' but the
laws themselves are 'guessed . . . extrapolations, not something that observations insist upon . . . Every scientific law, every scientific principle, every
statement of the results of an observation is some kind of a summary which
leaves out details, because nothing can be stated precisely' (Feynman, *The
Meaning of It All* [1998], pp, 24–5). Newton's Law of Gravitation, for
example,

> . . . is not exact; Einstein had to modify it, and we know it is not quite
> right yet, because we have still to put the quantum theory in. That is the
> same with all our other laws – they are not exact. There is always an edge of
> mystery, always a place where we have some fiddling around to do yet. This
> may or may not be a property of Nature, but it certainly is common to all
> the laws as we know them today. It may be only a lack of knowledge.
>
> Ibid., p. 33.

And whether or not the laws themselves are exact, their application is as
confused as their status: 'Nature . . . seems to be so designed that the most
important things in the real world appear to be a kind of complicated
accidental result of a lot of laws' (ibid., p. 122).

David Deutsch, a quantum physicist who is among those attempting to
design the first quantum computer, seems to agree in some ways with the
earlier of Feynman's views – but goes much further. He believes that we may
be able to construct a 'Theory of Everything' – and even that we may
already have constructed the first version of such a theory (because, unlike
Feynman in his perfectibilist mode – and like Popper – he believes that any
theory is bound to be superseded):

> If knowledge is to continue its open-ended growth . . . eventually our
> theories will become so general, deep and integrated with one another that
> they will effectively become a single theory of a unified fabric of reality.
> This theory will still not explain every aspect of reality: that is unattain-
> able. But it will encompass all known explanations, and will apply to the
> whole fabric of reality in so far as it is understood. Whereas all previous
> theories related to particular subjects, this will be a theory of all subjects:
> a *Theory of Everything*.
>
> It will not of course be the last such theory, only the first. In science we
> take it for granted that even our best theories are bound to be imperfect
> and problematic in some ways, and we expect them to be superseded in
> due course by deeper, more accurate theories . . .

The attainment of a Theory of Everything will be the last great unification, and at the same time it will be the first across-the-board shift to a new world-view. I believe that such a unification and shift are now under way.

David Deutsch, *The Fabric of Reality* (1997), pp. 17–18.

Deutsch makes clear that he is envisaging something much more universal than the 'theory of everything', or 'Grand Unified Theory', with which particle physicists hope to unify the various fundamental forces and describe all the possible particles. His proposed super-theory would comprehend quantum physics, but also take in evolution, epistemology, and the theory of computation. 'The four [subjects] taken together form a coherent explanatory structure that is so far-reaching, and has come to encompass so much of our understanding of the world, that in my view it may properly be called the first real Theory of Everything' (ibid., p. 28). Poor Casaubon, who was hoping merely to unify all the world's mythologies, has been comprehensively outflanked.

The American philosopher of science Nancy Cartwright, however, takes something like Kuhn's view – and goes much further in the same direction. She proposes an even more pragmatic and humanistic view of the fundamental laws, an approach that is dramatised by the provocative title of her book, *How the Laws of Physics Lie* (1983; the quotations that follow are drawn from different essays in the book). The fundamental scientific laws, she argues, are all drastic simplifications of a world that is indissolubly heterogeneous, so that no particular case is ever precisely described by them. And the more fundamental the law, the falser it is – the less true of any actual situation in the real world.

This is not to say there are no true laws. There are vast numbers of what she calls 'phenomenological' laws – specific laws relating to limited classes of phenomena – that are precise and predictive. Nor is she attempting to discredit the notion of fundamental laws. But the point of them, she says, is to bring out some aspect of the indescribable complexity of any real situation – and they achieve this by ignoring most of its aspects.

The data relating to any specific real situation, she says, always has to be 'prepared' to bring it within reach of any fundamental law, by adjustment, approximations, and the introduction of fictitious entities:

If the fundamental laws are true, they should give a correct account of what happens when they are applied in specific circumstances. But they do not. If we follow out their consequences, we generally find that the fundamental laws go wrong; they are put right by the judicious corrections of the applied physicist or the research engineer . . . There are no rigorous solutions for real life problems. Approximations and adjustments are required whenever theory treats reality . . . To get from a detailed factual knowledge of the situation to an equation, we must prepare the description of the situation to meet the mathematical needs of the theory . . . The lesson for the truth of fundamental laws is clear:

fundamental laws do not govern objects in reality; they govern only objects in models.

We have to *choose* a particular explanation strategy to suit the circumstances and our particular purposes. The common supposition is that

> ... if the right kinds of descriptions are given to the phenomena under study, the theory will tell us what mathematical description to use and the principles that make this link are as necessary and exceptionless in the theory as the internal principles themselves. But the 'right kind of description' for assigning an equation is seldom, if ever, a 'true description' of the phenomenon studied; and there are few formal principles for getting from 'true descriptions' to the kind of description that entails an equation. There are just rules of thumb, good sense, and, ultimately the requirement that the equation we end up with must do the job.

She makes an analogy between scientific theory and theatre, where we understand that historical truth has to be conventionalised and simplified to make it dramatically workable. She makes an even bolder analogy between natural objects and human beings, whose 'behaviour is constrained by some specific laws and by a handful of general principles, but it is not determined in detail, even statistically. What happens on most occasions is dictated by no law at all.' And she rejects the whole aim of so much of fundamental science, to reduce the universe to a series of abstract principles which explain everything. 'Our knowledge of nature,' she writes, 'nature as we best see it, is highly compartmentalised. Why think nature itself is unified?'

12 It would be more accurate to call it the Copenhagen/Göttingen Interpretation, since it was Max Born at Göttingen who first interpreted Schrödinger's wave-function as being not the representation of an actual wave but of the probability of finding a particle in any particular place.

13 Feynman, *The Character of Physical Law*, p. 58.

14 Popper, *The Logic of Scientific Discovery*, p. 230.

15 Popper, *Quantum Theory and the Schism in Physics* (1982).

16 Murray Gell-Mann, *The Quark and the Jaguar* (1994), pp. 136–7.

17 Ilya Prigogine, *The End of Uncertainty*, p. 51.

18 Ibid., p. 151. The reasons why Prigogine believes that he can dispense with the observer are mathematical, and I cannot pretend to understand them: 'We arrive at a realistic interpretation of quantum theory because the transition from wave functions to ensembles can now be understood as the result of Poincaré resonances without the mysterious intervention of an "observer" or the introduction of other uncontrollable assumptions' (ibid., p. 131). He argues that it is possible, and indeed necessary, to combine the indeterminism of quantum mechanics with the view that probability is an objective quality of the universe, and not of our statements about it. I cannot understand this.

19 Feynman, *The Meaning of It All*, p. 24.

20 Feynman, *The Character of Physical Law*, p. 168.

21 Popper, *The Logic of Scientific Discovery*, p. 18.

22 Ibid., p. 86.

23 Ibid., p. 91.

24 And perhaps even more oddly with Popper's claim to have been converted by Alfred Tarski to the possibility of 'absolute truth' – i.e., truth as 'correspondence to the facts', so that, for example, 'the statement, or the assertion, "*Snow is white*" corresponds to the facts if, and only if, snow is, indeed, white' (Popper, *Conjectures and Refutations*, pp. 302ff).

25 Popper, *The Logic of Scientific Discovery*, p. 93. Many physicists – Einstein, Feynman, and modern string theorists, for example – judge theories not only by their utility but by their aesthetic qualities – their 'beauty' or 'elegance'. This seems to me to imply even more strongly the indispensability of the observer, even if you see theories as being not human artefacts but natural objects. Oceans and mountains are not created by human beings, but if they have beauty it can only be in a universe where there are eyes to perceive it and minds to appreciate it.

26 Gell-Mannn, op. cit., pp. 140ff.

27 John C. Taylor, *Hidden Unity in Nature's Laws* (2001), p. 240.

28 Ibid., p. 374.

29 Gell-Mannn, op. cit., p. 138.

30 The starting point of both the Copenhagen Interpretation and multiple universe theory is the famous problem of the interference patterns. Bohr and Heisenberg proposed a 'thought experiment' in which a single photon reached a screen by way of a diffraction grating with two slits open in it. An interference pattern is created on the screen characteristic of the interference pattern produced by the combination of waves arriving from two different sources – until you attempt to observe the photon as it apparently passes simultaneously through the two slits, when it notoriously passes through only one of them, and the interference vanishes.

(This remained a 'thought experiment' when it was first proposed in the 1920s because it was not technologically possible at the time to mount it as a real experiment. A team at the University of Koblenz have now done an actual experiment with a real particle – though not a photon – and confirmed both parts of the prediction.)

Deutsch points out that the phenomenon is not limited to two slits:

Since different interference patterns appear when we cut slits at other places in the screen [i.e. the grating], provided that they are within the beam, shadow photons must be arriving all over the illuminated part of the screen whenever a tangible photon arrives. Therefore there are many more shadow photons than tangible ones. How many? Experiments cannot put an upper bound on the number, but they do set a rough lower bound. In a laboratory the largest area that we could conveniently illuminate with a laser might be about a square metre, and the smallest manageable size for the holes might be about a thousandth of a millimetre. So there are about 10^{12} (one trillion) possible hole-locations on the screen. Therefore there must be at least a trillion shadow photons accompanying each tangible one.

Deutsch, op. cit., p. 44.

439

His solution for this paradox is to assume that each of these 'shadow' photons really exists – but each of them in a separate universe of its own. (There is a parallel here with the 'modal realism' in philosophy, upheld by David Lewis – see *On the Plurality of Worlds* [1986] – which attempts to treat various philosophical problems by postulating that every stateable possibility is true, each in its own parallel world. See Chapter I/4.) So that in the experiment described the existence of a trillion virtual photons demonstrates the existence of a trillion other universes, all co-existing in parallel with this one. Since the scope of the experiment is restricted to the square metre that might be conveniently illuminated in the laboratory, it must be assumed that the trillion universes it reveals are only a tiny sample of the whole. Every photon – and every other sort of particle subject to quantum rules – must be accompanied by further hosts of virtual counterparts, some of them, presumably, existing in one or other of the trillion universes we have just stumbled across, others, presumably, in trillions of trillions more universes.

The only interaction between these other universes and our own is the interference patterns that appear when a particle and its virtual alter egos reassemble on the laboratory screen. This means that we can know only that much more about what's going on in the other universes than we do about the rather more modest number of elephants, and of tortoises on whose backs the elephants were standing, that were required in Greek cosmology to explain why the world didn't fall down. It appears to have a remarkable resemblance to flying saucer theory. An ambiguous phenomenon is reported which it is difficult to account for – and the explanation offered involves postulating a complete extra-terrestrial civilisation with an ability to suspend the laws of physics.

Reading Deutsch has encouraged me to adapt his approach to solving another mystery which has vexed theorists for many years – the single sock problem. This is the converse of quantum interference, where a single particle apparently acquires virtual partners. It is a matter of common knowledge that pairs of socks, while passing unobserved through the closed system of a washing machine, are repeatedly reduced to single socks, or at any rate to pairs consisting of one actual sock and one virtual sock undetectable by observation. The most likely explanation for this, it now dawns on me, by analogy with Deutsch, is that the missing socks are abstracted by fairy folk, and taken off to fairyland to be unwoven and reworked into garments for elves and pixies. This does, it's true, involve postulating an extensive fairy economic system about which we know little, but it does solve the problem of the virtual socks – and it does actually explain another mystery, which is where elves and pixies get their clothes from, a question that neither classical nor quantum physics has ever been able to answer satisfactorily.

And at least it requires the existence of only one fairyland, not trillions of fairylands. I can't help feeling, in fact, that Deutsch's figures are a considerable underestimate, if we are to take seriously what he says about these other universes. Although we can know nothing about what's going on in them, he suggests that things in each of them must be going their different ways. The particles are pursuing courses that are different from that of the one avatar we can observe. So are the physical objects that the particles make up. So, too, are the people who inhabit these different universes, our virtual other selves. 'When I introduced tangible and shadow photons,' writes Deutsch, 'I apparently distinguished them by saying that we can see the former, but not the latter. But who are "we"? While I was writing that, hosts of shadow Davids were writing it too . . . Many of those Davids are at this moment writing these very words. Some are putting it better. Others have gone for a cup of tea.'

Quite what it means to say that David Deutsch might have gone for a cup of tea in another universe even while he was still sitting at his desk in this one it's difficult to conjecture, since there is no way in which we can ever check the assertion or be affected by it. I suppose it suggests a certain picture, a certain imaginative tone and colour, just as the complex battlements of fairyland castles do. But, in so far as it has some notionally factual form, it introduces a notionally factual complication. Various photons in the light reflected from the David Deutsch labouring at his desk in this world have presumably interfered satisfactorily with various virtual counterparts in the light reflected from the David Deutsch who has been labouring at his desk in various parallel universes. Now we're told that some of these other David Deutsches have knocked off work. So that the next lot of photons emanating from our own familiar David Deutsch are no longer being interfered with by the photons from *that* David Deutsch. And yet the interference patterns have presumably remained unaltered. So another working David Deutsch must have come into existence in another universe to keep the system in balance. And for that matter another tea-drinking David Deutsch.

In other words, for every universe in the trillions of trillions of universes postulated by the theory, there must be trillions and trillions more to mirror the endlessly changing situations in each of them. I'm not clear whether the trillions and trillions of universes required by the original postulate would in fact be infinite, but the extra universes required to keep track of even one of their citizens, the ubiquitous Mr Deutsch himself, would surely be.

Not even flying-saucer theory and the vanishing-sock postulate require metaphysical hardware on this stupendous scale, even though they are both attempting explanations for phenomena that bulk a lot larger than those puzzling but faint shadows on the laboratory screen. (Silver cigars in the sky above Arizona! Local citizens abducted! A black-tie occasion, and only one black sock to go with the tie!) And since everything about this vast array, except the interference patterns themselves, remains entirely indeterminable, you might think that the relatively modest level of uncertainty introduced into the universe by Heisenberg and Bohr make the Copenhagen Interpretation seem by comparison almost as positivistic as the teaching of Mr Gradgrind.

It's only fair to say that, having introduced the idea of multiple universes to explain the single phenomenon of interference, Deutsch does go on to make use of it to explain various other problems of a more philosophical or conceptual nature, most notably involving time but also the questions of free will and counterfactuals, all of which we shall come to later. (Though of course in other universes other versions of ourselves have considered them ages ago. And resolved them, very satisfactorily.)

Multiple universe theory is not, however, much of an explanation, if it's one at all, for the phenomenon that it was introduced to explain, the interference patterns in the single particle/multiple slit experiment – because it doesn't explain why the shadow photons (an infelicitous oxymoron, incidentally, since a shadow in normal usage is precisely an absence of photons) have an effect on the tangible photon when they don't on anything else in the tangible universe.

And we still end up, just as we do in the Copenhagen Interpretation, with a reality which is accessible only partially to the observer, and which is expressible only through his participation in the world.

31 Most importantly, of energy, charge, and momentum; but also of many other quantities such as electron and baryon number.

32 They include in particular Einstein's principle of equivalence, adapted from Galileo, which states that no internal experiment can reveal whether a system is in constant motion in relation to another system, and gauge invariance, which applies to the forces involved in quantum mechanics.

33 Taylor, op. cit., p. 176.

34 Ibid., p. 204.

35 Mirror symmetry and charge conjugation invariance. Taylor describes the discovery by Chien-Shiung Wu of the breaking of the former in weak interactions as 'surely one of the great crucial experiments in the history of physics. The result astonished most physicists' (ibid., pp. 333–4).

36 'So far as we know there are no real units [of energy] . . . It is abstract, purely mathematical . . . there is a number such that whenever you calculate it it does not change. I cannot interpret it any better than that.' Richard Feynman, *The Character of Physical Law*, p. 70.

37 All quotations from the *Philosophiae Naturalis Principia Mathematica* (1687) are from the 1999 edition (entitled simply *The Principia*), translated from the Latin by I. Bernard Cohen and Anne Whitman.

38 Cf. D.T.Whiteside, *Introduction to the Mathematical Papers of Isaac Newton* (1974), pp 3–4.

39 Quoted in the article on Descartes in the 1911 edition of the *Encyclopaedia Britannica*.

40 James Gleick in his biography of Newton (Gleick, *Isaac Newton*, 2003) stresses the huge amount of practical experimenting that Newton performed.

Einstein, by his own account, was first prompted to construct the abstractions of both special and general relativity by imagining, if not observing, a concrete physical situation. In the case of special relativity he pictured to himself what he would experience if he were travelling with a ray of light. In the case of general relativity, 'I was sitting in my chair at the patent office at Bern when all of a sudden a thought occurred to me: "If a person falls freely he will not feel his own weight"' (quoted in Taylor, op. cit., p. 192).

41 The beginnings of these ideas go back even further: to Parmenides, about a century earlier, who held that nothing came into existence or ceased to be, but who derived this from his doctrine that nothing ever changed; and to Empedocles, who was influenced by Parmenides, and who also held that nothing comes into being or is destroyed, but taught that the elements of fire, air, water, and earth are combined by the power of love to form compounds (which are broken down into elements again by strife). Democritus, however, starts with the atomic hypothesis that he had learnt in the first place from his teacher Leucippus, who is credited by Aristotle as its inventor. The atoms hook on to each other to form new compounds in some way that doesn't involve the anthropomorphic principles uniting the elements in Empedocles. Like the elements themselves, the uniting principles are eternal and unalterable.

The basis from which Democritus derives this unalterability, says Louis Löwenheim (*Die Wissenschaft Demokrits*, Berlin [1914], p. 196), we can only speculate about. He suggests that it was the same as Helmholtz's two thousand and more years later – the axiom that every change in nature must have a sufficient cause, and that the quest for this cause must be continued until one has reached something permanently unchangeable. Löwenheim argues that 'the turning-point of unscientific alchemy into scientific chemistry is marked precisely by the moment when Galileo abandoned the Platonic–Aristotelian doctrine of the qualitative transformability of matter to return to Democritus's view of its qualitative untransformability.'

The principle of the conservation of matter is also propounded by Epicurus in his letter to Herodotus a century after Democritus, and it is to Epicurus that Lucretius credits it when he restates it in *De Rerum Natura* about two hundred years later still, in the first century BC. Some commentators assert that either Democritus or Epicurus or both also proposed the conservation of motion. If for 'motion' you not unreasonably read 'energy', then this would seem to foreshadow the most fundamental of all the modern conservation laws (under which, since Einstein, the conservation of matter has been subsumed). This is rather less certain territory, though. The history of Greek scientific thinking is difficult to piece together because only fragments survive from the works of the early philosophers who developed it, together with sometimes contradictory references in the writings of Aristotle, Plato, Cicero, and others. The first extant reference to anything that might be described as the conservation of motion seems to be in Lucretius (*De rerum natura*, Book II, pp. 303–7):

No powers whatsoever are capable of changing the totality of things. For neither is there any place whither matter could flee away out of the universe, it would be the same as ever, nor yet one where new power could collect and could break into the universe and change the whole nature of things and change no less their motion also.

According to Löwenheim:

It is clear that the poet Lucretius cannot be the discoverer of the principle, and even less the superficial Epicurus, especially since it is in the highest degree improbable that the same man who ascribed voluntary motion to atoms should have taught that the whole universe was governed by such lawfulness that no energy could ever appear which was not available from the beginning of the world. Since Cicero states that Epicurus took nothing from any predecessor other than Democritus, it seems reasonable to suppose that Democritus first expressed the principle of the conservation of energy and based it in the same way as his predecessor Empedocles did the principle of the conservation of matter.

It was specifically from Epicurus that the French philosopher and scientist Pierre Gassendi introduced the atomic hypothesis into modern science in the seventeenth century. You might also trace in Epicurus (in spite of his 'superficiality', as Löwenheim sees it) the first sketches of some of the other foundations of modern science. If you understand the transmission of what he calls 'images' as being by light, and his 'atoms' as including photons, then he has something like the modern conception of the speed of light as a constant which can never be exceeded:

The images have an unsurpassed fineness; and that is why they have unsurpassed speed too, since they find every passage suitably sized for there being no or few [bodies] to resist their flow . . . It is necessary that the atoms move at equal speed, when they move through the void and nothing resists them.

The Epicurus Reader, translated and edited by
Brad Inwood and L. P. Gerson (1994), p. 9.

He also seems to foreshadow quantum uncertainty and randomness:

. . . when the atoms move straight down through the void by their own weight, they deflect a bit in space at a quite uncertain time and in uncertain places, just enough that you could say that their motion had changed.
Epicurus, as later formulated by Lucretius, ibid., p. 65.

Even multiple universe theory:

There is an unlimited number of cosmoi, and some are similar to this one and some are dissimilar . . . For atoms of the sort from which a world might come to be or by which it might be made are not exhausted [in the production] of one world or any finite number of them, neither worlds like this one nor worlds unlike them. Consequently, there is no obstacle to the unlimitedness of worlds.

Ibid., p. 8.

Löwenheim credits Democritus with introducing early versions of causality, Newton's principle of action and reaction, and evolution. By 'evolution' he means not selection, but the understanding that one species developed out of another (and that, since the world was originally all water, the fauna were all in the first place marine animals, so that men are descended from worms). This he derives (op. cit., p. 165) from a reference in Pseudoplutarch (Plut. Epit. V, 19), who attributes it to the followers of Epicurus. 'But such a superficial thinker as Epicurus', says Löwenheim, 'cannot possibly have come upon a theory like evolution by himself. If he learned it he can only have taken it from Democritus.'

42 In a paper published in 1863, collected in Rudolf Clausius, *Abhandlungen über die Mechanische Wärmetheorie* (1864).

43 Clausius, incidentally, is always said to have introduced the Second Law, in the form 'Heat cannot of its own accord move from a colder to a hotter body,' in 1850. If you actually read the paper, though, you see that this isn't quite true, and that the origins of the law are (characteristically) rather more circumstantial. In the 1850 paper (see ibid.) he merely remarks, more or less in passing, that the idea of getting heat 'out of a *cold* body into a *hot* one . . . contradicts the behaviour of heat otherwise in everywhere demonstrating the endeavour to equalise any differences in temperature that arise and thus to move from the *hotter* body into the *colder*.' It was only later that he saw the full force of this. The familiar formulation appears first in a footnote added in 1863, where he says that 'the proposition assumed' in the original paper was to be regarded as a principle of the same importance as the one that work or heat cannot be got out of nothing. But in the first place it was a passing *assumption*.

44 Gell-Mannn, op. cit., p. 218. In other words, there are many disordered states but relatively few ordered ones; and disordered states, precisely because of their disorder, tend to be effectively indistinguishable from one another, so that a multiplicity of particular disordered states can be seen as general disorder. However, if we specified particular states of disorder they would be as rare as specified particular states of order, so that the Second Law might be rewritten (once again!) as stating that the likelihood, not of disordered states, but of *unspecified* ones, was increasing. This makes entropy dependent upon a human context. It is increasing because the states that we are interested in are ordered ones, and we are interested in them both because their order makes them discriminable and because they have the potential to change and to effect change.

45 The quotation from Boyle, for instance, is: 'The Wisdome of God does confine the creatures to the establish'd Laws of Nature' (Boyle, *Occasional Reflections* [1665], iv. vi).

46 Edgar Zilsel, 'The genesis of the concept of physical law'. *Philosophical Revew* 51 (1942), edited Cornell, published New York. I haven't been able to trace any reprint.

47 Descartes, *Discourse on Method* (1637; translated by John Veitch, 1912), Part V.

48 Galileo's translators use the term, says Zilsel, but Galileo himself doesn't. He casts the many generalisations that he offers from his experiments in the language of geometry, as theorems, propositions, lemmata and corollaries.

49 According to Kepler's translator, William H. Donahue, he does elsewhere in his work use the expression 'leges motuum' (laws of motion). The idea that 'inanimate objects "obey" the "laws of motion"' much as rational beings obey laws laid down by their ruler Donahue dismisses as preposterous, but in this, says Donahue, Kepler seems merely to be following current usage. 'The notion of "laws of motion" was ubiquitous in sixteenth-century astronomical works, and . . . it was plausible in this context because the positions of the planets were thought to be governed by angels' (Johannes Kepler, *New Astronomy* [1609, trans Donahue 1992]).

Nicholas Jardine, in *The Birth of History and Philosophy of Science* (1984), p. 240, also believes that Zilsel overlooks the 'ubiquity' of the notion of the laws of nature in sixteenth-century astronomical works.

50 Descartes, op. cit., p. 34.

51 Ibid., p. 35. Jonathan Bennett, a philosopher who is very familiar with the work of Descartes, tells me he believes that this view, in the *Discourse*, is at odds with the 'more considered and central views' that Descartes expressed elsewhere.

52 Zilsel (in op. cit., pp. 258ff) mentions earlier English uses of the term: by Bacon, where he considers it really to be synonymous with form, and by Hooker, who in *The Laws of Ecclesiastical Polity* discusses the natural law which is kept 'unwittingly by the heavens and elements'. But this Zilsel dismisses as teleological and anthropocentric.

53 Or at any rate to be anticipated. A regulatory law, like a constitutive one, can be antecedent to the event, so as to pre-empt any irregularity. The commune of Châteauneuf-du-Pape famously issued an ordinance in the 1950s forbidding flying saucers from landing on its territory, and for all I know the law remains in force, even though no instance of this particular nuisance had been observed at the time, or has been since.

54 According to Zilsel, though, this is precisely what the laws of nature were understood to be doing when the concept was first introduced. 'Let us suppose . . . the government to be omnipotent and the police to be omniscient. In this ideal case the behaviour of the citizens would completely conform to the demands of the lawgiver and laws would be always observed. With such an ideal state nature was compared in the seventeenth century' (op. cit., p. 246).

55 Popper, *The Logic of Scientific Discovery*, p. 19.

56 There is a parallel here with the 'deep grammar', supposed by some linguists to be hard-wired into our brains, and supposed not just to regularise our linguistic performance but to make it possible; a subject considered in a later section.

57 The concept of energy itself, which is what the Laws of Thermodynamics 'govern', and which began life as a quite straightforward and familiar metaphor, has long since developed so many shifting avatars, including unrealised potential forms and of course mass, that it is now too abstract and ambiguous to be grasped at all by reference to the ordinary physical world. Cf. Feynman: 'So far as we know there are no real units [of energy] . . . It is abstract, purely mathematical, that there is a number such that

whenever you calculate it it does not change. I cannot interpret it any better than that' (*The Character of Physical Law* [1965], p. 70).

58 Feynman, *The Character of Physical Law*, pp. 51ff.

59 'One of the arguments (non-scientific but perhaps "transcendental" . . . in favour of this . . . is: if no regularities were apparent in nature then neither observations nor language could exist: neither a descriptive nor an argumentative language' (Popper, *The Logic of Scientific Discovery*, p. 282).

This is part of an addendum dated to the 1972 edition of the book. His description of the argument as 'transcendental' aligns it with Kant's arguments for the a priori truth of causality, and sits oddly with his attack on Heisenberg, in another later addendum, for implying a Kantian view of physical objects as unknowable 'things-in-themselves', whose knowable appearances arise 'from a kind of interaction between the things in themselves and ourselves' (ibid., p. 476). This seems to me pretty much what Popper himself is saying about the 'regularities' in nature and their expression in scientific laws (and to be a reasonable representation of the truth in both cases).

60 See also chapter IV/1 for a discussion of what we tell ourselves in dreams.

61 Richard Feynman, *Six Easy Pieces* (1995), p. 69.

62 According to Aristotle things simply fell towards their 'natural place', which was the centre of the universe. It was Copernicus's realisation that the earth was not the centre of the universe that first made gravitation seem in need of explanation.

For the various sources of the traditional story of Newton and the apple see Gleick, op. cit. It's no doubt mythical – and in more ways than one. The phenomenon that required analysis was not, in the first place, understood to be the fall of anything – it was the curved orbits of the planets, established by Kepler, and it seems to have been Hooke who first suggested (in a letter to Newton) a 'hypothesis' he had formed that this could be explained by breaking the curve down into a combination of two rectilinear motions – one along the tangent to the curve and one of falling inward toward a centre. (See Bernard Cohen's introduction to *Newton*, op. cit., p. 15.)

63 Cf. Taylor and Chandrasekhar, in note 9 to chapter I/1, on the generalised nature of our knowledge of any particular physical object, and Feynman, in notes 8 and 9 to the present chapter, on 'the infinite amount of logic' needed to figure out what's going on in even the tiniest space.

64 Popper, *Conjectures and Refutations* (1963), p. 61.

65 D. W. Winnicott, 'The relationship of a mother to her baby at the beginning,' in *The Family and Individual Development* (1965), and elsewhere.

I/3 Events and their Ancestry

1 This use of causality implies that all causal relations are between discrete sequential events, an idea which John Searle says underlies 'our official theories of causation', and which he believes to be 'a flawed conception'

(The Mystery of Consciousness [1997], pp. 7–8). Although many cause-and-effect relations are like that, he argues, by no means all are.

> Look around you at the objects in your vicinity and think of the causal explanation of the fact that the table exerts pressure on the rug. This is explained by the force of gravity, but gravity is not an event. Or think of the solidity of the table. It is explained causally by the behaviour of the molecules of which the table is composed. But the solidity of the table is not an extra event, it is just a feature of the table.

This seems to me to be a most strained and misleading use of 'cause'. The pressure of the table on the rug may be *explained* by gravitation, in the sense that the general truth of objects being attracted by the earth (or, even more generally, by other objects) brings out what this particular state of affairs has in common with other similar states of affairs, but to say it's *caused*, or explained *causally*, by gravitation is to say that the table is attracted by the earth because it's attracted by the earth. (Unless by gravitation you mean indeed a series of events – the effects of as yet undetected gravitational waves or particles.) The solidity of the table, likewise, isn't *caused* by the behaviour of the molecules – it *is* the behaviour of the molecules, as viewed at the macroscopic level.

Even if you agree that 'cause' can be used to mean a kind of generalising exegesis of a concept, it's not the use that Searle needs for the point he is making. He is proposing these examples of 'non-event causation' to offer a model for understanding what he sees as the causal relationship between neurobiological processes and consciousness (more on this in chapter IV/3) and he is drawing a distinction between his view of consciousness and various other views. If by 'cause' he means merely an exegesis then no distinction is being made. Everyone, whatever his views of consciousness, would surely agree that it in some way *involves* neurobiological processes. No brain – no experiences. But this isn't the question at issue. What Searle is trying to show is that these processes are the cause of mental events in the usual sense of causing – that they bring the events into being.

I might extend Searle's usage a little to account for the table itself. You ask me why there is a table in my living-room. There are all kinds of possible answers I might give you. 'Because I want something to dance on . . . Because it was given me by visiting Martians . . . Because the joiner made it . . . Because there always has been . . .' They all tell you something about the table's provenance. But if I say, 'Because of its molecular structure,' what kind of answer would that be? What would it tell you? 'What caused the red ball to go into the middle pocket?' – 'The diameter of the white ball.'

2 There *are* people in the world who manage to give practical expression to philosophical doubt of this sort. I once spent tedious hours in a jury room with a fellow juryman who refused to accept *any* of the evidence that had been put before us. We had to be convinced 'beyond reasonable doubt', he reminded us self-righteously at frequent intervals, and the only thing beyond his capacity for reasonable doubt was the reasonable doubt itself.

3 Carl Friedrich von Weizsäcker, 'A Reminiscence from 1932', in *Niels Bohr: A Centenary Volume*, ed. A. P. French and P. J. Kennedy (1985). This was during a conversation in a Berlin taxi, when Weizsäcker was nineteen. 'In that moment,' says Weizsäcker, 'I decided to study physics to understand this.'

4 Heisenberg, in *Quantum Theory and Measurement*, ed. Wheeler and Zusek (1983).

5 In fact the very first introduction of probability into quantum mechanics dates from 1926, the year before Heisenberg's paper, and occurs in a foot-note to a paper by Max Born (see Abraham Pais, *Niels Bohr's Times* [1991], p. 286).

6 John Wheeler described the Big Bang as confronting us 'with the greatest crisis in physics' (quoted in Prigogine, *The End of Uncertainty* [1996 in French, 1997 in English], p. 164).

7 See next chapter, Grand Theatre.

8 Max Born, the pioneer of quantum mechanics who first introduced the idea that it had a statistical interpretation, and that Schrödinger's wave formulation represented not a physical wave but a distribution of probabilities, was careful to distinguish between causality and determinism. Quantum theory, he agreed, in a lecture he gave in Oxford late in his life, was 'frankly and shamelessly statistical and indeterministic'. He asked if we could be content with accepting chance, not cause, as the supreme law of the physical world.

> To this last question I answer that not causality, properly understood, is eliminated, but only a traditional interpretation of it consisting in its identification with determinism . . . Causality in my definition is the postulate that one physical situation depends on the other, and causal research means the discovery of such dependence. This is still true in quantum physics, though the objects of observation for which a dependence is claimed are different: they are the probabilities of elementary events, not those single events themselves . . .
>
> Nancy Greenspan, *The End of the Certain World* (2005), p. 271.

9 Oliver Sacks, *Awakenings* (1973).

10 Jonathan Bennett, *Events and their Names* (1988), p. 44.

11 In fact indeterminacy already had a foothold even in the theory of Newtonian mechanics partly through the many-body problem (see I/2, note 1), but also through the concept of infinity, because infinitely remote causes can produce effects which are not fully determinable within the system. And almost twenty years before the uncertainty principle Poincaré anticipated the failure of predictability in the kind of system studied in chaos theory. It is not always possible, he wrote in *Science and Method* (1908), even if we have a perfect understanding of the natural laws involved, to predict a forthcoming situation with the same degree of accuracy as in specifying the situation that gives rise to it. 'It may happen that small differences in the initial conditions produce very great ones in the final phenomena. A small error in the former will produce an enormous error in the latter. Prediction becomes

impossible.' (A remarkable prediction in itself, of difficulties that would be systematically formulated and studied only half a century later.)

12 Such constructions can only ever be relatively random, because in mathematics there is no absolute way of generating a sequence of random numbers. *Any* sequence of numbers is in theory necessarily part of some possible series, i.e. it can be systematically generated by an algorithm, however obscure.

> Modern mathematical definitions of random sequences have been constructed based on Kolmogorovs's definition of complexity in information theory: a random sequence is one with maximal complexity. In other words a sequence is random if the shortest formula which computes it is extremely long.
>
> Deborah J. Bennett, *Randomness*, p. 163.

Again:

> Kolmogorov thought that only approximate randomness applied to finite populations. A perfectly random sequence may be merely an abstraction, but approximate randomness may be good enough. Of course, we must add here that the definition of 'good enough' may depend on the situation.
>
> Ibid., p. 172.

So randomness, in this sense, is a function (once again!) of the use to which it is being put. It also may have to be manipulated artificially to make it fit the task in hand:

> Several authors have pointed out that a sequence obtained by a perfect random selection process may be very unrandom in appearance . . . For example, Hacking points out that an enormously large table of random digits might very well contain an extensive string of zeros. If it did not contain such a string, the table's randomness might be suspect. Yet if the table contained such a string, this portion of the table would be unsuitable for small samples of digits. Hacking says he could be dissatisfied to exclude the string, since one would be accused of arbitrarily tampering with the table's randomness. On the other hand, he would be loath to include the string, since its use would be unsuitable for many purposes.'
>
> Ibid., p. 167.

In effect sequences of random numbers are defined into being by declaring the existence of any apparently non-random features irrelevant to the purposes in hand; as if a fish fillet were to be labelled: 'Every care has been taken to ensure that this product is free of bones. Please ignore any that remain.'

13 We say that these events are accidental, due to chance. And since they constitute the *only* possible source of modifications in the genetic text, itself the *sole* respository of the organism's hereditary structures, it necessarily follows that chance *alone* is at the source of every innovation, of all creation in the biosphere. Pure chance, absolutely free but blind, at

the very root of the stupendous edifice of evolution: this central concept of modern biology is no longer one among other possible or even conceivable hypotheses. It is today the *sole* conceivable hypothesis, the only one compatible with observed and tested fact.

Jacques Monod, *Chance and Necessity,* 1972, p. 110.

More on this later in the section.

14 Quoted in Gustav Born, *Freedoms and Limits of Science*, p. 9 in Alexander von Humboldt Stiftung, *Mitteilungen*, Heft Nr 72.

15 Op. cit., p. 189.

16 Ibid., p. 132. 'In this new formulation, the basic quantity is no longer the wave function corresponding to a probability *amplitude*, but *probability itself*.'

17 Ibid., p. 155. (Democritus and the other classical atomists believed that atoms were in a state of constant vibration – a remarkable foreshadowing, based upon nothing that could then be observed, of the constant internal activity ascribed to atoms in modern physics.)

18 Ibid., p. 132.

19 There is another parallel here with the relationship between the expression of a scientific law and its objective content, and between a sentence and the proposition that it states, etc. See previous chapter.

20 Max Perutz, *I Wish I'd Made You Angry Earlier* (1998), p. 204.

21 Max Born's insistence, referred to earlier, that quantum mechanics, while not deterministic, remains causal, is also based on the claim that 'observable events obey laws of chance'. But I think he is making a slightly different point from Perutz, because he goes on to say that 'the probability for these events itself spreads according to laws which are in all essential features causal laws' (Greenspan, op. cit., p. 271). What I think he is arguing is that in particular quantum situations you can limit and define the distribution of probability (so that, for instance, a particle is likelier to be in some places rather than in other places), even though within these distributions exact positions are not causally determined. The rules of the game were causal, he said, the events themselves probabilistic. But the 'rules of the game' that express the probabilities are not the laws of chance – they are the laws of physics.

22 It may also be the case that, statistically, as it happens, the judges have tended to favour works by authors whose surname begins with a letter in the first half of the alphabet. This seems unappealing as a guide to the future. Nevertheless, had we used it as a guide in the past, we can see now from the record that it might have proved effectively predictive.

23 Michael Barnsley, quoted in James Gleick, *Chaos* (1988), p. 239.

24 James Crutchfield, ibid., p. 306.

25 Roderick V. Jensen, ibid., p. 306.

26 Joseph Ford, ibid., p. 306.

27 Joseph Ford, ibid., p. 306

28 John Hubbard, ibid., p. 239

29 The history and implications of indeterminism are admirably summed up by
Gustav Born, himself an experimental pharmacologist, and the son of Max
Born, quoted above, and strikingly illustrated in relation to cosmology by
the work of Victor Weisskopf that he quotes:

> The introduction of *indeterminism* into particle physics through quantum
> mechanics can be seen as the beginning of the realisation that all sorts of
> other natural processes are describable only in probabilistic terms. This
> applies, for example, to amplification processes in which very small causes
> can have very large, entirely indeterminate effects. Weisskopf illustrates
> this by the development of star systems. In the hot gases of the Primal
> Universe, minute statistical fluctuations in density distribution increase
> through the gravitational attraction of neighbouring gas molecules, so that
> clusters form. This becomes a positive feedback, inducing the originally
> almost uniform gases to separate into ever increasing clusters, which ulti-
> mately become the galaxies. So although one *can* predict that gravitational
> amplication *must* produce clusters it is impossible to predict how and
> when the individual events occur.
>   Thus, nature creates new shapes that cannot be predicted from the laws
> of physics, except in very general terms. We may be able to understand the
> occurrence of spiral arms and the like, but not the immense variety of
> detail that we admire when we look at pictures of galaxies. This is one of
> many examples in the physical sciences where complete predictability is *in
> principle* impossible, and thus a demonstration of the existence of absolute
> limits due to the nature of nature itself.
> Gustav Born, *Freedoms and Limits of Science*, Alexander von Humboldt
>   Stiftung, *Mitteilungen*, Sonderdruck aus Heft Nr 72, p. 9.

30 Entropy, it occurs to me, might be seen (to adapt a psychiatric phrase) as a
loss of causal affect.
31 Spinoza, according to Jonathan Bennett (*A Philosophical Guide to
Conditionals* [2003], p. 203) was led by his determinism into hostility to
subjunctive (including counterfactual) conditionals. Other philosophers, he
says, hold that such conditionals 'must have antecedents that could have
come about through a different legal outcome of an indeterminist
transaction'. Bennett himself, though rejecting determinism, believes that any
proposed analysis of counterfactuals 'must survive being combined with the
hypothesis that [the actual world] is deterministic'. He thinks this can be
done by means of David Lewis's 'modal realism' (of which more later).
32 Cf. Leibniz, 'The whole universe with all its parts would have been different
from the beginning if the least thing in it had happened differently from how
it did' (Leibniz, 1686, reprinted in *Philosophical Essays* [1989], p. 73).
33 David Lewis, *On the Plurality of Worlds* (1986).
34 Jonathan Bennett describes this as 'extreme realism'. His own version, which
he calls 'abstract realism', is the view that the various worlds are 'propositions,
states of affairs, or ways-things-could-be – abstract objects of some kind'
(Bennett, *A Philosophical Guide to Conditionals* [2003], pp. 153ff). I can't

really see any difference between the two. In Lewis's terms, 'If I'd known you were coming I'd have baked a cake,' means that in some other world (or 'at some other world' as philosophers awkwardly say – the 'at' being an acronym for 'according to'), there is an actual physical cake that I (or some counterpart of me) did in fact bake. In Bennett's terms, it means that in, or according to, some other world the proposition that I baked a cake is true. But if it's true that I baked a cake, then somewhere there must be a cake, a real, eatable cake with sultanas in it, that I baked (not to mention an 'I' who baked it). Unless it means that the proposition is true only in the sense of being compatible with a set of propositions about this alternative world which *all* happen to be false. But in such a world there are no sultanas in the cake even on Lewis's analysis.

35 Bennett, op. cit., p. 185.

36 David Deutsch (*The Fabric of Reality*, 1997) believes, like Lewis, that counterfactuals are one of the problems that can be solved by multiple universe theory. Deutsch's version, which seems to differ from Lewis's in allowing some kind of interaction with the actual universe at the quantum level, whereas Lewis's excludes causal interaction of any sort, is very robust on the subject of alternative David Deutsches. They are out there propounding multiple universe theory, only better, and drinking cups of tea (see chapter I/2).

I'm encouraged to risk a counterfactual of my own about David Deutsch: 'If David Deutsch had put forward his theory of counterfactuals in the year 1453, it would by now have saved the world 73,956 philosopher-hours.' And yes, it's suddenly all become very clear. We are not saying anything cloudy or vague. We are asserting the simple factual proposition that in some other universe or universes David Deutsch (not *this* David Deutsch, of course, but one or more of his counterparts) *did* put forward such a theory in 1453, and that it *has* saved the world, in that universe or universes, 73,956 philosopher-hours. And if he (or they) actually did, and it actually has, then the statement is quite simply a true statement of fact.

A small practical problem remains. The factual statement that one or more David Deutsches did his or their stuff in 1453, and that this did have the effect claimed, may, like any other factual statements, be false; and, although there are techniques for verifying factual statements about events in *this* universe it's hard to see how we could use them for verifying factual statements about events in other universes, since by definition we can never know anything whatsoever about the contents of them (except that the photons in them somehow interfere with the photons in this universe).

Look into it a little more deeply, though, and you realise that this is to underestimate the power of multiple universe theory. The statement, we can confidently assert, *is* true. It *must* be true – one or more David Deutsches *did* put forward such a theory in 1453, and it *did* have such an effect. Because if it were *not* true then I should be free to propose another counterfactual: that if it *had* been true then David Deutsch *would* have done his stuff then, and it would have had the effect claimed; which of course means that in one or more universes it *was* true, and David Deutsch *did* do his stuff then. And if you say that it might in its turn be false that this was true in any universe, then I propose another counterfactual: that if it *had* been true that this was true . . . And so on, through all the unending wealth of different universes, and of universes upon universes.

Let me try another counterfactual: 'If David Deutsch had thought for two minutes about what he was suggesting he would have laughed at his own idiocy.' David Deutsch will not be offended at my saying this. He will, if he is consistent, placidly agree that, however long and seriously he pondered the question in this universe, in at least one other universe a version of himself did indeed think for two minutes about it and fall off his chair laughing. He would go on to agree, I hope, that in yet another universe or so yet another David Deutsch or so had suggested that the problem of counterfactuals could be solved by the application of ginseng extract to the soles of the feet.

37 Many-world theorists also talk about worlds that differ from this one as a result of miracles, small or large. But a world in which miracles are possible is surely, *ipso facto*, an indeterministic world; a miracle is precisely an event which is not determined by normal causal laws. David Lewis, in his celebrated paper 'Counterfactual Dependence and Time's Arrow', explains that what he means when he says that a miracle takes place is that there is a violation of the laws of nature:

> But note that the violated laws are not laws of the same world where they are violated. That is impossible: whatever else a law may be, it is at least an exceptionless regularity. I am using 'miracle' to express a relation between different worlds. A miracle at w_1 [one possible world], relative to w_0 [another possible world], is a violation at w_1 of the laws of w_0, which are at best the almost-laws of w_1. The laws of w_1 itself, if such there be, do not enter into it.

This is surely the Leibnizian route once again.

38 Even Gell-Mann, who is so dismissive of the 'anthropocentric' interpretation of quantum mechanics referred to here, agrees that 'nothing can ever be measured with perfect accuracy', which 'gives rise to effective indeterminacy at the classical level over and above the indeterminacy in principle of quantum mechanics' (*The Quark and the Jaguar*, p. 26).
See also chapter I/1.

39 Benoit Mandelbrot, *Fractal Geometry*, p. 18, quoted in Gleick, *Chaos*, pp. 95–7, 105.

40 I have made very free in this section with the supposed views of a supposed determinist. *Are* there actually any determinists in the world these days, though? Jonathan Bennett believes that among philosophers, at any rate, there are not – only ones who investigate what the implications would be if determinism *were* true. Among scientists, though, I get the impression that there are quite a lot who would *like* to be determinists, if only they could see how to manage it.

41 Op. cit., p. 108.

42 According to Monod, some other aspects of biological structure are also random. The law that governs the sequence of amino acids in a polypeptide chain, he says, is that of chance. 'To be more specific: these structures are "random" in the sense that, even knowing the exact order of 199 residues [amino acids] in a protein containing 200, it would be impossible to formulate any rule, theoretical or empirical, enabling us to predict the nature of the one residue not yet identified by analysis.' The precisely deterministic process of genetic replication (except in so far as it is undermined by quantum randomness) then reproduces that same random pattern in each new example of the protein (op. cit., p. 93).

43 Ibid., p. 111.

44 Or in these days, of course, by people dropping cigarette ends, or lighting barbecues. But, whether you regard such events as determined, and, if so, which of Monod's categories you regard them as falling into, they remain random within the life-cycle of the plants concerned.

I/4 Grand Theatre

1 A. S. Eddington, *The Nature of the Physical World* (1958).

2 Ilya Prigogine, *The End of Uncertainty* (1996 in French; 1997 in English), p. 19.

3 Ibid., p. 154.

4 Prigogine finds the unidirectional nature of time so fundamental that he believes it was established even before the Big Bang, as a feature of the 'meta-universe, the medium in which invididual universes are born'. Even before *our* universe was created, he says, 'there was an arrow of time, and this arrow will go on forever' (op. cit., pp. 181–2).

5 Clausius, *Abhandlungen über die Mechanische Wärmetheorie*, 1864, p. 50. But see note 42 to chapter I/3, on his introduction of the concept.

6 See Chapter I/3.

7 Ibid., p. 26.

8 First published in 1988.

9 Hawking has managed to get himself confused by his own proposal. They wouldn't die before they were born. They would rise from their graves and come to life. And then, a lifetime later, disappear back into their mother's womb, and separate into ovum and spermatazoon.

10 No life at all, in fact:

> Evolution in the biosphere is therefore a necessarily irreversible process defining a direction in time; a direction which is the same as that enjoined by the law of increasing entropy, that is to say, the second law of thermodynamics. This is far more than a mere comparison: the second law is founded upon considerations identical to those which establish the irreversibility of evolution. Indeed, it is legitimate to view the irreversibility of evolution as an expression of the second law in the biosphere.
>
> Jacques Monod, *Chance and Necessity*, p. 118.

11 The recession of the galaxies doesn't imply that we have a special position at the centre of the universe; any observer anywhere in the universe would have the same impression. Nevertheless, since all observations are from a particular point of view, for each of these observers (the observers on the Earth among them) this is the objective situation.

12

> The maximum speed through space occurs if *all* of an object's motion through time is diverted to motion through space. This occurs when all of its previous light-speed motion through time is diverted to light-speed motion through space. But having used up all of its motion through time, this is the *fastest* speed through space that the object – any object – can possibly achieve . . . . Something travelling at light speed through space will have no speed left for motion through time. Thus light does not get old; a photon that emerged from the big bang is the same age today as it was then. There is no passage of time at light speed.
>
> Brian Greene, The Elegant Universe (1999), p. 50.

13 The idea of further time dimensions is mentioned by Brian Greene in *The Elegant Universe* (pp. 204–5). 'Some theorists have been exploring the possibility of incorporating extra time dimensions into string theory,' he says, 'but as yet the situation is inconclusive.' Various mystical ideas of multi-dimensional time were fashionable in the twenties and thirties of the last century. I believe that J. B. Priestley's 'time' plays (*Time and the Conways*, for instance) were based on a notion put forward by J. W. Dunne in *An Experiment with Time*; a notion which as I recall was based on a misunderstanding of phenomenalism. If we can perceive a physical object only by experiencing some kind of internal perceptual entities, he argued, then presumably we can perceive those internal entities only by experiencing some further layer of entities, and so on, through an infinite series of super-imposed worlds. But in the phenomenalist analysis of perception we don't *perceive* the sense-data (or qualia, or whatever). We entertain them in some (even more mysterious) way that doesn't involve an infinite regress.

14 The possibility of any real straight line, as a smoothly continuous physical entity without lateral extension, finally disappeared, if it had ever existed, with the introduction of the atom, and was made even less imaginable by the indeterminacies revealed in quantum mechanics.

15 Daniel Dennett, *Consciousness Explained*, p. 382.

16 Ibid., p. 376.

17 This, of course, is the system that is generalised in the lock and key, and in the way in which cells recognise each other by nesting (stereospecific complementarity – cf. Monod, *Chance and Necessity*, pp. 61ff).

18 Jonathan Bennett, 'Substance, Reality and Primary Qualities' (*American Philosophical Quarterly*, 1965). Bennett adduces the 'phenol argument' as an analogy with what he regards as the 'mis-expressed' distinction drawn by Locke between primary and secondary qualities. This is invalid, he says, because the assertion that the phenol-thio-urea does not change is false; it changes in respect of its taste.

19 Contrast, though, a similarly brief glimpse of the real world. When you try to read the name of a passing station from the window of a high-speed train you see not the first letter or two before it's gone – you see the whole word, already *there,* but unreadable because unresolvable.

20 Cf. Philip Larkin, on the 'sparkling armada of promises' forever heading towards us, each of which 'never anchors; it's / No sooner present than it turns to past' ('Next Please'). Even if it did anchor, though, what cargo could it unload that would be as rich as the one we saw in anticipation?

21 Abraham Pais, *Niels Bohr's Times,* p. 22.

I/5 Fingerhold

1 The first two chains, for instance (discovered by Poulet in 1918) were one with five links:

> 12496, 14288, 15472, 14536, 14264. The next number is 12496 again, so that the same sequence continues ad infinitum

and one with twenty-eight links:

> 14316, 19116, 31704, 47616, 83328, 177792, 295488, 629072, 589786, 294896, 358336, 418904, 366556, 274924, 275444, 243760, 376736, 381028, 285778, 152990, 122410, 97946, 48976, 45946, 22976, 22744, 19916, 17716; which then starts again with 14316.
>   See: http://primes.utm.edu/glossary/page.php?sort=SociableNumbers

Only one further chain was located (by Borho in 1969, with four links) – this is how rare and unpredictable they are – until computers opened the way to another fifty or so.

2 This is what the linguistician Derek Bickerton, in *Roots of Language* (1981), calls a 'Flintstone scenario' – an unverifiable explanation for the origins of some activity by imagining some simple everyday scene from prehistoric life. This particular scenario is a fiction, certainly; but fiction is a path to the truth – provided you understand that it's fiction.

3 My preposterous piece of Flintstoniana actually echoes the entirely sober procedure by which Gottlob Frege, the father of mathematical logic, derived number from a non-arithmetical source. He took as the primary given the comparability of different groups of objects in respect to the multiplicity of their members; he saw that the membership of one group can be mapped on to another group without reference to any pre-existing concept of number. If you watch a vast army marching past you may have no idea of how many soldiers it comprises, or how many rifles they are carrying, but if you observe that each soldier has a rifle you know at any rate one thing – that the number of rifles is the same as the number of soldiers. The class of soldiers on parade and the class of rifles they are carrying have this feature in common, and this common feature identifies the two classes as jointly forming a further class – a class of two classes. This class probably has other members – the class of dog-tags round the soldiers' necks, the class of triggers

457

on their rifles – and other members yet, if only we knew about them – particular groups of sheep and turnips that could all be matched up, member by member, with the soldiers and rifles. All this is very like Abel matching up his sheep and his fingers. Frege, however, then goes on to use this insight to provide a logical, non-arithmetic definition of number as an abstract concept. Any particular number is the class of all classes with matching memberships. So twelve, for example, is the class of all groups of twelve; and, in general, x is the class of all classes with x members.

4 The Pirahã people, in the Amazon region of Brazil, may seem to belie this. They famously have almost no concept of number. According to some reports they have only a single number word, which vaguely signifies 'one' but may also mean 'small'. Other accounts suggest that they have a term for a second number, which may be 'two' or 'not many', with perhaps another signifying some kind of undifferentiated multiplicity for any grouping larger than two. Perhaps the latter provides enough of a background contrast against which the analogy between all groups with one-ish or two-ish members becomes apparent. All reports seem to agree that the Piraha have no words for any specific grouping larger than two, and without the words, interestingly, they are unable even wordlessly to match sets of more than two or three members accurately. (*Guardian*, 20 August 2004, and *The Independent*, 6 May 2006)

   You might ask why, if my Flintstones scenario of Cain and Abel has any substance at all, this particular people has never grasped the fiveness of their fingers. Perhaps they lack the other side of the equation – objects they needed to quantify. They are reported to be 'imprecise about quantities of fish and manioc'. Perhaps their methods of fishing and cultivation, or their economic relations, never gave them any really crucial interest in being precise.

5 Compare the complexity and elusiveness of the blue in a blue sky, discussed in chapter I/1.

6 Cf. Gell-Mann, as quoted in chapter I/3: 'Nothing can ever be measured with perfect accuracy' (*The Quark and the Jaguar*, p. 26).

7 Cf. Cartwright, in chapter I/2: 'Approximations and adjustments are required whenever theory treats reality' (*How the Laws of Physics Lie*, p. 13). 'To get from a detailed factual knowledge of the situation to an equation, we must prepare the description of the situation to meet the mathematical needs of the theory' (ibid., p. 15).

8 Bart Kosko, *Fuzzy Thinking* (1994), p. 155.

9 Quoted by Kosko in ibid., p. 146.

10 Ibid., p. 149.

11 Ibid., p. 172.

12 Ibid., p. 176.

13 I imagine that this is what Nietzsche is referring to when he talks about the 'continual falsification of the world by means of numbers' (*Beyond Good and Evil* [1886], p. 35).

14 The constants are c, the velocity of light; G, the gravitational constant; and h, the Planck constant.

15 *c*, the velocity of light, and *G*, the gravitational constant, are determined by measurement. There are suggestions that *c* may have had a different value in the earlier stages of the universe, and that *G* may vary over distance and time. *h*, the Planck constant, sometimes defined as the constant of action, is the value required to bring the energy of the quantum into relationship with radiation frequency.

16 Nietzsche again: 'the fictions of logic' – ibid., p. 35.

17 By Heron of Alexandria, in the first century, but apparently not taken seriously as entities until the nineteenth century – a slow burn almost as long as that of the jet engine, a forerunner of which he also invented.

18 This is the so-called 'correspondence theory' of truth, propounded by Tarski and warmly endorsed by Popper, who believes that it establishes contingent truth as 'objective or absolute', and that it 'appears to be accepted today with confidence by all who understand it'.(See I/2, note 23).

19 Neal Ascherson, *Games with Shadows* (1988).

20 Sortov had an earlier and more distinguished follower – William James, who attempts to rescue from 'the stream of thought' the elusive elements overlooked by traditional psychologists, such as instantaneous glimpses of someone's meaning, one's own intentions to say something, and the ability to anticipate the sense of the sentence one is reading. 'One may admit that a good third of our psychic life', he writes, 'consists in these rapid premonitory perspective views of schemes of thought not yet articulate.' He describes them as 'in very large measure constituted of *feelings of tendency*, often so vague that we are unable to name them at all.' But he insists that they are 'among the *objects* of the stream', and says that one of his goals is 'the re-instatement of the vague to its proper place in our mental life' (William James, *The Principles of Psychology* [1890], Vol. I, pp. 253–4).

II/I Why the Marmalade?

1 In any case man is not the only creature that perceives and categorises the world around it. The lion and the antelope, the cockroach and the worm, also have to identify food, danger, homes, and mates – and thereby also begin to give it form. (See chapters IV/4 and V/2.)

2 Motive and intention are overlapping but not identical concepts. My motive in writing that very sentence, for instance, and my intention in writing it *might* be the same. I might simply want to make clear how the two concepts are related. But even if my intention was to make the concepts clear, you might feel that my motive was different – mere pedantry, perhaps; in other words, something of which I was myself unaware. It's perfectly natural for you to claim to know more about my motives than I do. On the other hand, while it's natural for you to think you know more about my intentions than I choose to reveal, it's quite odd for you to know more about them than I *know*. Odd, but perhaps not impossible; see what follows.

3 More on all this in Chapter III/2.

4 G. E. M. Anscombe, *Intention* (1963).

5 And we shall be back at the breakfast table, looking at some of these things that might be going through my mind, in the next chapter.

II/2 How the Marmalade?

1 William James gives a very good account of the problem of how anyone ever decides to get out of bed in the morning, and finds that it contains 'in miniature form the data for an entire psychology of volition' (James, *The Principles of Psychology* [1890], Vol. II, pp. 524–5).

2 Daniel Dennett quotes Paul Valéry on the need 'to produce future'. Dennett comments:

> In order to cope, an organism must either armour itself (like a tree or a clam) and 'hope for the best', or else develop methods of getting out of harm's way and into the better neighbourhoods in its vicinity. If you follow this latter course, you are confronted with the primordial problem that every *agent* must continually solve: Now what do I do?
>
> Daniel Dennett, *Consciousness Explained*, p. 177.

3 The topic of attention, and how we thereby select our experience, is explored by William James (*The Principles of Psychology* [1890], Vol. I, pp. 402ff.) – he claims for the first time in psychology, the idea having been ignored by the empiricist philosophers who believed that the world simply flowed in upon us. He raises the question of whether we can attend to more than one object at a time, and quotes various authorities as finding it possible to have four or even six simultaneous objects of attention, while Julius Caesar is said to have dictated four letters at the same time as writing a fifth. James records wonderful nineteenth-century experiments in which he and others wrote one calculation while performing a different one in the head, or wrote out one poem while reciting another. A certain M. Paulhan, for example, in the *Revue scientifique* of 1887:

> I write the first four verses of [Racine's] *Athalie*, whilst reciting eleven of Musset. The whole performance occupies 40 seconds. But reciting alone takes 22 and writing alone 31, or 53 altogether, so that there is a difference in favour of the simultaneous operations.

James's own answer to how many 'entirely disconnected systems or processes of conception can go on simultaneously' is: 'not easily more than one, unless the processes are very habitual; but then two, or even three, without very much oscillation of the attention.' I suspect that 'oscillation of the attention' – more familiar now in the performance of computers as multi-tasking – probably accounts for feats such as M. Paulhan's mental athletics.

4 See particularly Gilbert Ryle (though he does not specifically mention deciding):

> When a person is described as having fought and won, or as having journeyed and arrived, he is not being said to have done two things, but to have done one thing with a certain upshot. Similarly a person who has

aimed and missed has not followed up one occupation by another; he has done one thing, which was a failure. So, while we expect a person who has been trying to achieve something to be able to say without research what he has been engaged in, we do not expect him necessarily to be able to say without research whether he has achieved it. Achievements and failures are not occurrences of the right type to be objects of what is often, if misleadingly, called 'immediate awareness'. They are not acts, exertions, operations, or performances, but, with reservations for purely lucky achievements, the fact that certain acts, operations, exertions, or performances have had certain results.

> *The Concept of Mind* (1949), p. 144.

5 St James the Great or St Claude of Besançon, both of whom saved the lives of hanged men, one by supporting the victim, one by cutting him down (see Réau, *L'Iconographie chrétienne*).

6 I suppose a determinist might try to argue that a decision was the product not of the state of affairs as perceived, but of the state of affairs itself, so that the perception of it which seemed (even to him) to be motivating the decision was merely a non-functional psychological decoration and not a part of the causal link at all. This would seem to make the mechanics of the link even less accessible to observation, in the case of his own decisions just as much as the decisions of others – perhaps beyond the range of imagination.

7 Some economists now recognise that economic systems are chaotic, and therefore probabilistic in the same way that chaotic physical systems are. Cf. Paul Ormerod, *Butterfly Economics* (1998).

8 In the first place by your decisions, conscious or unconscious, about what you will take in of the world around you. William James, claiming to be the first psychologist to direct attention to attention, is superb on the subject:

> Millions of items of the outward order are present to my sense which never properly enter into my experience. Why? Because they have no *interest* for me. *My experience is what I agree to attend to.* Only those items which I *notice* shape my mind – without selective interest, experience is an utter chaos. Interest alone gives accent and emphasis, light and shade, background and foreground – intelligible perspective, in a word. It varies in every creature, but without it the consciousness of every creature would be a gray chaotic indiscriminateness, impossible for us even to conceive.
>
> > Op. cit., p. 402.

> Suffice it meanwhile that each of us literally *chooses*, by his ways of attending to things, what sort of a universe he shall appear to himself to inhabit.
>
> > Ibid., p. 424.

9 Cf. Isaiah Berlin, *The Proper Study of Mankind* (1997), p. 109:

> The assumption of common speech is that freedom is what distinguishes

man from all that is non-human. But if you look at a dog, dodging back and forth, looking for scents, realising that it's time to catch up with its master; or at an ant, endlessly changing direction as it scurries about its business, it seems clumsy, unworkable, to have to suppose that in theory each of these changes of direction could be accounted for deterministically, by varying currents of scent, by built-in alarms that trigger anxiety about the whereabouts of the master, or of continuing too far in the same direction.

10 The economic parallels of the experiments are drawn by Alan Kirman, in his chapter in *The Economy as an Evolving Complex System II*, ed. Arthur, Durlauf and Lane, (US, 1997), and referred to in Paul Ormerod, *Butterfly Economics* (1998).

11 Isaiah Berlin, *The Sense of Reality*, p. 110.

12 The pharmacologist Susan Greenfield offers a similar argument to resolve what she sees as the contradiction between free will and determinism, by suggesting what is in effect a version of Niels Bohr's complementarity principle: that the subjective and objective accounts of the event are mutually exclusive yet equally valid.

13 Deutsch uses his multiple universe theory, introduced to explain the superposition of states at the quantum level, as a solution to combining free will and determinism. Free will survives in a deterministic universe, he suggests, although your choices are determined, because some copies of you in some universes choose to do one thing, while other copies of you in other universes choose to do other things. (By 'choosing' I assume he means seeming to choose, while in fact being entirely constrained by the history of that particular universe.) The theory, happily, not only solves the problem of free will, but decides which 'choice' is the right one: the right 'choice' is one which is made by more copies than other 'choices'.

So we know that, while the David Deutsch in this familiar universe of ours was compelled by his genetic inheritance, his upbringing, and the general course of intellectual history as we know it, to put forward this particular theory, this doesn't in any way detract from his personal achievement in doing so, because in other universes other David Deutsches are putting forward (have put forward, will put forward) quite other theories, some of them explaining free will in astrological terms, others in terms of cold fusion. If only we knew what proportion of David Deutsches was putting forward each of these theories we should be able to judge which of them was correct.

Sadly, since we can't know anything whatsoever about any of these other universes the question must remain forever undecidable; though I have a sneaking private suspicion, unsupportable by any possible evidence, that however many trillions of David Deutsches are putting forward however many trillions of trillions of theories, not a single one of those theories can be as daft as this one.

III/1 A Cast of Characters

1 The difficulty (perhaps the impossibility) of introspection has been much remarked upon. For example:

> Sometimes I am, sometimes I think.
>
> Paul Valéry, quoted in Daniel Dennett,
> *Consciousness Explained*, p. 423.

> Life is like standing on a chess board where you can see every square except the one you're standing on.
>
> Richard Feynman; the source of this I cannot locate.

Niels Bohr suggested that the difficulty of introspection was a particular instance of his general principle of complementarity, which proposed that some states of affairs can be fully described and understood only through the application of mutually incompatible theories (the prime example being the behaviour of quantum entities, which have to be viewed both as particles and waves).

William James strikes off two characteristically brilliant similes in quick succession:

> The attempt at introspective analysis . . . is in fact like seizing a spinning top to catch its motion, or trying to turn up the gas quickly enough to see how the darkness looks.
>
> James, *The Principles of Psychology* (1890), Vol I, p. 244.

James himself, however, manages over and over again to get the gas turned up remarkably fast. He is superb (and so far as I know unsurpassed) in observing and describing what he calls 'the stream of thought' – particularly in suggesting its continuity, and in capturing the elusive, formless interstices between the more definite perceptions, decisions, etc., which are usually studied by psychologists (and philosophers, although he doesn't say so). The metaphor of the bird in what follows is unforgettable:

> As we take, in fact, a general view of the wonderful stream of our consciousness, what strikes us first is this different pace of its parts. Like a bird's life, it seems to be made of an alternation of flights and perchings. The rhythm of language expresses this, where every thought is expressed in a sentence, and every sentence closed by a period. The resting-places are usually occupied by sensorial imaginations of some sort, whose peculiarity is that they can be held before the mind for an indefinite time, and contemplated without changing; the places of flight are filled with thoughts of relations, static or dynamic, that for the most part obtain between the matters contemplated in the periods of comparative rest.
>
> Ibid., p. 243.

2 Cf. David Burke's account, in a book that he and I wrote together, *Celia's Secret* (2000), pp. 21-2, of what goes through his head when he's on the stage acting:

Acting is mostly a twin-track mental activity. In one track runs the role, requiring thoughts ranging from, say, gentle amusement to towering rage. Then there is the second track that is monitoring the performance: executing the right moves, body language, and voice level; taking note of audience reaction and keeping an eye on fellow actors; coping with emergencies such as a missing prop or a faulty lighting cue. These two tracks run parallel, night by night. If one should go wrong, then it is likely that the other will misbehave, too . . .

But there is a third and wholly subversive track which intrudes itself at intervals, full of phantom thoughts and feelings that come and go of their own volition. This ghost train of random musings is, of course, to be discouraged, but it can never be entirely denied . . . I have been guilty during a performance of dwelling on everything from shopping lists to food fantasies, and I have one particularly alarming *idée fixe* that afflicts me from time to time. It is the temptation to do or say something so outrageous that it would stop the play, empty the house, and end my career. The spectre will appear without warning like the Ghost itself [of Hamlet's father] and beckon me to follow it over the beetling brow of the cliff, urging me to drop my trousers, or shout obscenities at the leading lady.

3 Bruce Chatwin, for example, in his account of the mythology of the native Australian peoples in *The Songlines*, describes how the Ancestors, awakened by the primordial warmth of the sun, called first themselves and then the rest of the world into being by finding words for them:

The mud fell from their thighs, like placenta from a baby. Then, like the baby's cry, each Ancestor opened his mouth and called out, 'I AM!' 'I am – Snake . . . Cockatoo . . . Honey-ant . . . Honeysuckle . . .' And this first 'I am!', this primordial act of naming, was held, then and forever after, as the most secret and sacred couplet of the Ancestor's song.

Each of the Ancients (now basking in the sunlight) put his left foot forward and called out a second name. He put his right foot forward and called out a third name. He named the waterhole, the reedbeds, the gum trees – calling to right and left, calling all things into being and weaving their names into verses.

Chatwin, *The Songlines* (1987), p. 81.

4 Frigyes Karinthy, *A Voyage Round My Skull* (1938), pp. 121–2.
5 Some novels model this by telling the story at second hand, purely through depositions, documents, gossip, conjectures, etc.
6 Robbe-Grillet, *Towards a New Novel* (1962), p. 91.
7 Robbe-Grillet, *Le Voyeur* (1955), pp. 122–3.
8 Robbe-Grillet, op. cit., p. 117.
9 Cf. Daniel C. Dennett, *Consciousness Explained*, p. 79:

A novel tells a story, but not a true story, except by accident. In spite of our knowledge or assumption that the story told is not true, we can, and do, speak of what is *true in the story* . . . What is true in the story is much,

much more than what is explicitly asserted in the text. It is true that there are no jet planes in Holmes's London (though this is not asserted explicitly or even logically implied in the text), but also true that there are piano tuners (though – as best I recall – none is mentioned, or again, logically implied.) In additon to what is true and false in the story, there is a large indeterminate area: while it is true that Holmes and Watson took the 11:10 from Waterloo station to Aldershot one summer's day, it is neither true nor false that that day was a Wednesday. (*The Crooked Man.*)

More on this in the next chapter.
10 Dennett, op. cit., p. 417.

III/2 Is it True about Lensky?

1 Though I gather that there is now some doubt about whether David was a historical figure at all. He is a wonderfully complex and vivid character (I particularly like the story of his making a fool of himself in his wife's eyes by dancing with such enthusiasm around the Ark of the Covenant, as it finally entered Jerusalem, that he exposed himself), and if he is invented there must have been a natural novelist among the authors of the Book of Kings.

2 Or is entirely independent of it, as in the case of tautologies, which are always true, and contradictions, which are always false. Even with non-tautologous and non-contradictory propositions, either truth-value in the input statements can be equally well accommodated. They can all be false; logic will operate upon them and produce further statements with a known truth-value, some of them false, some of them, such as the negations of the false propositions, true. It doesn't even matter if the truth-value of the input statements is unknown. It simply means that the truth-value of the output statements is also unknown, so that they are waiting as hypotheticals for the initial values to be inserted. For the output statements to have a definite truth-value all we need to know is that the input propositions have definite truth-values as well.

3 If it's true that Onegin didn't shoot Lensky then this also seems to imply, in any ordinary use of language, that there were an Onegin and a Lensky for it to be true of. Russell introduced a way of analysing factual propositions so as to get round this difficulty (he specifically mentioned fictitious entities such as the present King of France). Propositions are to be broken down into an existential quantifier ($\exists$) and a qualifier, so that 'Charles I was beheaded' is to be translated as '$\exists \ x$ (x is Charles I) (x was beheaded)', which can be true only if '$\exists \ x$ (x is Charles I)' is true – ie, if there is or was such a person as Charles I. So 'Charles XXI was beheaded' is quite simply false. 'Onegin didn't shoot Lensky' would presumably be translated as '$\exists \ x \exists \ y$ (x is One-gin) (y is Lensky) ~ (Onegin shot Lensky)', meaning: 'There is an x and there is a y such that x is Onegin and y is Lensky and it's not the case that Onegin shot Lensky.' Since the two existential clauses fail, the whole proposition fails, in exactly the same way as all propositions about Charles XXI. But, since 'Onegin shot Lensky' also fails in the same way, this avoids the false

implication at the cost of (once again) obliterating the distinction between 'Onegin shot Lensky' and 'Onegin didn't shoot Lensky'.

4 This absurdity is avoided once again by subjecting the proposition to Russell's analysis, since the two existential quantifiers, '∃ x∃ y (x is Onegin) (y is Lensky)', remain false, so we are at least in no sillier a position than we were by starting with 'Onegin shot Lensky'.

5 You might think that a way out of the dilemma was found by Lukasiewicz and other modern logicians who demonstrated that logic can in theory work with more than two values. The first extra value to be introduced was sometimes interpreted as 'probable' – which is unconvincing enough in the case of factual propositions, and is plainly irrelevant to the problems raised by fictitious ones. Could it be reinterpreted as 'fictitious'? Assigning all fictitious propositions this value is open to the same objection as labelling them all false (that it doesn't differentiate between 'Onegin shot Lenksy' and 'Lensky shot Onegin'). Lukasiewicz and his followers went on to demonstrate the possibility of a logic that had an infinite number of different truth-values. Could we try to solve the problem of differentiation by supposing that this infinity of values represents a graduated scale of truth, so that 'Onegin shot Lensky' is truer than 'Lensky shot Onegin', but not as true as 'd'Anthès shot Pushkin'? Apart from the ridiculous unreality of this, because it obliterates the obvious distinction between statements about fictitious characters, Onegin and Lensky, and statements about real people, d'Anthès and Pushkin, it also surely means (if it means anything at all), that the truth-values of 'Onegin shot Lensky' and 'Onegin didn't shoot Lensky', even if they are not absolutes, are still in inverse proportion to each other, so that if one is not very true then it implies that the other is quite considerably true.

6 Many factual statements, of course, are also copies of other statements, or derivations from them, and are justified in practice by their consistency with them. (See chapter III/1, also later in the present chapter.) Others are derived from some fresh observation of the world, with all the indeterminacy that this so often involves. All the same, the truth-conditions in both cases go back eventually to some kind of relationship to an event or situation in the real world.

7 Covert counterfactuals, for instance. Then at least all the problems would be ones that philosophers were familiar with. 'Had it been the case that, at some unspecified time past, there was a young woman called Rapunzel, or a young man called Onegin, then this is what would have happened . . .' But this is just as absurd as our other analyses. For a start, to describe something as a counterfactual is, *ipso facto*, to assert that its premiss is false, which opens up all the old difficulties. Then again, a counterfactual at any rate appears to be asserting something which might be, or might have been, true or false. It is notoriously difficult to say what the criteria for its truth-value are, but at any rate it is asserting something that is subject to debate, to the production of evidence and parallels; it is asserting (truly, falsely, or indeterminably) that, in different circumstances, had the conditional premiss been

true, then the statement that follows would also be true. None of this is the case with the story. The storyteller is not for a moment asserting, in however dilute a form, that *if* there really had been some girl called Rapunzel, then a case could be argued for saying that she had long hair which she let down for her lover to climb.

8 Unless you're a photon, of course, or some other particle subject to quantum conditions. In the macroscopic world, though, is it *logically* impossible to be in two places at the same time? 'I am in London' is not, in terms of formal logic, a contradiction of 'I am in Paris'. Not, at any rate, unless 'I am in London' is taken as implying, among other things, 'I am not in Paris' – though why it should, in terms of formal logic, I can't quite see. However, it's plainly more than a simple physical impossibility, like the impossibility of stretching your arm from London to Paris. You can imagine someone developing an arm five hundred kilometres long; but to imagine someone in two places at once is surely to enter a fairy-tale world. In practice, of course, people *do* manage to accept more than one invitation for the same time. And they *do* manage to entertain the idea that Lensky shot Onegin as well as Pushkin's version. You were doing it then, even as you read the previous sentence! ('We are all doublethinkers,' says David Lewis, in *On the Plurality of Worlds* [1986], p. 30. We are able not only to entertain contradictory ideas, but to *believe* them, or both to believe and not to believe them.)

9 What he actually says in the poem is not the bald statement by which I have paraphrased it, but a vivid moment-by-moment narrative, culminating in a bleak and perfect account of what it is to be dead. It comes in the following three stanzas in chapter 6 (for those lucky enough to be able to read Pushkin's Russian):

XXX
«Теперь сходитесь».

Хладнокровно,
Еще не целя, два врага,
Походкой твердой, тихо, ровно
Четыре перешли шага,
Четыре смертные ступени.
Свой пистолет тогда Евгений,
Не преставая наступать,
Стал первый тихо подымать.
Вот пять шагов еще ступили,
И Ленский, жмуря левый глаз,
Стал также целить - но как раз
Онегин выстрелил ... Пробили
Часы урочные: поэт
Роняет молча пистолет,

XXXI

На грудь кладет тихонько руку
И падает. Туманный взор
Изображает смерть, не муку.
Так медленно по скату гор,
На солнце искрами блистая,
Спадает глыба снеговая.
Мгновенным холодом облит,
Онегин к юноше спешит,
Глядит, зовет его ... напрасно:
Его уж нет. Младой певец
Нашел безвременный конец!
Дохнула буря, цвет прекрасный
Увял на утренней заре,
Потух огонь на алтаре!..

XXXII

Недвижим он лежал, и странен
Был томный мир его чела.
Под грудь он был навылет ранен;
Дымясь, из раны кровь текла.
Тому назад одно мгновенье
В сем сердце билось вдохновенье,
Вражда, надежда и любовь,
Играла жизнь, кипела кровь:
Теперь, как в доме опустелом
Все в нем и тихо и темно;
Замолкло навсегда оно.
Закрыты ставни, окны мелом
Забелены. Хозяйки нет.
А где, бог весть. Пропал и след.

My best shot at translating this, while preserving Pushkin's metre and rhyme scheme:

XXX

'Walk forward!'
        Each man – calm, collected,
His pistol still held pointing low –
Began to pace the path directed
With measured tread towards his foe.
Four steps they'd taken, quiet and steady,
Four fatal paces – when already
Onegin chose to be the one
Who softly lifted first his gun.
Five paces more with stride unbroken,
And Lensky, squinting, did the same.
But, even as he took his aim,

Onegin fired. And fate had spoken.
The poet, still without a sound,
Lets drop his pistol on the ground,

XXXI

Then lays his hand upon his breast
And falls. The clouding of his eye
Proclaims him now by death possessed.
So, slowly and in silence, high
Upon a sunlit mountain summit,
A snowy mass will shift and plummet.
Onegin, touched by sudden chill,
Makes haste to where his foe lies still.
He looks, he calls. He cannot wake him.
The youthful singer's song is done,
His little race already run.
The storm is spent, and grey dawn breaking.
The lovely flower has dropped. And cold
The altar flame that burned of old.

XXXII

So strange, the way he lay unmoving,
A languid peace upon his brow,
While from the cruel wound torn through him
The blood flowed dark and smoking now.
One minute past, but one short minute,
That breast had harboured still within it
A heart by poet's vision fired,
By love, and hate, and hope inspired.
As in a house that's been deserted,
There reigned in every shuttered room,
A never-ending silent gloom
Where all things to grey dust reverted.
The owner's gone. But who knows where?
No sign that he was ever there.

10 David Lewis, *Philosophical Papers* (1983), Vol I, p. 261.
11 E.g. Kendall L. Walton, 'Pictures and Make Believe', *Philosophical Review* 82 (1973); and 'Fearing Fiction', *Journal of Philosophy* 75 (1978).
12 Though Coleridge used it in connection not with fiction as such but with poetry in general.
13 See chapter III/1.
14 So of course does literary narrative at moments of tension – including the passage from *Eugene Onegin* quoted above. Pushkin approaches the moment of the shot in the past tense, but, immediately after 'Onegin fired', the action shifts into the present, and Lensky 'lets drop his pistol', to die not then but now. Fourteen lines later, as Onegin registers the fact of his friend's

death, the tense returns to the past.

15 Daniel C. Dennett, *Consciousness Explained*, p. 78 (footnote).

16 The Apostolic Creed is no longer thought to have been composed by the twelve Apostles, but is believed (in the more pragmatic sense of the word) to be the work of bishops preparing candidates for baptism a century or two later. The Nicene Creed, similarly, is now thought to have originated not with the Council of Nicaea in 325 but the Council of Constantinople in 381; and the Athanasian Creed not to have been composed by Athanasius in the first century but by hands unknown in the fifth century. Not that this changes their status as sources, any more than it would if *Eugene Onegin* turned out to have been written, not by Pushkin in the 1820s, but by Lermontov in the 1830s.

17 The latter is a paradoxical expression, on the face of it. If hope were sure and certain it wouldn't be hope, in any normal usage. I assume that the phrase is intended to convey something like 'confident anticipation'; or perhaps mysteriously to combine the reassuring but inert character of fore-knowledge with hope's more bracing uncertainty.

18 Usually quoted as 'I believe because it is impossible.' A little over the top, perhaps, either way, but the striking phrase is something Tertullian had a gift for. In any case he was a convert, and converts, like teenage girls who have just discovered make-up, tend to lay it on a bit thick.

19 Ferdinand de Saussure, *Le Cours de linguistique générale* (1916), p. 110.

20 Goffman, Erving, *The Presentation of Self in Everyday Life* (1959), pp. 20ff.

21 Genesis 18, v. 23ff.

22 Cf. David Lewis, once again, 'We are all doublethinkers.' (*On the Plurality of Worlds* [1986], p. 30.)

IV/1 Ricefiring

1 Jared Diamond, *The Rise and Fall of the Third Chimpanzee* (1991), p. 46.

2 The epitome of this traditional view is to be found in the logical atomism of Wittgenstein's *Tractatus*, which suggests that the fundamental stuff of the world is elementary facts, and that these facts are expressed in the elementary propositions into which language could, at any rate in principle, be broken down. Modern philosophy began when he abandoned this view later in life, and started to look at the ways in which language is actually used.

3 Daniel C Dennett, *Consciousness Explained*, pp. 300ff.

4 Justin Leiber, *Invitation to Cognitive Science* (1991), p. 8.

5 Noam Chomsky, *Men of Ideas*, p. 218.

6 Bryan Magee, *Confessions of a Philosopher* (1997), p. 81.

7 Even if you take the view that mathematical equations express propositions, mathematics can in theory be deconstructed (even if never completely) into the axioms of logic, and therefore into language. And if equations are to express *substantive* propositions – i.e. to say something about the contents of the world rather than showing forth purely formal relationships – then they have to be supplemented with language: the names of units, classes, forces, etc.

8 The question arises as to what exactly constitutes language, and whether it is restricted to what can be expressed in symbolic codes of sounds, marks, gestures, etc., but this we shall come to later.

9 In Douglas Adams, *The Hitchhiker's Guide to the Galaxy* (1979).

10
This process of creolization is a natural experiment in language evolution that has unfolded independently dozens of times in the modern world . . . What is striking is that the linguistic outcomes of all these independent natural experiments share so many similarities, both in what they lack and in what they possess. On the negative side, creoles are simpler than normal languages in that they usually lack conjugations of verbs for tense and person, declensions of nouns for case and number, most prepositions, distinctions between events in the past and present, and agreement of words for gender. On the positive side, creoles are advanced over pidgins in many respects: consistent word order; singular and plural pronouns for the first, second, and third persons; relative clauses; indications of the anterior tense (describing actions occurring before the time under discussion, whether or not that time is the present); and particles or auxiliary verbs preceding the main verb and indicating negation, anterior tense, conditional mood, and continuing as opposed to completed actions. Furthermore, most creoles agree in placing a sentence's subject, verb, and object in that particular order, and also agree in the order of particles or auxiliaries preceding the main verb.

Jared Diamond, op. cit., p. 143.

11 Daniel Dennett, who accepts that some of the syntactical structure of language in our brains may be innate, thinks that other parts of it may be memes (Richard Dawkins's term for self-replicating ideas), and argues that it doesn't really matter where the dividing line is drawn, because whether the structures are real or virtual they lay down some of the tracks on which thoughts can then travel. This seems to me slightly back-to-front. Tracks are made by people or vehicles passing over the ground. If there are tracks they have evolved as part of the thinking, not as preconditions of it. But I agree that a division into real and virtual doesn't really affect the argument.

12 Derek Bickerton, *Roots of Language* (1981); and *Language and Species*, (1990).

13 Opposition is at last being expressed by some linguists to the idea of a universal grammar. Andrew Wedel is reported to have proposed that language has an innate property of self-organisation, so that its structures have emerged from repeated small-scale interactions between its elements. 'I think there is a big shift,' Wedel is reported as saying (*The Independent*, 2 March 2005), 'from the explanation from a single level, advocated by Chomsky, that one grammar algorithm is coded in our genes, to a more layered set of explanations where structure gradually emerges, over time, through many cycles of talking and learning.' There seem to be analogies here, if I understand it aright, with the self-organising systems studied by Prigogine. Some linguists also now believe that the theory is undercut by the

striking lack of syntax in the language of the Piraha people of the Amazon basin, who also lack any sense of number (see I/5, note 4). Professor Dan Everett, who has studied the Pirahã for many years, is reported as saying that hypotheses such as universal grammar are inadequate to account for Pirahã usage. These theories assume that language evolution has ceased to be shaped by the social life of species. Pirahã grammar, he argues, arises from their culture, not from any pre-existing template. (*The Independent*, 6 May 2006)

14 And of course there are various dialects of ice skating. Compare figure-skating and speed-skating (different forms reflecting different purposes), and the old-fashioned Fenland style, still practised by elderly dons at Cambridge within my memory, with hands tucked behind back, or the style adopted by the Reverend Robert Walker (arms folded across chest) in Raeburn's painting.

15 The extension of innate grammars from language to other human activities is dismissed by Steven Pinker (in *The Language Instinct*, pp. 419–20) as 'a common academic parody of Chomsky'. The particular activities he mentions are 'bicycling, matching ties with shirts, rebuilding carburettors, and so on'. He proposes various criteria for distinguishing them from language (and not only language, but a long list of other activities, such as the recognition of faces, foods, and various dangers, that he hypothesises also require what he calls 'modules' in the brain). The clearest of the criteria is historical: 'Using biological anthropology, we can look for evidence that a problem is one that our ancestors had to solve in the environments in which they evolved, so language and face recognition are at least candidates for innate modules, but reading and driving are not.'

This would seem to deal with my extension of walking to ice-skating, though not with walking itself. But the question of whether the activity is venerable enough to suggest a grammar fixed in the neurological hardware is a secondary one. If you need rules for talking then you need them for walking, and if you need them for walking then you need them for skating – even if in the case of skating we acquire the rules as we go along.

Another of his criteria I find obscure: 'Using engineering analyses, we can examine what a system would need, in principle, to do the right kind of generalizing for the problem it is solving . . .'

A third criterion seems to me highly dubious: 'Using data from psychology and ethnography, we can test the following prediction: when children solve problems for which they have mental modules, they should look like geniuses, knowing things they have not been taught; when they solve problems that their minds are not equipped for, it should be a long hard slog.' On this basis three of my grandchildren must be endowed with mental modules for bicy-cling, since they all stayed upright at the first attempt, and two more have mental modules for swimming and ski-ing.

'Modules' are presumably parts of the brain that specialise in particular activities. There may well be such specialised sections, and they may well have developed adaptively, but this is not to say that they operate through codes of *rules*. Whatever the neurological apparatus is that keeps the sole of the foot aligned with the ground, for instance, it must be infinitely more responsive, flexible, and adaptable than a rulebook. What keeps an aircraft aligned with the ground? Do we have to imagine that the auto-pilot has to refer each incipient change in the aircraft's attitude to a rule to find out what instructions to send to the elevators?

16 Except for German, where the former is *gehen* and the latter is *fahren*.

17 The structural complications in the deep grammar implied by this last small anomaly in Russian (but otherwise lost to us) are more considerable than might at first appear.

All Russian verbs are categorised by their *aspect*, so that for each verb there is usually a pair of forms, one perfective and one imperfective, differentiated most frequently by a prefix, sometimes a suffix or a completely different root, corresponding approximately to the familiar perfect and imperfect tenses of verbs in other Indo-European languages. Verbs of motion, however, constitute what the authoritative work on the subject, *An Introduction to Russian Aspectology* (2000), by A. Zaliznyak and A. Shmelev, calls 'a special aspectual subsystem from the point of view of their relationship to aspectual correlativity'. This is because the imperfective form of each verb is divided again to form a further pair, depending upon whether the motion it designates is 'determinate' (roughly speaking a single action in a single direction which is or was in the course of happening) or 'indeterminate' (where the action is repeated or general).

The pairs formed by the verbs in these two sub-classes, explain Zaliznyak and Shmelev, are not aspectual, since both members of each have the same aspect. Nevertheless, the semantic relationship between the verbs of each pair is 'often reminiscent of the relationship between verbs of different aspect entering into an aspectual pair, and morphologically verbs of indeterminate motion are formed from verbs of determinate motion by means of the same mechanisms as in the formation of the imperfective correlates in aspectual pairs.' (And there are two particular verbs in the subsystem that form a pair on yet another semantic level: the ones that designate the basic act of going, depending on whether the going is self-powered or not. Each verb in this pair has the two aspects, perfective and imperfective, and the two forms of the imperfective aspect, determinate and indeterminate.)

To muddy the waters a little further: according to the *Introduction* (А Зализняк и А Шмелев: *Введение в русскую аспектологию* for those who would prefer to see all this in the appropriate language), where this особая аспектуальная подсистема с точки зрения их отношения к видовой коррелятивности is described, there are alternative nomenclatures. The 'verbs of determinate motion' (глаголы определенного движения) such as идти are also called 'verbs of directional or uni-directional motion, or motor-non-iterative verbs' (глаголы направленного – однонаправленного – движения или моторно-некратные) while the 'verbs of indeterminate motion' (глаголы неопределенного движения) such as ходить are also called 'verbs of non-directional or vari-directional motion, or motor-iterative verbs' (глаголы неопределенного – разнонаправленного – движения, или моторно-кратные). To confuse things further still, incidentally, in other sources the two sub-classes of imperfectives are called 'mono-iterative,' and 'multi-iterative', or 'frequentative' – однократные и многократные глаголы.

18 Bickerton, op. cit., p. 296.

19 Ibid., pp. 30–33.

20 Jared Diamond, op. cit., pp. 142 and 148.

21 You might say that stage directions are not assertions but instructions to the performers, a polite equivalent to (Rise!), (Think!), (Exit!), but this is scarcely plausible. The performer may himself have to rise and exit to enact the character's rising and exiting, but he doesn't have to think to enact the thinking. If the stage direction were addressed to the performer rather than the character it should read: Appear to think!

22 E.g. Bickerton, *Roots of Language*, p. 51.

IV/2 The Rule of Rules

1 A brief general statement of Chomsky's position:

> Given the evidence available to us today, it seems to me reasonable to propose that in every human language surface structures are generated from structures of a more abstract sort, which I will refer to as 'deep structures', by certain formal operations of a very special kind generally called 'grammatical transformations'. Each transformation is a mapping of labeled bracketings on to labeled bracketings. Deep structures are themselves labeled bracketings. The infinite class of deep structures is specified by a set of 'base rules'. Transformations applied in sequence to deep structures in accordance with certain fixed conventions and principles ultimately generate the surface structure of the sentences of the language. Thus a set of base rules defining an infinite class of deep structures and a set of grammatical transformations can serve to generate the surface structures.
>
> Noam Chomsky, *Language and Mind* (1968), p. 162.

2 The term 'rules', unlike 'laws' in the scientific sense, seems always to have been used to refer quite straightforwardly to measures originating with human agencies.

3 The *historical* derivation of the French double negative is presumably single – a negative particle, as in English, negating a series of positive words: 'not . . . a step, a point, a person' etc. But *pas, point*, and *personne* have long since acquired, in tandem with their positive sense, a negative one, in the appropriate context, even when used without a verb to be negated with a *'ne'*: *Pas de chance. Aucun espoir. 'Qui est là?'* – *'Personne.'* Or the shortest and most celebrated diary entry in history – by Louis XVI for 14 July 1789, the day that the mob stormed the Bastille and set in motion the revolution that was going to bring him to the guillotine four years later: *'Rien.'*

4 Just as many speakers of demotic French do sometimes drop the particle (*Sais pas*, etc.).

5 English-speakers also often indicate a negative by means of a construction with *no* negative constituent ('bugger all', etc.). The attempt to combine colloquial usage with grammar-book rules leads to nonsense, as in the joke about the angry foreigner who shouts: 'You think you know fuck all! I tell you – you know fuck *nothing*!'

6 John R. Searle, *Speech Acts* (1969), p. 33.

7 Ibid., p. 34.

8 Chomsky, op. cit., p. 29.

9 Searle's whole thesis seems curiously at odds with his well-known views on consciousness. The Chinese Room argument, deployed against the idea that the computer is a model of the human mind, is a demonstration that the ability to operate a set of rules doesn't in itself imply understanding, and that any account of our consciousness which attempts to reduce it to a code of rules cannot be complete. The same argument plainly applies to language; indeed the Chinese Room argument seems to take this as a given.

IV/3 Mailing a Cat

1 Cf. once again the two wonderful passages from William James quoted in note 8 to chapter II/2.

2 Bryan Magee, *Confessions of a Philosopher* (1997), p. 79.

3 More on this in chapter V/1.

4 Cf. William James, at his brilliant best again in a much-quoted passage:

> Suppose we try to recall a forgotten name. The state of our consciousness is peculiar. There is a gap therein, but no mere gap. It is a gap that is intensely active. A sort of wraith of the name is in it, beckoning us in a given direction, making us at moments tingle with the sense of our close-ness, and then letting us sink back without the longed-for term. If wrong names are proposed to us, this singularly definite gap acts immediately so as to negate them. They do not fit into its mould and the gap of one word does not feel like the gap of another, all empty of content as both might seem necessarily to be when described as gaps. When I vainly try to recall the name of Spalding my consciousness is far removed from what it is when I vainly try to recall the name of Bowles.
>
> William James, *The Principles of Psychology* (1890), Vol. I, pp. 251–2.

5 Pinker, *The Language Instinct*, chapter 3, p. 81.

6 Ibid. p. 82.

7 Cf. Chapter I/2.

8
> Pre-linguistic thinking is an observing, a slow collecting of similarities, an attending, an exercising of the memory-traces, which is continued until the need arises to fix the new knowledge in a sign.
>
> Fritz Mauthner, *Beiträge zu einer Kritik der Sprache*, Vol 1, p. 217.
> Quoted in Gershon Weiler, *Mauthner's Critique of Language* (1970), p. 46.

9   C. S. Lewis, in *Bluspels and Flalansferes* (1939), reminds us of the
philologist's aphorism that our language is full of dead metaphors. In his
1844 essay 'The Poet', the philosopher and poet Ralph Waldo Emerson
said, 'Language is fossil poetry.' If not all of our words, certainly a great
number of them, began as metaphors. Lewis mentions 'attend' as having
once meant 'stretch'. Etc. . . . Philological scholars will delve deeper (see
what I mean?) and show that even words whose origins are less obvious
were once metaphors, perhaps in a dead (get it?) language.
                    Richard Dawkins, *Unweaving the Rainbow* (1998), p. 310.

10 Or, as Stendhal put it somewhat earlier, '*Un roman est comme un archet, la
caisse du violon qui rend les sons c'est l'âme du lecteur.*' (A novel is like a
violinist's bow; the belly of the violin that gives back the sounds is the
reader's soul.) *La Vie de Henri Brulard* (1890), p. 180.

IV/4 Likeness

1 In *Constructions* (1974).
2 The relationship between the infinitude of possibilities before us, and the
finite selection we make from them, is (once again) brilliantly captured in
this extended metaphor of William James's:

> The mind, in short, works on the data it receives very much as a sculptor
> works on his block of stone. In a sense the statue stood there from eterni-
> ty. But there were a thousand different ones beside it, and the sculptor
> alone is to thank for having extricated this one from the rest. Just so the
> world of each of us, howsoever different our several views of it may be, all
> lay embedded in the primordial chaos of sensations, which gave the mere
> matter to the thought of all us indifferently. We may, if we like, by our
> reasoning unwind things back to that black and jointless continuity of
> space and moving clouds of swarming atoms which science calls the only
> real world. But all the while the world *we* feel and live in will be that
> which our ancestors and we, by slowly cumulative strokes of choice, have
> extricated out of this, like sculptors, by simply rejecting certain portions of
> the given stuff. Other sculptors, other statues from the same stone! Other
> minds, other worlds from the same montonous and inexpressive chaos!
> My world is but one in a million alike embedded, alike real to those who
> may abstract them. How different must be the worlds in the consciouness
> of ant, cuttle-fish, or crab!
>   James, *The Principles of Psychology* (1890), Vol. I, p. 288.

3   Man perceives in the world only what already lies within him; but to perceive what lies within him man needs the world; for this, however, activity
and suffering are indispensable.
                        Hugo von Hoffmansthal, *The Book of Friends.*

4 Oliver Sacks, *A Leg to Stand On* (1984), pp. 51–3.

5 M. Merleau-Ponty, *Phenomenology of Perception* (1945, 1958), p. 18. (Cf. Chapter I/1.)

6 To put this in terms of James's sculptural metaphor: our making of perceptual sculptures is facilitated because so much of the time the physical material has been pre-sculpted for us.

7 There is a parallel in the way that we form discrete elements to make up the language that describes the world. As Saussure remarks, the sounds of speech are classified by, among other things, the degree of openness of the buccal cavity, as measured on an arbitrary scale of 0–6 – and there is even an element of arbitrariness in the discrete classification of the parts of the mouth. (Ferdinand de Saussure, *Le Cours de linguistique générale* [1916], pp. 71 and 720.) And yet out of this barely differentiable material we make words which are (mostly) as definite and clearly-outlined as things (mostly) seem to be.

8 The fact that so many simple comparatives and superlatives are formed by inflection – bigger, biggest – or even by becoming entirely new words – bad/worse, *bon/meilleur* – suggests how fundamental they are, how deeply embedded in the structure of our thinking.

9 See also the discussion of 'fuzzy logic' in chapter I/5.

10 And which are remarkably different in different countries. I was amused, comparing the translations of one of my novels, to see that in Italian the protagonist and his neighbour's wife were *tu* to each other from the first moment of meeting, and that in German they switched from *Sie* to *du* with the first kiss, while in French they remained *vous* even in bed together.

11 William James makes a somewhat similar point by likening our understanding of language in general to the particular case of a child's hearing a story:

> We think it odd that young children should listen with such rapt attention to the reading of stories expressed in words half of which they do not understand, and of none of which they ask the meaning. But their thinking is in form just what ours is when it is rapid. Both of us make flying leaps over large portions of the sentences uttered and we give attention only to substantive starting points, turning points, and conclusions here and there. All the rest, 'substantive' and separately intelligible as it may *potentially* be, actually serves only as so much transitive material . . . The children probably feel no gap when through a lot of unintelligible words they are swiftly carried to a familiar and intelligible terminus.
>
> Op. cit., p. 264, footnote.

This is another of James's wonderful pieces of observation. It seems to me, though, that our ability to grasp the sense of language goes way beyond this. In the appropriate context we can dispense even with the 'substantive starting points, turning points, and conclusions' – even with the 'familiar and intelligible terminus' itself – and piece together a meaning out of scraps of the 'transitive material'.

V/1 Off-Line

1 In chapter IV/3.

2 Some people, I believe, claim not to dream at all. Dreams are such a noticeable part of most people's experience that dream-blindness must be as restrictive as colour-blindness, or tone-deafness. Or perhaps it would imply a view of the world that differed even more radically from the one that most people have. One of the scales on which we locate our experience – coherence, perhaps – would be different, because one of its extremities would have vanished, as if freezing point or boiling point had been removed from the thermometer.

3 Our journey across France didn't continue, though, in spite of my intervention. The story was sidetracked by the garageman's offering me a conductor's baton with a handle in the form of a beautifully exact model railway engine, for three pounds. When I demurred, he reduced the price to two pounds. Trying not to let my wife overhear, I told him I would give him two pounds ten. By that I think I meant not £2.10, but £2.10s. in the old currency – £2.50. Or perhaps I meant both at once – £2.50 to make the offer sound better to the man, and £2.10 to ease my conscience for spending the money.

4 Cf. Pepys, in the entry in his Diary dated 15 August 1665:

> Up by 4 a-clock and walked to Greenwich, where called at Captain Cockes and to his chamber, he being in bed – where something put my last night's dream into my head, which I think is the best that ever was dreamed – which was, that I had my Lady Castlemayne in my armes and was admitted to use all the dalliance I desired with her, and then dreamed that this could not be awake but that it was only a dream.

5 Sometimes, though, beneath the surface of the apparently random particularities of a dream, one can detect (when one thinks about it afterwards, in the waking world) an innate identity that somehow asserts itself in spite of all evidence to the contrary. Amsterdam obstinately persists in being Amsterdam even though you can see Notre-Dame and the Brandenburg Gate. I was trying in a dream to buy orange juice in Blackheath, and was told in the shop that it was no longer obtainable there – I would have to go down the road to Epsom for it. There seemed nothing surprising about this at the time, even though Epsom is on the other side of London, and I can see the structural connection now I'm awake. Blackheath is the suburban village where I lived for many years as an adult, and it adjoins the suburban town of Lewisham; Epsom is the suburban town adjoining Ewell, the suburban village where I grew up. It was perfectly reasonable to be referred from the small shops of Ewell/Blackheath to the larger ones of Lewisham/Epsom.

   Did I picture Epsom when the Blackheath shopkeeper told me to go there for orange juice? Not exactly; but I felt it as Epsom rather than Lewisham.

6 There are translation problems here (once again!). If '*cogito*' or '*je pense*' means what English-speakers mean by 'I am thinking', then it may be plausible to accept that my present thinking includes the mental activity

involved in intentionally and meaningfully uttering the statement that I am now uttering. The usual English translation of the original, though, is 'I think'. This seems to imply not just my present activity but a regular and characteristic propensity. Even if my knowledge that I am doing what I am doing at this present moment is privileged in some special way, my knowledge about what I do at other moments seems no more secure than any other knowledge about the world.

7 The possibility is discussed, with great caution, by Roger Penrose in *The Emperor's New Mind* (1989), pp. 516ff.

V/2 Home Address

1 William James, in his admirable efforts to steer clear of special mental stuff or ineffable spirit, makes a heroic effort to dissolve the self out into the stream of thought that he has so perceptively described – and makes a rare but spectacular departure from sense in the process. He begins by asserting (*The Principles of Psychology*, Vol I, pp. 300ff) that mental activities such as attending, asserting, negating, etc, are 'felt as movements of something in the head' which can often be exactly described. The sensation of remembering or reflecting, for instance, he finds as being 'due to an actual rolling outwards and upwards of the eyeballs, such as I believe occurs in me in sleep', while consenting and negating are largely a matter of the opening and closing of the glottis.

I can certainly feel my eyeballs rolling as I reflect on this, and my glottis closing. The constriction of my glottis becomes even more noticeable at what follows. James tries to persuade us that what he calls the 'self of selves' (the essential self, the self in its most inward manifestation) 'when carefully examined, is found to consist mainly of the collection of these peculiar motions in the head or between the head and throat'. This is not *all* it consists of, he concedes – or at any rate not in our present state of knowledge. But 'if the dim portions which I cannot yet define should prove to be like unto these distinct portions in me, and I like other men, it would follow that our entire feeling of spiritual activity, or what commonly passes by that name, is really a feeling of bodily activities whose exact nature is by most men overlooked.'

As the chapter goes on, permanent paralysis of the glottis begins to threaten. If the self is simply a stream of thoughts, who or what is thinking them? James's answer is that *they* are. The thoughts themselves are thinking them. Or, to be precise, they are thinking each other. This is how they form a continuous entity, by 'knowing' their immediate predecessors and successors.

Each pulse of cognitive consciousness, each Thought, dies away and is replaced by another. The other, among the things it knows, knows its own predecessor, and finding it 'warm', in the way we have described, greets it, saying: 'Though art *mine*, and part of the same self with me.' Each later Thought, knowing and including thus the Thoughts which went before, is the final receptacle – and appropriating them is the final owner – of all that they contain and own. Each Thought is thus born an owner, and dies owned, transmitting whatever it realized as its Self to its own later proprietor.

Ibid. p. 339.

2 E.g. Derek Bickerton, who seems to be talking about the self as consciousness in the following passage:

And what constitutes the 'I' that activates? Analogy from observed conspecifics, use of mirrors and other reflecting substances, plus the higher-order 'traffic-control' neurons which must exist to establish priorities in brain activity if the whole thing isn't to degenerate into electrochemical chaos.

The Roots of Language, p. 224.

3 The most notable, perhaps, being by Daniel Dennett, in *Consciousness Explained*. A bold title, and a brilliant, subtle, and sympathetic analysis in terms of loops in the neural circuitry etc, and 'overwritings of content' which (more interestingly) 'yield over the course of time something rather like a narrative stream or sequence . . .' (p. 135).

Our tales are spun, but for the most part we don't spin them; they spin us. Our human consciousness, and our narrative selfhood, is their product, not their source. These strings or streams of narrative issue forth *as if* from a single source – not just in the obvious physical sense of flowing from just one mouth, or one pencil or pen, but in a more subtle sense: their effect on any audience is to encourage them to (try to) posit a unified agent whose words they are, about whom they are: in short, to posit a *center of narrative gravity*.

Ibid., p. 417.

Dennett is at his best on the analogy with stories – of describing another person's mental world on the basis of what they report, and treating it as one does a work of fiction, which has value in its own right, without any necessary connection with the 'real' world; though this only works when we are dealing with creatures able to communicate in language – articulate human beings (ibid., pp. 78–82). But he goes on to suggest (ibid., p. 85) that there may be 'real', objectively ascertainable, events in people's brains from the inspection of which we could 'reasonably suppose that we had discovered what they were *really* talking about . . .' This seems to me to be the crucial reductionist slide from experience as experience to the neurological phenomena associated with it – the step across the category boundary. Could you 'reasonably suppose' that you knew what the symphony was 'really' about if you looked at the score, or the graph of the sounds on an oscilloscope?

The same step is involved in Dennett's parallel with the discovery by anthropologists of a real human source of stories about a god (ibid., pp. 82ff). In this case, however, we are relating like to like – fictitious person to real person. The parallel in the case of neurological events/experience would be quite different. It would be the discovery that whenever the tribe reported sightings of the god there were certain objective phenomena that seemed to explain them – high barometric pressure, particular algae in the water supply, or whatever. This might well be true, but it would be preposterous to think that what the tribe were *really* reporting, when they reported experiences of the god, was barometric readings and biochemical analyses. Whisky makes you drunk, but a bottle of whisky is not the experience of drunkenness.

4 See John R. Searle, *The Mystery of Consciousness* (1997), for a league table of his fellow philosophers' approaches – and his demolitions of all of them.

Searle himself argues for a *causal* connection between neural activity and what he and other modern philosophers call 'qualia' (the elements of our conscious experience). This seems to me at least as misleading as all the other approaches, and in two connected but separate ways. Interpreting perception as the creation of a second lot of things inside ourselves (qualia, sense data, sensibilia, etc.), comes from our longing to pluck solid objects out of our traffic with the world, and fix the unfixable flux of our experience, and it obscures the nature of that traffic by moving the problem (of how we and things relate) one stage further back into the shadows. And even if you try to analyse perception in this way, the idea that neural activity is the *cause* of these supposed phenomena is ridiculous. The neural activity *is* the experience. The experience *is* the neural activity. You can trace a causal chain from external object to brain, via the light travelling to the eye, its refraction by the lens, its conversion into electrical signals in the retina, the transmission of the signals to the brain, the resultant electro-chemical processes there . . . then what? A sudden leap from physical events in topographically real space to non-physical objects located nowhere?

A wound causes pain; a wound causes certain activities in the pain centres of the brain. These are parallel statements. To say that certain activities in the pain centres of the brain cause pain is a confusion of two different languages. The scoring of a goal may be said by one sports commentator to cause the crowd to roar, while a more precise observer may report that it causes 27,433 people to utter sounds indicative of approval. But the roar of the crowd doesn't cause 27,433 people to utter sounds indicative of approval, and nor does the accumulation of individual sounds cause the noise of the crowd. Saying that events in the brain are the *cause* of consciousness is like saying that the functioning of the leg muscles is the cause of walking. 'Why are you walking from Land's End to John O'Groats?' – 'Because of the alternating contraction and relaxation of the quadriceps and the gluteus maximus.'

Searle tries to make his account more palatable by his argument that causality doesn't necessarily have to be sequential. For a discussion of this see chapter I/4.

5 Yet the cruel truth is that the central objective of the now majestic research program in neuroscience remains beyond reach: there is only the most shaky understandng of how the brain, and the human brain particu-

larly, engenders mind – the capacity to reflect on past events, to think and to imagine.

> John Maddox, *What Remains to be Discovered* (1998), p. 276.

6 You can make a three-dimensional model of the world, of course, and it can be viewed from here, from there, from infinitely many places, just as what it represents can be. But, like what it represents, it has to be viewed from *somewhere*.

7 Cf. William James on attention, quoted in note 8 to chapter II/2.

8 According to Maxine Hong Kingston, in *China Men* (1977), p. 48, Chinese immigrants into the USA used to sell their identities in this way. Once they had established US citizenship they supplied other would-be immigrants with the details of their lives, and the lives of the various 'paper sons' whose fictitious births they had registered over the years, and the purchasers would expend prodigious efforts in studying them and learning them by heart. The US immigration officials, unable to tell one Chinese from another, could establish that it was the right person returning from a supposed visit home, or a genuine member of his family arriving to join him, only by interrogating him about biographical details that they already had on record – 'trap questions about how many pigs did they own in 1919, whether the pig house was made out of bricks or straw, how many steps on the back stoop, how far to the outhouse . . .' (ibid., p. 60).

9 So do people of all ages. Jean Piaget, the child psychologist, recalled in great detail an incident in his childhood when a man had tried to kidnap him from his pram, and his nurse was injured in protecting him. Years later his parents received a letter from the nurse confessing that she had invented the story, and returning the watch she had been given as a reward. Piaget realised that he had manufactured his recollection from the verbal accounts he had heard of the incident. (Piaget, *The Child's Construction of Symbols* [1945], quoted in Douwe Draaisma, *Why Life Speeds Up As You Get Older* [2001/2004], p. 23.)

At once, as I read this, I began vaguely to recall similar incidents in my own life . . . only to wonder if I am confecting them in my turn by pure suggestion.

10 In *The Man Who Mistook his Wife for a Hat*, for instance.

11 Some computer engineers, who have realised that artificial intelligence won't work without artificial feelings to drive it, are reported to be doing something rather like this already. Kim Jong-Hwan, the director of an intelligent-robot research centre, is reported to have developed a computer code that will determine 'a robot's propensity to "feel" happy, sad, angry, sleepy, hungry or afraid'. His method of doing this is the one proposed here: an imitation of human genetics – 'a series of artificial chromosomes . . . modelled on human DNA, though equivalent to a single strand of genetic code rather than the complex double helix of a real chromosome' (*Guardian*, 2 February 2005).

12 Creativity, whether in machines or in human beings, is often treated as an abstract quality or virtue, a criterion of intellectual worth in its own right, as

if it were the *point* of art, science, and everything else, quite detached from the nature and value of what is being created. The primary question, though, is not about the number and novelty of works produced, but about their utility, enjoyability, wisdom, marketability, etc. – about their relation to human (or computer) tastes and purposes.

Acknowledgments

My cousin Dr Keith Frayn, Professor of Human Metabolism at Oxford, for the estimate of the total number of cells in the human body, and of atoms per cell, in *Prospectus*.

My stepdaughter Jo Tomalin for finding the explanation and history of sociable numbers.

Zoya Anderson, for finding me *An Introduction to Russian Aspectology*, the authoritative work on the subject.

Graham Farmelo, Head of Science Communication at the Science Museum, London, for advice at various points and in particular for reading Chapter I/2 and making a number of helpful suggestions.

Finn Aaserud, the Director of the Niels Bohr Institute in Copenhagen, for telling me about N. David Mermin's paper.

John Cole, a sharp-eyed reader, for pointing out my gross mis-estimate, in the Prospectus to earlier editions of this book, of the number of our forebears on earth, and referring me to the estimate quoted in this edition.

Alan Lightman, who followed up his review of the first edition in *Nature* by correcting my derivation of indeterminacy, and attempting to introduce me to calculus.

My wife Claire Tomalin, who interrupted work on her own book to read an early draft of this one, and who made many helpful comments on style, tone and presentation.

And above all to Jonathan Bennett, formerly Professor of Philosophy at the Universities of British Columbia and Syracuse, who fifty years ago taught me philosophy at Cambridge, who has remained a friend, and who has over the years given me much help and support. He read an early draft of the present book, saved me from a number of errors, and suggested various improvements. His work is much quoted in the text (though I don't always agree with him). I should make it clear that he remains critical of many things, and dissents from many of the views expressed (most of all the central argument, which he regards as 'anthropocentrism run amok').

Index

William James on, 479
self-interest, 191–2
Seurat, Georges, 221
sexual arousal, 25
Shakespeare, William, 243
Shmelev, A., 473
shoes, 156–7
Short History of Nearly Everything, A
   (Bill Bryson), 423
signs, 262
silver bullets, 76
Simenon, Georges, 229
singularites, 4
Six Easy Pieces (Richard Feynman), 447
size, 114–15, 116
skating, 282–3
sky, 26, 29–30
Slavonic languages, 355
snooker, 73–5, 78, 83, 93, 96, 99,
   105–6, 174
sociable numbers, 140–1
Sodom and Gomorrah, 266
Songlines, The (Bruce Chatwin), 464
space, 23
   conceptions of, 113, 114
   creation of, 111, 114, 116
   dimensions of, 128–32
   speed in, 456
   time and, 111–12
space-time, 23, 47, 52, 349
Speech Acts (John R. Searle), 475
Spinoza, Baruch, 452
standardisation, 28
starlings, 416
stars, 30
Stendhal, 476
story *see also* fiction: narrative; novels
   dreams as, 383–4, 388–9

history and, 192
nature and simplicity of, 178
novels and, 464–5
part of the human condition, 234
philosophy as, 424
planning of, 357
sources of, 374
time as, 223
straight lines, 53, 128–32, 456
string theory, 52, 111, 123
structuralism, 345, 350, 352, 353, 360
Structure of Scientific Revolutions, The
   (Thomas S. Kuhn), 435
'Substance, Reality and Primary
   Qualities' (Jonathan Bennett), 456
Sun, the, 11, 49
supernovae, 11
Süsskind, Patrick, 318
Sweden, 355–6
symmetry, 50, 51
syntax, 285, 287, 315 *see also* grammar; language; linguistics
   Chomsky on, 280, 296
   development of, 296–7
   structure of, 471
   universal grammar and, 282, 284,
     336, 471–2
   verbs and, 294, 296–7, 336, 471–2

Talmud, 268
Tarski, Alfred, 439, 459
tautology, 465
taxonomy, 48, 66
Taylor, John C.
   on grand unified theories, 52
   matter, our knowledge of, 427, 447
   on quantum histories, 47
   quoted, 425, 439, 442